D1172026

Handbook of
Structural Steel
Connection Design
and Details

Handbook of
Structural Steel
Connection Design
and Details

Akbar R. Tamboli, P.E. FASCE   Editor

Consulting Engineer
Princeton, New Jersey

McGraw-Hill

New York   San Francisco   Washington, D.C.   Auckland   Bogotá
Caracas   Lisbon   London   Madrid   Mexico City   Milan
Montreal   New Delhi   San Juan   Singapore
Sydney   Tokyo   Toronto

Library of Congress Cataloging-in-Publication Data

Tamboli, Akbar R.
  Handbook of structural steel connection design and details.
    p.   cm.
  ISBN 0-07-061497-0
  1. Steel, Structural--Handbooks, manuals, etc.   2. Building, Iron
and steel—Handbooks, manuals, etc.
  TI472 .H26 1999
  624.1'821—dc20
                                                          98-46680
                                                              CIP

McGraw-Hill

A Division of The McGraw-Hill Companies

  567890 BKM BKM   0987

P/N 0-07-134442-X
Part of 0-07-061497-0

ISBN-13: 978-0-07-134442-5

ISBN-10: 0-07-134442-X

The sponsoring editor for this book was Larry Hager, the editing supervisor was Frank Kotowski, Jr., and the production supervisor was Pamela A. Pelton. It was set in Century Schoolbook by Estelita F. Green of McGraw-Hill's Professional Book Group composition unit.

McGraw-Hill books are available at special quantity discounts to use as premiums and sales promotions, or for use in corporate training programs. For more information, please write to the Director of Special Sales, McGraw-Hill, 11 West 19th Street, New York, NY 10011. Or contact your local bookstore.

Contents

Preface

The need for the *Handbook of Structural Steel Connections Design and Details* with an LRFD approach was recognized at the time the *Steel Design Handbook: LRFD Method* was published.

This handbook was developed to serve as a comprehensive reference source for the design of steel connections using the LRFD method. Each topic is written by leading experts in the field. Emphasis is given to provide examples from actual practice. Examples are focused to give a cost-effective approach. The theory and criteria are explained and cross-references to equations to AISC are given where applicable.

The book starts with a discussion of fasteners for structural connections. It then goes into the design of connections for axial, moment, and shear forces. Detailed connection design aspects are covered in this chapter.

Welded joint design and production are treated as a separate topic, and state-of-the-art information on welding is given for use in daily practice. How to control weld cracking and joint distortion is explained for use in general consulting practice. Partially restrained connection design is explained with practical examples.

Recent seismic activity has created the need for the design of connections for seismically resistant structures. These types of connections are covered with detailed examples. Commonly used connection details are shown for use in daily practice by fabricator, detailer, and consulting engineers.

Sometimes fabricators and engineers need to design connections for special structures. Actual examples of how to approach these needs are given from real projects which are built.

To ensure quality of connection, construction inspection and quality control are vital. Therefore, detailed information on these aspects is given to achieve desired goals.

Most steel structures have steel decking. To ensure good quality and interaction, steel deck details are explained thoroughly.

The latest trend in composite construction has created the need for the design of composite construction connections. Steel-to-concrete shear wall and composite column connections are explained in detail to achieve proper interaction and strength.

The editor gratefully acknowledges the efforts of contributors in preparing excellent manuscripts. Thanks are due to the management and staff at CUH2A, Inc.

The editor and authors are indebted to several sources for the information presented. Space considerations preclude listing all, but credit is given wherever feasible, especially in references throughout the book.

Users of this handbook are urged to communicate with the editor regarding all aspects of this book, particularly any error or suggestion for improvement.

Akbar R. Tamboli

Contributors

Abulhassan Astaneh, Ph.D., P.E. *Professor, Department of Civil Engineering, University of California* (CHAP. 10)

Atorod Azizinamini, Ph.D., P.E. *Associate Professor, Civil Engineering Department, University of Nebraska-Lincoln* (CHAP. 10)

Ahmed El-Remaily, P.E. *Ph.D. candidate, Civil Engineering Department, University of Nebraska-Lincoln* (CHAP. 10)

Scott Funderburk *The Lincoln Company* (CHAP. 3)

Richard B. Heagler, P.E. *Director of Engineering, Nichols J. Bouras, Inc.* (CHAP. 9)

Thomas Kane, C. Eng., M.I. Struct. E. *Technical Manager, Cives Steel Company* (CHAPS. 1 AND 2)

L. A. Kloiber *President, Lejeune Steel* (CHAP. 7)

Robert T. Leon *Professor, School of Civil and Environmental Engineering, Georgia Institute of Technology* (CHAP. 4)

James O. Malley *Senior Principal, Degenkolb Engineers* (CHAP. 5)

Duane K. Miller, P.E. *The Lincoln Electric Company* (CHAP. 3)

Shane Noel, *Designer, Degenkolb Engineers* (CHAP. 5)

Raymond S. Pugliesi *Associate, Degenkolb Engineers* (CHAP. 5)

Research Engineers, Inc. *Software Development Group* (CD-ROM DISK)

Bahram Sharooz, Ph.D. *Associate Professor, Civil Engineering Department, University of Cincinnati* (CHAP. 10)

Robert E. Shaw, P.E. *President, Steel Structures Technology Centers, Inc.* (CHAP. 8)

Akbar Tamboli, P.E. FASCE *Consulting Engineer, CU2HA, Inc.* (APP. A, B, C, D)

William A. Thornton, Ph.D., P.E.   *Chief Engineer, Cives Steel Company* (CHAPS. 1 AND 2)

David R. Williams, P.E.   *Principal, Williams Engineering Associates* (CHAP. 6)

Acknowledgments

The editor would like to acknowledge the input and help received from the many people, specifically those listed below for the time and encouragement they provided:

- Omer Blodgett, The Lincoln Electric Company
- Theodore Galmbos, University of Minnesota
- Lynn Beedle, Lehigh University

Appreciation is expressed to Al Perry of CUH2A, Inc., Princeton, New Jersey for his encouragement during the handbook preparation, and Irwin Cantor and Ysrael Seinuk of Cantor-Seinuk Group of New York City for encouraging the use of LRFD approach in major projects like Seven World Trade Center, New York City; Newport Office Tower, Jersey City, New Jersey; and Chase Metrotech Complex, Brooklyn, New York.

The editor would also like to acknowledge the help and assistance provided by Larry S. Hager, sponsoring editor of this handbook, who had put forth invaluable support during the process of preparing the manuscript. Also thanks to the many other individuals at McGraw-Hill responsible for bringing the book to press, including Frank Kotowski, Jr., editing supervisor; Margaret Webster-Shapiro and Howard Grossman, in preparing cover artwork; and Pamela A. Pelton, production supervisor.

The editor wishes to extend his thanks and appreciation to his wife, Rounkbi; and his children Tahira, Ajim, and Alamgir for their patience and understanding during the preparation of this handbook.

1

Fasteners for Structural Connections

William A. Thornton
Thomas Kane
Cives Steel Company
Roswell, GA

1.1 Introduction

There are two ways to connect structural steel members—using bolts or welds. Rivets, while still available, are not currently used for new structures and will not be considered here. This chapter will present the basic properties and requirements for bolts and welds.

(Courtesy of The Steel Institute of New York.)

Connections are an intimate part of a steel structure and their proper treatment is essential for a safe and economic structure. An intuitive knowledge of how a system will transmit loads (the art of load paths) and an understanding of structural mechanics (the science of equilibrium and limit states) are necessary to achieve connections which are both safe and economic. Chapter 2 will develop this material. This chapter is based on the bolting and welding requirement specifications of the American Institute of Steel Construction (AISC), "Load and Resistance Factor Design Specification for Structural Steel Buildings," 1994, and the American Welding Society *Structural Welding Code,* D1.1 (1998).

1.2 Bolted Connections

1.2.1 Types of bolts

There are three kinds of bolts used in steel construction. These are high-strength structural bolts manufactured under the American Society for Testing and Materials (ASTM) Specifications A325 and

A490, Fig. 1.1, and common bolts manufactured under ASTM A307, Fig. 1.2. The A325 and A490 bolts are *structural bolts* and can be used for any building application. A307 bolts, which were referred to previously as *common bolts,* are also variously called *machine bolts, ordinary bolts,* and *unfinished bolts.* The use of these bolts is limited primarily to shear connections in nonfatigue applications.

Structural bolts (A325 and A490) can be installed fully tensioned or snug tight. *Fully tensioned* means that the bolt is torqued until a tension force approximately equal to 70 percent of its tensile strength is produced in the bolt. *Snug-tight* is the condition that exists when all plies are in contact. It can be attained by a few impacts of an impact wrench or the full effort of a person using an ordinary spud wrench. Common bolts (A307) can be installed only to the snug-tight condition. There is no recognized procedure for tightening these bolts beyond this point.

Fully tensioned structural bolts must be used in certain locations. Section J1.11 of the AISC specification requires that they be used for the following joints:

Column splices in multistory framing, if it is more than 200 ft high, or when it is between 100 and 200 ft high and the smaller horizontal dimension of the framing is less than 40 percent of the height, or when it is less than 100 ft high and the smaller horizontal dimension is less than 25 percent of the height.

Connections, in framing more than 125 ft high, on which column bracing is dependent as well as connections of all beams or girders to columns.

Crane supports, as well as roof-truss splices, truss-to-column joints, column splices and bracing, in framing supporting cranes with capacity exceeding 5 tons.

Connections for supports for impact loads, loads causing stress reversal, or running machinery.

Any other connections stipulated on the design plans.

Also, section J3.1 of the AISC specification requires that bolts subject to tension loads be fully tensioned. In all other cases, A307 bolts and snug-tight A325 and A490 bolts can be used.

In general, the use of high-strength structural bolts should conform to the requirements of the Research Council on Structural Connections (RCSC) "Specification for Structural Joints Using ASTM A325 and A490 Bolts, LRFD," 1988. This document which is specific to A325 and A490 bolts contains all of the information on design, installation, inspection, washer use, compatible nuts, etc., for these bolts. There is no comparable document for A307 bolts. The RCSC "bolt spec" was developed in the 1950s to allow the replacement of rivets with bolts.

High-strength bolts are available in many sizes, as shown in Table 1.1. Most standard connection tables, however, apply primarily to ¾- and ⅞-in-diameter bolts. Shop and erection equipment is generally set up for these sizes, and workers are familiar with them.

1.2.2   Washer requirements

The RCSC specification requires that design details provide for washers in connections with high-strength bolts as follows:

1. A hardened beveled washer should be used to compensate for the lack of parallelism where the outer face of the bolted parts has a greater slope than 1:20 with respect to a plane normal to the bolt axis.

2. For A325 and A490 bolts for slip-critical connections and connections subject to direct tension, hardened washers are required as specified in items 3 through 7 listed here. For bolts that can be tightened only snug-tight, if a slotted hole occurs in an outer ply, a flat hardened washer or common plate washer should be installed over the slot. For other connections with A325 and A490 bolts, hardened washers are not generally required.

3. When the calibrated wrench method is used for tightening the bolts, hardened washers should be used under the element turned by the wrench.

4. For A490 bolts tensioned to the specified tension, hardened washers should be used under the head and nut in steel with a specified yield point less than 40 ksi.

TABLE 1.1   Thread Lengths for High-Strength Bolts

Bolt diameter, in	Nominal thread, in	Vanish thread, in	Total thread, in
½	1.00	0.19	1.19
⅝	1.25	0.22	1.47
¾	1.38	0.25	1.63
⅞	1.50	0.28	1.78
1	1.75	0.31	2.06
1⅛	2.00	0.34	2.34
1¼	2.00	0.38	2.38
1⅜	2.25	0.44	2.69
1½	2.25	0.44	2.69

5. A hardened washer conforming to ASTM F436 should be used for A325 or A490 bolts 1 in or less in diameter tightened in an over-sized or short-slotted hole in an outer ply.

6. Hardened washers conforming to F436 but at least $\frac{5}{16}$ in thick should be used instead of washers of standard thickness under both the head and nut of A490 bolts more than 1 in in diameter tightened in oversized or short-slotted holes in an outer ply. This requirement is not met by multiple washers even though the com-bined thickness equals or exceeds $\frac{5}{16}$ in.

7. A plate washer or continuous bar of structural-grade steel, but not necessarily hardened, at least $\frac{5}{16}$ in thick and with standard holes, should be used for an A325 or A490 bolt 1 in or less in diameter when it is tightened in a long slotted hole in an outer ply. The washer or bar should be large enough to cover the slot completely after installation of the tightened bolt. For an A490 bolt greater than 1 in in diameter in a long slotted hole in an outer ply, a single hardened washer (not multiple washers) conforming to F436, but at least $\frac{5}{16}$ in thick should be used instead of a washer or bar of structural-grade steel.

The requirements for washers specified in items 4 and 5 are satis-fied by other types of fasteners meeting the requirements of A325 or A490 and with a geometry that provides a bearing circle on the head or nut with a diameter at least equal to that of hardened F436 wash-ers. Such fasteners include "twist-off" bolts with a splined end that extends beyond the threaded portion of the bolt. During installation, this end is gripped by a special wrench chuck and is sheared off when the specified bolt tension is achieved.

The RCSC specification permits direct tension-indicating devices, such as washers incorporating small, formed arches designed to deform in a controlled manner when subjected to the tightening force. The specification also provides guidance on use of such devices to ensure proper installation.

1.2.3 Fully tensioned and snug-tight bolts

As pointed out in a previous section, fully tensioned bolts must be used for certain connections. For other locations, snug-tight bolts should be used because they are less expensive with no reduction in strength. The vast majority of shear connections in buildings can be snug-tight, and shear connections are the predominate connection in every building. Also, if common bolts provide the required strength, they should be used because they are less expensive than structural bolts. There is no danger of interchanging the two types because all bolts are required to have clear identifying marks (see Fig. 1.1 for structural bolts and Fig. 1.2 for common bolts).

1.2.4 Bearing-type versus slip-critical joints

Connections made with high-strength bolts may be slip-critical (that is, the material joined is clamped together by the tension induced in the bolts by tightening them) or bearing-type (that is, the material joined is restricted from moving primarily by the bolt shank). In bearing-type connections, bolt threads may be included in, or excluded from, the shear plane. Different design strengths are used for each condition. Also, bearing-type connections may be either fully tensioned or snug-tight, subject to the limitations already discussed. Snug-tight bolts are much more economical to install and should be used where permitted. The slip-critical connection is the most expensive because it requires that the faying surfaces be free of paint, grease, and oil, or that a special paint be used. Hence this type of connection should be used only where required by the governing design specification, for example, where it is undesirable to have the bolts slip into the bearing or where stress reversal could cause slippage. Slip-critical connections, however, have the advantage in building construction that, when used in combination with welds, the fasteners and welds may be considered to share the stress.

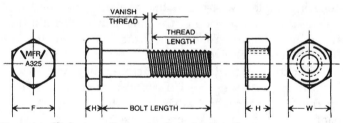

Figure 1.1 High-strength structural-steel bolt and nut.

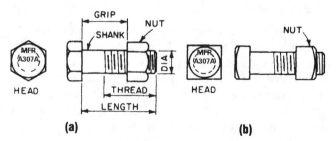

Figure 1.2 Unfinished (machine) or common bolts.

Threads included in shear planes. The bearing-type connection with threads in shear planes is most frequently used. Since the location of the threads is not restricted, the bolts can be inserted from either side of a connection. Either the head or the nut can be the element turned. Paint of any type is permitted on the faying surfaces.

Threads excluded from shear planes. The bearing-type connection with threads excluded from shear planes is the most economical high-strength bolted connection, because fewer bolts are generally needed for a given required strength. But this type should be used only after careful consideration of the difficulties involved in excluding the threads from the shear planes. The location of the thread runout or vanish depends on which side of the connection the bolt is entered and whether a washer is placed under the head or the nut. This location is difficult to control in the shop but even more so in the field.

Total nominal thread lengths and vanish thread lengths for high-strength bolts are given in Table 1.1. It is common practice to allow the last $\frac{1}{8}$ in of vanish thread to extend across a single shear plane.

In order to determine the required bolt length, the value shown in Table 1.2 should be added to the grip (that is, the total thickness of all connected material, exclusive of washers). For each hardened flat washer that is used, add $\frac{5}{32}$ in and for each beveled washer, add $\frac{5}{16}$ in. The tabulated values provide appropriate allowances for manufacturing tolerances and also provide for full thread engagement with an installed heavy hex nut. The length determined by the use of Table 1.2 should be adjusted to the next longer $\frac{1}{4}$-in length.

1.2.5   Bolts in combination with welds

In new work, ASTM A307 bolts or high-strength bolts used in bearing-type connections should not be considered as sharing the stress in combination with welds. Welds, if used, should be provided to carry

TABLE 1.2   Lengths to Be Added to Grip

Nominal bolt size, in	Addition to grip for determination of bolt length, in
$\frac{1}{2}$	$\frac{11}{16}$
$\frac{5}{8}$	$\frac{7}{8}$
$\frac{3}{4}$	1
$\frac{7}{8}$	$1\frac{1}{8}$
1	$1\frac{1}{4}$
$1\frac{1}{8}$	$1\frac{1}{2}$
$1\frac{1}{4}$	$1\frac{5}{8}$
$1\frac{3}{8}$	$1\frac{3}{4}$
$1\frac{1}{2}$	$1\frac{7}{8}$

the entire stress in the connection. High-strength bolts proportioned for slip-critical connections may be considered as sharing the stress with welds.

In welded alterations to structures, existing rivets and high-strength bolts tightened to the requirements for slip-critical connections are permitted for carrying stresses resulting from loads present at the time of alteration. The welding needs to be adequate only to carry the additional stress.

1.2.6 Standard, oversized, short-slotted, and long-slotted holes

In general, a connection with a few large-diameter fasteners costs less than one of the same capacity with many small-diameter fasteners. The fewer the number of fasteners, the fewer the number of holes to be formed and the less installation work needed. Larger-diameter fasteners are particularly favorable in connections where shear governs because the load capacity of a fastener in shear varies with the square of the fastener diameter. For practical reasons, however, $\frac{3}{4}$- and $\frac{7}{8}$-in-diameter fasteners are preferred.

The AISC specification requires that holes for bolts be $\frac{1}{16}$ in larger than the nominal fastener diameter. In computing net area or a tension member, the diameter of the hole should be taken $\frac{1}{16}$ in larger than the hole diameter.

Holes can be punched or drilled. Punching usually is the most economical method. To prevent excessive damage to material around the hole, however, the specifications limit the maximum thickness of material in which holes may be punched full size. These limits are summarized in Table 1.3.

In *buildings,* holes for thicker material may be either drilled from the solid or subpunched and reamed. The die for all subpunched holes and the drill for all subdrilled holes should be at least $\frac{1}{16}$ in smaller than the nominal fastener diameter.

TABLE 1.3 Maximum Material Thickness, in, for Punching Fastener Holes*

Type of steel	AISC
A36 steel	$d + \frac{1}{8}$†
High-strength steels	$d + \frac{1}{8}$†
Quenched and tempered steels	$\frac{1}{2}$‡

*Unless subpunching or subdrilling and reaming are used.
†d = fastener diameter, in.
‡A514 steel.

Oversize holes can be used in slip-critical connections, and the oversize hole can be in some or all the plies connected. The oversize holes are $\frac{3}{16}$ in larger than the bolt diameter for bolts $\frac{5}{8}$ to $\frac{7}{8}$ in in diameter. For bolts 1 in in diameter, the oversize hole is $\frac{1}{4}$ in larger and for bolts $1\frac{1}{8}$ in in diameter and greater, the oversize hole will be $\frac{5}{16}$ in larger.

Short-slotted holes can be used in any or all the connected plies. The load has to be applied 80 to 100° normal to the axis of the slot in bearing-type connections. Short slots can be used without regard to the direction of the applied load when slip-critical connections are used. The short slots for $\frac{5}{8}$- to $\frac{7}{8}$-in-diameter bolts are $\frac{1}{16}$ in larger in width and $\frac{1}{4}$ in larger in length than the bolt diameter. For bolts 1 in in diameter, the width is $\frac{1}{16}$ in larger and the length $\frac{5}{16}$ in larger and for bolts $1\frac{1}{8}$ in in diameter and larger, the slot will be $\frac{1}{16}$ in larger in width and $\frac{3}{8}$ in longer in length.

Long slots have the same requirement as the short-slotted holes, except that the long slot has to be in only one of the connected parts at the faying surface of the connection. The width of all long slots for bolts is $\frac{1}{16}$ in greater than the bolt diameter, and the length of the long slots for $\frac{5}{8}$-in-diameter bolts is $\frac{15}{16}$ in greater; for $\frac{3}{4}$-in-diameter bolts, $1\frac{1}{8}$ in greater; for $\frac{7}{8}$-in-diameter bolts, $1\frac{5}{16}$ in greater; for 1-in-diameter bolts, $1\frac{1}{2}$ in greater; and for $1\frac{1}{8}$-in-diameter and larger bolts, $2\frac{1}{2}$ times the diameter of bolt.

When finger shims are fully inserted between the faying surfaces of load transmitting parts of the connections, this is not considered as a long-slot connection.

1.2.7   Edge distances and spacing for bolts

Minimum distances from centers of fasteners to any edges are given in Table 1.4.

The AISC specification has provisions for minimum edge distance: The distance from the center of a standard hole to an edge of a connected part should not be less than the applicable value from Table 1.4.

Maximum edge distances are set for sealing and stitch purposes. The AISC specification limits the distance from the center of the fastener to the nearest edge of the parts in contact to 12 times the thickness of the connected part, with a maximum of 6 in. For unpainted weather steel, the maximum is 7 in or 14 times the thickness of the thinner plate. For painted or unpainted members not subject to corrosion, the maximum spacing is 12 in or 24 times the thickness of the thinner plate. Pitch is the distance, in, along the line of principal stress between centers of adjacent fasteners. It may be measured along one or more lines of fasteners. For example, suppose bolts are staggered along two parallel lines. The pitch may be given as the distance between successive bolts in each line separately. Or it may be

TABLE 1.4 Minimum Edge Distances, in, for Fastener Holes in Steel for Buildings

Fastener diameter, in	At sheared edges	At rolled edges of plates, shapes, or bars or gas-cut edges*
½	⅞	¾
⅝	1⅛	⅞
¾	1¼	1
⅞	1½†	1⅛
1	1¾†	1¼
1⅛	2	1½
1¼	2¼	1⅝
Over 1¼	1¾ d‡	1¼d ‡

*All edge distances in this column may be reduced ⅛ in when the hole is at the point where stress does not exceed 25 percent of the maximum allowed stress in the element.

†These may be 1¼ in at the ends of beam connection angles.

‡d = fastener diameter, in.

SOURCE: From AISC "Specification for Structural Steel Buildings."

given as the distance, measured parallel to the fastener lines, between a bolt in one line and the nearest bolt in the other line.

Gage is the distance, in, between adjacent lines of fasteners along which pitch is measured or the distance, in, from the back of an angle or other shape to the first line of fasteners.

The minimum distance between centers of fasteners should be at least 3 times the fastener diameter. (The AISC specification, however, permits 2⅔ times the fastener diameter).

Limitations also are set on maximum spacing of fasteners for several reasons. In built-up members, *stitch fasteners,* with restricted spacings, are used between components to ensure uniform action. Also, in compression members such fasteners are required to prevent local buckling.

Designs should provide ample clearance for tightening high-strength bolts. Detailers who prepare shop drawings for fabricators generally are aware of the necessity for this and can, with careful detailing, secure the necessary space. In tight situations, the solution may be to stagger the holes (Fig. 1.3), variations from standard gages (Fig. 1.4), use of knife-type connections, or use of a combination of shop welds and field bolts.

Minimum clearances for tightening high-strength bolts are indicated in Fig. 1.5 and Table 1.5.

1.2.8 Installation

All parts of a connection should be held tightly together during installation of fasteners. Drifting done during assembling to align holes

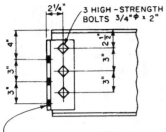

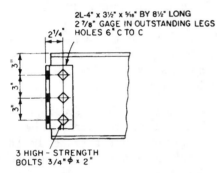

Figure 1.3 Staggered holes provide clearance for high-strength bolts.

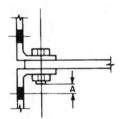

Figure 1.4 Increasing the gage in framing angles.

Figure 1.5 The usual minimum clearances.

TABLE 1.5 Clearances for High-Strength Bolts

Bolt diameter, in	Nut height, in	Usual minimum clearance, A, in	Minimum clearance for twist-off bolts, A, in	
			Small tool	Large tool
⅝	⅝	1	1⅝	—
¾	¾	1¼	1⅝	1⅞
⅞	⅞	1⅜	1⅝	1⅞
1	1	1⁷⁄₁₆		1⅞
1⅛	1⅛	1⁹⁄₁₆		—
1¼	1¼	1¹¹⁄₁₆		—

should not distort the metal or enlarge the holes. Holes that must be enlarged to admit fasteners should be reamed. Poor matching of holes is cause for rejection.

For connections with high-strength bolts, surfaces, when assembled, including those adjacent to bolt heads, nuts, and washers, should be free of scale, except tight mill scale. The surfaces also should be free of defects that would prevent solid seating of the parts, especially dirt, burrs, and other foreign material. Contact surfaces within slip-critical joints should be free of oil, paint (except for qualified paints), lacquer, and rust inhibitor.

High-strength bolts usually are tightened with an impact wrench. Only where clearance does not permit its use will bolts be hand-tightened.

Each high-strength bolt should be tightened so that when all fasteners in the connection are tight it will have the total tension, kips, for its diameter. Tightening should be done by one of the following methods, as given in the RCSC Specifications (1988).

Calibrated-wrench method.   When a calibrated wrench is used, it must be set to cut off tightening when the required tension has been exceeded by 5 percent. The wrench should be tested periodically (at least daily on a minimum of three bolts of each diameter being used). For this purpose, a calibrating device that gives the bolt tension directly should be used. In particular, the wrench should be calibrated when bolt size or length of air hose is changed. When bolts are tightened, bolts previously tensioned may become loose because of compression of the connected parts. The calibrated wrench should be reapplied to bolts previously tightened to ensure that all bolts are tensioned to the prescribed values.

Turn-of-the-nut method.   When the turn-of-the-nut method is used, tightening may be done by impact or hand wrench. This method involves three steps:

1. *Fit up of connection:*   Enough bolts are tightened a sufficient amount to bring contact surfaces together. This can be done with fit-up bolts, but it is more economical to use some of the final high-strength bolts.

2. *Snug tightening of bolts:*   All high-strength bolts are inserted and made snug-tight (tightness obtained with a few impacts of an impact wrench or the full effort of a person using an ordinary spud wrench). While the definition of snug-tight is rather indefinite, the condition can be observed or learned with a tension-testing device.

3. *Nut rotation from snug-tight position:*   All bolts are tightened by the amount of nut rotation specified in Table 1.6. If required by bolt-entering and wrench-operation clearances, tightening, including by

TABLE 1.6 **Number of Nut or Bolt Turns from Snug-Tight Condition for High-Strength Bolts***

Bolt length (Fig. 1.1)	Both faces normal to bolt axis	Slope of outer faces of bolted parts	
		One face normal to bolt axis and the other sloped†	Bolt faces sloped†
Up to 4 diameters	⅓	½	⅔
Over 4 diameters but not more than 8 diameters	½	⅔	⅚
Over 5 diameters but not more than 12 diameters‡	⅔	⅚	1

*Nut rotation is relative to the bolt regardless of whether the nut or bolt is turned. For bolts installed by ½ turn and less, the tolerance should be ±30°. For bolts installed by ⅔ turn and more, the tolerance should be ±45°. This table is applicable only to connections in which all material within the grip of the bolt is steel.

†Slope is not more than 1:20 from the normal to the bolt axis, and a beveled washer is not used.

‡No research has been performed by RCSC to establish the turn-of-the-nut procedure for bolt lengths exceeding 12 diameters. Therefore, the required rotation should be determined by actual test in a suitable tension-measuring device that simulates conditions of solidly fitted steel.

the calibrated-wrench method, may be done by turning the bolt while the nut is prevented from rotating.

Direct-tension-indicator and alternative design bolts. The direct tension indicator (DTI) hardened-steel load-indicator washer has dimples on the surface of one face of the washer. When the bolt is torqued, the dimples depress to the manufacturer's specification requirements, and proper torque can be measured by the use of a feeler gage. Special attention should be given to proper installation of flat hardened washers when load-indicating washers are used with bolts installed in oversize or slotted holes and when the load-indicating washers are used under the turned element.

The alternative design bolt (also called a twist-off or TC bolt) is a bolt with an extension to the actual length of the bolt. This extension will twist off when torqued to the required tension by a special torque gun. A representative sample of at least three bolts and nuts for each diameter and grade of fastener should be tested in a calibration device to demonstrate that the device can be torqued to 5 percent greater tension than that required.

Since both of these devices involve an irreversible mechanism, that is, yielding on the dimples or fracture of the twist-off element, bolts should be installed in all holes and brought to the snug-tight condition. All fasteners should then be tightened, progressing systematically from the most rigid part of the connection to the free edges in a manner that will minimize relaxation of previously tightened fasteners prior to final twist-off or yielding of the control or indicator element of the individual devices. In some cases, proper tensioning of the bolts may require more than a single cycle of systematic tightening.

An excellent source of information on bolt installation is the *Structural Bolting Handbook* (1995).

1.3   Welded Connections*

Welded connections often are used because of their simplicity of design, fewer parts, less material, and decrease in shop handling and fabrication operations. Frequently, a combination of shop welding and field bolting is advantageous. With connection angles shop-welded to a beam, field connections can be made with high-strength bolts without the clearance problems that may arise in an all-bolted connection.

Welded connections have a rigidity that can be advantageous if properly accounted for in design. Welded trusses, for example, deflect less than bolted trusses, because the end of a welded member at a joint cannot rotate relative to the other members there. If the end of a beam is welded to a column, the rotation there is practically the same for column and beam.

A disadvantage of welding, however, is that shrinkage of large welds must be considered. It is particularly important in large structures where there will be an accumulative effect.

Properly made, a properly designed weld is stronger than the base metal. Improperly made, even a good-looking weld may be worthless. Properly made, a weld has the required penetration and is not brittle.

Prequalified joints, welding procedures, and procedures for qualifying welders are covered by AWS D1.1, *Structural Welding Code— Steel*, American Welding Society (1998). Common types of welds with structural steels intended for welding when made in accordance with AWS specifications can be specified by note or by symbol with assurance that a good connection will be obtained.

In making a welded design, designers should specify only the amount and size of weld actually required. Generally, a $\frac{5}{16}$-in weld is considered the maximum size for a single pass.

The cost of fit-up for welding can range from about one-third to several times the cost of welding. In designing welded connections, therefore, designers should consider the work necessary for the fabricator and the erector in fitting members together so they can be welded.

---

*Section 1.3 is from A. Tamboli, *Steel Design Handbook: LRFD Method*, McGraw-Hill, NY, 1997.

1.3.1   Types of welds

The main types of welds used for structural steel are fillet, groove, plug, and slot. The most commonly used weld is the fillet. For light loads, it is the most economical, because little preparation of material is required. For heavy loads, groove welds are the most efficient, because the full strength of the base metal can be obtained easily. Use of plug and slot welds generally is limited to special conditions where fillet or groove welds are not practical.

More than one type of weld may be used in a connection. If so, the allowable capacity of the connection is the sum of the effective capacities of each type of weld used, separately computed with respect to the axis of the group.

Tack welds may be used for assembly or shipping. They are not assigned any stress-carrying capacity in the final structure. In some cases, these welds must be removed after final assembly or erection.

Fillet welds have the general shape of an isosceles right triangle (Fig. 1.6). The size of the weld is given by the length of leg. The strength is determined by the throat thickness, the shortest distance from the root (intersection of legs) to the face of the weld. If the two legs are unequal, the nominal size of the weld is given by the shorter of the legs. If welds are concave, the throat is diminished accordingly, and so is the strength.

Fillet welds are used to join two surfaces approximately at right angles to each other. The joints may be lap (Fig. 1.7) or tee or corner (Fig. 1.8). Fillet welds also may be used with groove welds to reinforce

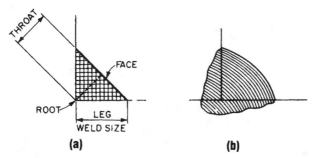

Figure 1.6   Fillet weld: (*a*) theoretical cross section and (*b*) actual cross section.

Figure 1.7   Welded lap joint.

Figure 1.8   (*a*) Tee joint and (*b*) corner joint.

corner joints. In a skewed tee joint, the included angle of weld deposit may vary up to 30° from the perpendicular, and one corner of the edge to be connected may be raised, up to ³⁄₁₆ in. If the separation is greater than ¹⁄₁₆ in, the weld leg should be increased by the amount of the root opening.

Groove welds are made in a groove between the edges of two parts to be joined. These welds generally are used to connect two plates lying in the same plane (butt joint), but they also may be used for tee and corner joints.

Standard types of groove welds are named in accordance with the shape given the edges to be welded: square, single V, double V, single bevel, double bevel, single U, double U, single J, and double J (Fig. 1.9). Edges may be shaped by flame cutting, arc-air gouging, or edge planing. Material up to ⅝ in thick, however, may be groove-welded with square-cut edges, depending on the welding process used.

Groove welds should extend the full width of the parts joined. Intermittent groove welds and butt joints not fully welded throughout the cross section are prohibited.

Groove welds also are classified as complete-penetration and partial-penetration welds.

In a *complete-joint-penetration weld,* the weld material and the base metal are fused throughout the depth of the joint. This type of weld is made by welding from both sides of the joint or from one side

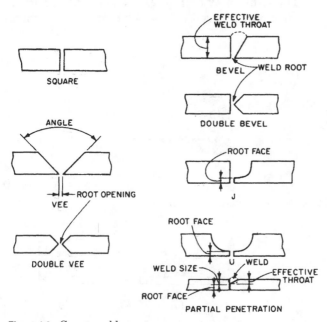

Figure 1.9 Groove welds.

to a backing bar or backing weld. When the joint is made by welding from both sides, the root of the first-pass weld is chipped or gouged to sound metal before the weld on the opposite side, or back pass, is made. The throat dimension of a complete-joint-penetration groove weld, for stress computations, is the full thickness of the thinner part joined, exclusive of weld reinforcement.

Partial-joint-penetration welds generally are used when forces to be transferred are small. The edges may not be shaped over the full joint thickness, and the depth of the weld may be less than the joint thickness (Fig. 1.9). But even if the edges are fully shaped, groove welds made from one side without a backing strip or made from both sides without back gouging are considered partial-joint-penetration welds. They are often used for splices in building columns carrying axial loads only. In bridges, such welds should not be used where tension may be applied normal to the axis of the welds.

Plug and slot welds are used to transmit shear in lap joints and to prevent buckling of lapped parts. In buildings, they also may be used to join components of built-up members. (Plug or slot welds, however, are not permitted on A514 steel.) The welds are made, with lapped parts in contact, by depositing weld metal in circular or slotted holes in one part. The openings may be partly or completely filled, depending on their depth. Load capacity of a plug or slot completely welded equals the product of hole area and allowable stress. Unless appearance is a main consideration, a fillet weld in holes or slots is preferable.

Economy in selection.   In selecting a weld, designers should consider not only the type of joint but also the type of weld that would require a minimum amount of metal. This would yield a saving in both material and time.

While the strength of a fillet weld varies with size, the volume of metal varies with the square of the size. For example, a ½-in fillet weld contains 4 times as much metal per inch of length as a ¼-in weld but is only twice as strong. In general, a smaller but longer fillet weld costs less than a larger but shorter weld of the same capacity.

Furthermore, small welds can be deposited in a single pass. Large welds require multiple passes. They take longer, absorb more weld metal, and cost more. As a guide in selecting welds, Table 1.7 lists the number of passes required for some frequently used types of welds.

Double-V and double-bevel groove welds contain about half as much weld metal as single-V and single-bevel groove welds, respectively (deducting effects of root spacing). Cost of edge preparation and added labor of gouging for the back pass, however, should be considered. Also, for thin material, for which a single weld pass may be sufficient, it is uneconomical to use smaller electrodes to weld from two

TABLE 1.7   Number of Passes for Welds

Weld size,* in	Fillet welds	Single-bevel groove welds (backup weld not included)		Single-V groove welds (backup weld not included)		
		30° bevel	45° bevel	30° open	60° open	90° open
³⁄₁₆	1					
¼	1	1	1	2	3	3
⁵⁄₁₆	1					
⅜	3	2	2	3	4	6
⁷⁄₁₆	4					
½	4	2	2	4	5	7
⅝	6	3	3	4	6	8
¾	8	4	5	4	7	9
⅞		5	8	5	10	10
1		5	11	5	13	22
1⅛		7	11	9	15	27
1¼		8	11	12	16	32
1⅜		9	15	13	21	36
1½		9	18	13	25	40
1¾		11	21			

*Plate thickness for groove welds.

sides. Furthermore, poor accessibility or less favorable welding position (Sec. 1.3.4) may make an unsymmetrical groove weld more economical, because it can be welded from only one side.

When bevel or V grooves can be flame-cut, they cost less than J and U grooves, which require planning or arc-air gouging.

For a given size of fillet weld, the cooling rate is faster and the restraint is greater with thick plates than with thin plates. To prevent cracking due to resulting internal stresses, the AISC Specification section J2-2 sets minimum sizes for fillet welds depending on plate thickness (Table 1.8).

To prevent overstressing of base material at a fillet weld, standard specifications also limit the maximum weld size. They require that allowable stresses in adjacent base material not be exceeded when a fillet weld is stressed to its design strength.

A limitation is also placed on the maximum size of fillet welds along edges. One reason is that edges of rolled shapes are rounded, and weld thickness consequently is less than the nominal thickness of the part. Another reason is that if weld size and plate thickness are nearly equal, the plate corner may melt into the weld, reducing the length of weld leg and the throat. Hence standard specifications require the following: *Along edges of material less than ¼ in thick, maximum size of fillet weld may equal material thickness. But along edges of materi-*

TABLE 1.8   Minimum Fillet-Weld Sizes and Plate-Thickness Limits

Size of fillet welds,* in		Maximum plate thickness, in‡	Minimum plate thickness for fillet welds on each side of the plate, in	
Buildings†	AWS D1.1		36-ksi steel	50-ksi steel
⅛§		¼		
3⁄16		½	0.43	0.31
¼		¾	0.37	0.41
5⁄16		Over ¾	0.72	0.52

*Weld size need not exceed the thickness of the thinner part joined, but AISC requires that care be taken to provide sufficient preheat to ensure weld soundness.
†When low-hydrogen welding is employed, AWS D1.1 permits the thinner part joined to be used to determine the minimum size of fillet weld.
‡Plate thickness is the thickness of the thicker part joined.
§Minimum weld size for structures subjected to dynamic loads is 3⁄16 in.

al ¼ in or more thick, the maximum size should be 1⁄16 in less than the material thickness.

Weld size may exceed this, however, if drawings definitely show that the weld is to be built out to obtain full throat thickness. AWS D1.1 requires that the minimum effective length of a fillet weld be at least 4 times the nominal size, or else the weld must be considered not to exceed 25 percent of the effective length.

Subject to the preceding requirements, intermittent fillet welds may be used in buildings to transfer calculated stress across a joint or faying surfaces when the strength required is less than that developed by a continuous fillet weld of the smallest permitted size. Intermittent fillet welds also may be used to join components of built-up members in buildings.

Intermittent welds are advantageous with light members where excessive welding can result in straightening costs greater than the cost of welding. Intermittent welds often are sufficient and less costly than continuous welds (except girder fillet welds made with automatic welding equipment).

Weld lengths specified on drawings are effective weld lengths. They do not include distances needed for start and stop of welding.

To avoid the adverse effects of starting or stopping a fillet weld at a corner, welds extending to corners should be returned continuously around the corners in the same plane for a distance of at least twice the weld size. This applies to side and top fillet welds connecting brackets, beam seats, and similar connections, on the plane about which bending moments are computed. End returns should be indicated on design and detail drawings.

Fllet welds deposited on opposite sides of a common plane of contact between two parts must be interrupted at a corner common to both welds.

If longitudinal fillet welds are used alone in end connections of flat-bar tension members, the length of each fillet weld should at least equal the perpendicular distance between the welds. The transverse spacing of longitudinal fillet welds in end connections should not exceed 8 in unless the design otherwise prevents excessive transverse bending in the connections.

In material ⅝ in or less thick, the thickness of plug or slot welds should be the same as the material thickness. In material greater than ⅝ in thick, the weld thickness should be at least half the material thickness but not less than ⅝ in.

The diameter of the hole for a plug weld should be at least equal to the depth of the hole plus ⁵⁄₁₆ in, but the diameter should not exceed 2¼ times the thickness of the member.

Thus, the hole diameter in ¾-in plate could be a minimum of ¾ + ⁵⁄₁₆ = 1¹⁄₁₆ in. The depth of metal would be at least ⅝ in > (½ × ¾ = ⅜ in).

Plug welds may not be spaced closer center-to-center than 4 times the hole diameter.

The length of the slot for a slot weld should not exceed 10 times the part thickness. The width of the slot should be at least equal to the depth of the hole pus ⁵⁄₁₆ in, but the width should not exceed 2¼ times the part thickness.

Thus, the width of the slot in ¾-in plate could be a minimum of ¾ + ⁵⁄₁₆ = 1¹⁄₁₆ in. The weld metal depth would be at least ⅝ in > (½ × ¾ = ⅜ in). The slot could be up to 10 × ¾ = 7½ in long.

Slot welds may be spaced no closer than 4 times their width in a direction transverse to the slot length. In the longitudinal direction, center-to-center spacing should be at least twice the slot length.

1.3.2 Welding symbols

These should be used on drawings to designate welds and provide pertinent information concerning them. The basic parts of a weld symbol are a horizontal line and an arrow:

Extending from either end of the line, the arrow should point to the joint in the same manner as the electrode would be held to do the welding.

Welding symbols should clearly convey the intent of the designer. For the purpose, sections or enlarged details may have to be drawn to show the symbols, or notes may be added. Notes may be given as part of welding symbols or separately. When part of a symbol, the note should be placed inside a tail at the opposite end of the line from the arrow:

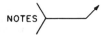

The type and length of weld are indicated above or below the line. If noted below the line, the symbol applies to a weld on the arrow side of the joint, the side to which the arrow points. If noted above the line, the symbol indicates that the other side, the side opposite the one to which the arrow points (not the far side of the assembly), is to be welded.

A fillet weld is represented by a right triangle extending above or below the line to indicate the side on which the weld is to be made. The vertical leg of the triangle is always on the left.

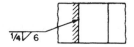

The preceding symbol indicates that a ¼-in fillet weld 6 in long is to be made on the arrow side of the assembly. The following symbol requires a ¼-in fillet weld 6 in long on both sides.

If a weld is required on the far side of an assembly, it may be assumed necessary from symmetry, shown in sections or details, or explained by a note in the tail of the welding symbol. For connection angles at the end of a beam, far-side welds generally are assumed:

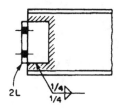

The length of the weld is not shown on the symbol in this case because the connection requires a continuous weld the full length of each angle on both sides of the angle. Care must be taken not to omit the length unless a continuous full-length weld is wanted. "Continuous" should be written on the weld symbol to indicate length when such a weld is required. In general, a tail note is advisable to specify welds on the far side, even when the welds are the same size.

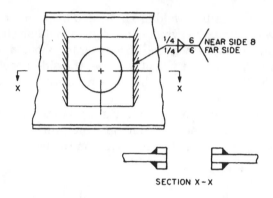

SECTION X – X

For many members, a stitch or intermittent weld is sufficient. It may be shown as

$$\diagup \overline{1/4 \, V \, 2\text{--}10}$$

This symbol calls for ¼-in fillet welds on the arrow side. Each weld is to be 2 in long. Spacing of welds is to be 10 in center-to-center. If the welds are to be staggered on the arrow and other sides, they can be shown as

$$\diagup \overline{\begin{array}{c} 1/4 \diagdown 2\text{--}10 \\ 1/4 \, V \, 2\text{--}10 \end{array}}$$

Usually, intermittent welds are started and finished with a weld at least twice as long as the length of the stitch welds. This information is given in a tail note:

$$\diagup \overline{\begin{array}{c} 1/4 \diagdown 2\text{--}10 \\ 1/4 \, V \, 2\text{--}10 \end{array}} \diagdown 4" \text{ at ends}$$

When the welding is to be done in the field rather than in the shop, a triangular flag should be placed at the intersection of arrow and line:

This is important in ensuring that the weld will be made as required. Often, a tail note is advisable for specifying field welds.

A continuous weld all around a joint is indicated by a small circle around the intersection of line and arrow:

Such a symbol would be used, for example, to specify a weld joining a pipe column to a base plate. The all-around symbol, however, should not be used as a substitute for computation of the actual weld length required. Note that the type of weld is indicated below the line in the all-around symbol, regardless of shape or extent of joint.

The preceding devices for providing information with fillet welds also apply to groove welds. In addition, groove-weld symbols must designate material preparation required. This often is best shown on a cross section of the joint.

A square-groove weld (made in thin material) without root opening is indicated by

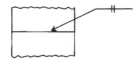

Length is not shown on the welding symbol for groove welds because these welds almost always extend the full length of the joint.

A short curved line below a square-groove symbol indicates weld contour. A short straight line in that position represents a flush weld surface. If the weld is not to be ground, however, that part of the symbol is usually omitted. When grinding is required, it must be indicated in the symbol.

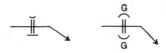

The root-opening size for a groove weld is written in within the symbol indicating the type of weld. For example, a ⅛-in root opening for a square-groove weld is specified by

and a ⅛-in root opening for a bevel weld, not to be ground, is indicated by

In this and other types of unsymmetrical welds, the arrow not only designates the arrow side of the joint but also points to the side to be shaped for the groove weld. When the arrow has this significance, the intention often is emphasized by an extra break in the arrow.

The angle at which the material is to be beveled should be indicated with the root opening:

A double-bevel weld is specified by

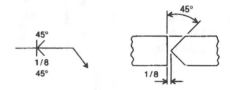

A single-V weld is represented by

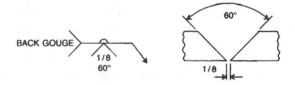

A double-V weld is indicated by

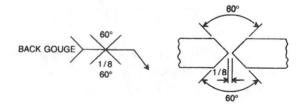

Summary. Standard symbols for various types of welds are summarized in Fig. 1.10. The symbols do not indicate whether backing, spac-

	BACK WELDS	FILLET WELDS	PLUG OR SLOT WELDS	GROOVE WELDS							WELD ALL AROUND	FIELD WELD
				SQUARE	VEE	BEVEL	U	J	FLARE VEE	FLARE BEVEL		
ARROW SIDE												
OTHER SIDE												
BOTH SIDES												

Figure 1.10 Summary of welding symbols.

er, or extension bars are required. These should be specified in general notes or shown in detail drawings. Preparation for J and U welds is best shown by an enlarged cross section. Radius and depth of preparation must be given.

In preparing a weld symbol, insert size, weld-type symbol, length of weld, and spacing, in that order from left to right. The perpendicular leg of the symbol for fillet, bevel, J, and flare-bevel welds should be on the left of the symbol. Bear in mind also that arrow-side and other-side welds are the same size unless otherwise noted. When billing of detail material discloses the identity of the far side with the near side, the welding shown for the near side also will be duplicated on the far side. Symbols apply between abrupt changes in direction of welding unless governed by the all-around symbol or dimensioning shown.

Where groove preparation is not symmetrical and complete, additional information should be given on the symbol. Also it may be necessary to give weld-penetration information, as in Fig. 1.11. For the weld shown, penetration from either side must be a minimum of $\frac{3}{16}$ in. The second side should be back-gouged before the weld there is made.

Welds also may be a combination of different groove and fillet welds. While symbols can be developed for these, designers will save time by supplying a sketch or enlarged cross section. It is important to convey the required information accurately and completely to the workers who will do the job. Actually, it is common practice for

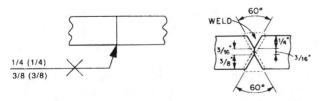

Figure 1.11 Penetration information is given on the welding symbol in (a) for the weld shown in (b). Penetration must be at least $\frac{3}{16}$ in. Second side must be back-gouged before the weld on that side is made.

designers to indicate what is required of the weld and for fabricators and erectors to submit proposed procedures.

1.3.3 Welding material

Weldable structural steels permissible in buildings are listed with the required electrodes in Table 1.9. Welding electrodes and fluxes should conform to AWS 5.1, 5.5, 5.17, 5.18, 5.20, 5.23, 5.25, 5.26, 5.28, or 5.29 or applicable provisions of AWS D1.1. Weld metal deposited by electroslag or electrogas welding processes should conform to the requirements of AWS D1.1 for these processes. For welded connections in buildings, the electrodes or fluxes given in Table 1.9 should be used in making complete-joint-penetration groove welds.

1.3.4 Welding positions

The position of the stick electrode relative to the joint when a weld is being made affects welding economy and quality.

The basic welding positions are as follows:

Flat with the face of the weld nearly horizontal. The electrode is nearly vertical, and welding is performed from above the joint.

Horizontal with the axis of the weld horizontal. For groove welds, the face of the weld is nearly vertical. For fillet welds, the face of the weld usually is about 45° relative to horizontal and vertical surfaces.

Vertical with the axis of the weld nearly vertical. (Welds are made upward.)

Overhead with the face of the weld nearly horizontal. The electrode is nearly vertical, and welding is performed from below the joint.

Where possible, welds should be made in the flat position. Weld metal can be deposited faster and more easily and generally the best and most economical welds are obtained. In a shop, the work usually is positioned to allow flat or horizontal welding. With care in design, the expense of this positioning can be kept to a minimum. In the field, vertical and overhead welding sometimes may be necessary. The best assurance of good welds in these positions is use of proper electrodes by experienced welders.

The AWS specifications require that only the flat position be used for submerged-arc welding, except for certain sizes of fillet welds. Single-pass fillet welds may be made in the flat or the horizontal position in sizes up to $\frac{5}{16}$ in with a single electrode and up to $\frac{1}{2}$ in with multiple electrodes. Other positions are prohibited.

TABLE 1.9   Matching Filler-Metal Requirements for Complete-Penetration Groove Welds in Building Construction

Base metal*	Shielded metal-arc	Submerged-arc	Gas metal-arc	Flux-cored arc
		Welding process		
A36,† A53 grade B A500 grades A and B A501, A529, and A570 grades 30 through 50	AWS A5.1 or A5.5§ E60XX E70XX E70XX-X	AWS A5.17 or A5.23§ F6XX-EXXX F7XX-EXXX or F7XX-EXX-XX	AWS A5.18 ER70S-X	AWS A5.20 E6XT-X E7XT-X (Except −2, −3, −10, −GS)
A242,‡ A441, A572 grade 42 and 50, and A588‡ (4 in and under)	AWS A5.1 or A5.5§ E7015, E7016, E7018, E7028 E7015-X, E7016-X, E7018-X	AWS A5.17 or A5.23§ F7XX-EXXX F7XX-EXX-XX	AWS A5.18 ER70S-X	AWS A5.20 E7XT-X (Except −2, −3, −10, −GS)
A572 grades 60 and 65	AWS A5.5§ E8015-X, E8016-X E8018-X	AWS A5.28§ F8XX-EXX-XX	AWS A5.28§ ER 80S-X	AWS A5.29§ E8XTX-X
A514 over 2½ in thick	AWS A5.5§ E10015-X, E10016-X, E10018-X	AWS A5.28§ F10XX-EXX-XX	AWS A5.28§ ER 100S-X	AWS A5.29§ E10XTX-X
A514 2½ in thick and under	AWS A5.5§ E11015-X, E11016-X, E11018-X	AWS A5.23§ F11XX-EXX-XX	AWS A5.28§ ER 110S-X	AWS A5.29§ E11XTX-X

*In joints involving base metals of different groups, low-hydrogen filler-metal requirements to the lower-strength group may be used. The low-hydrogen processes are subject to the technique requirements applicable to the higher-strength group.

†Only low-hydrogen electrodes may be used for welding A36 steel more than 1 in thick for dynamically loaded structures.

‡Special welding materials and procedures (for example, E80XX-X low-alloy electrodes) may be required to match the notch toughness of base metal (for applications involving impact loading or low temperature) or for atmospheric corrosion and weathering characteristics.

§Deposited weld metal should have a minimum impact strength of 20 ft-lb at 0°F when Charpy V-notch specimens are required.

When groove-welded joints can be welded in the flat position, submerged-arc and gas metal-arc processes usually are more economical than the manual shielded metal-arc process.

Designers and detailers should detail connections to ensure that welders have ample space for positioning and manipulating elec-

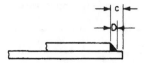

Figure 1.12   Minimum landing for a fillet weld.

trodes and for observing the operation with a protective hood in place. Electrodes may be up to 18 in long and ⅜ in in diameter.

In addition, adequate space must be provided for deposition of the required size of the fillet weld. For example, to provide an adequate landing c, in, for the fillet weld of size D, in, in Fig. 1.12, c should be at least D + ⁵⁄₁₆. In building column splices, however, $c = D$ + ³⁄₁₆ often is used for welding splice plates to fillers.

1.3.5   Weld procedures

Welds should be qualified and should be made only by welders, welding operators, and tackers qualified as required in AWS D1.1 for buildings. Welding should not be permitted under any of the following conditions:

When the ambient temperature is below 0°F

When surfaces are wet or exposed to rain, snow, or high wind

When welders are exposed to inclement conditions

Surfaces and edges to be welded should be free from fins, tears, cracks, and other defects. Also, surfaces at and near welds should be free from loose scale, slag, rust, grease, moisture, and other material that may prevent proper welding. AWS specifications, however, permit mill scale that withstands vigorous wire brushing, a light film of drying oil, or antispatter compound to remain. But the specifications require all mill scale to be removed from surfaces on which flange-to-web welds are to be made by submerged-arc welding or shielded metal-arc welding with low-hydrogen electrodes.

Parts to be fillet-welded should be in close contact. The gap between parts should not exceed ³⁄₁₆ in. If it is ¹⁄₁₆ in or more, the fillet-weld size should be increased by the amount of separation. The separation between faying surfaces for plug and slot welds and for butt joints landing on a backing should not exceed ¹⁄₁₆ in. Parts to be joined at butt joints should be carefully aligned. Where the parts are effectively restrained against bending due to eccentricity in alignment, an offset not exceeding 10 percent of the thickness of the thinner part joined, but in no case more than ⅛ in, is permitted as a departure from theoretical alignment. When correcting misalignment in such cases, the parts should not be drawn in to a greater slope than ½ in in 12 in.

For permissible welding positions, see Sec. 1.3.4. Work should be positioned for flat welding whenever practicable.

In general, welding procedures and sequences should avoid needless distortion and should minimize shrinkage stresses. As welding progresses, welds should be deposited so as to balance the applied heat. Welding of a member should progress from points where parts are relatively fixed in position toward points where parts have greater relative freedom of movement. Where it is impossible to avoid high residual stresses in the closing welds of a rigid assembly, these welds should be made in compression elements. Joints expected to have significant shrinkage should be welded before joints expected to have lesser shrinkage, and restraint should be kept to a minimum. If severe external restraint against shrinkage is present, welding should be carried continuously to completion or to a point that will ensure freedom from cracking before the joint is allowed to cool below the minimum specified preheat and interpass temperatures.

In shop fabrication of cover-plated beams and built-up members, each component requiring splices should be spliced before it is welded to other parts of the member. Up to three subsections may be spliced to form a long girder or girder section.

With too rapid cooling, cracks might form in a weld. Possible causes are shrinkage of weld and heat-affected zone, austenite-martensite transformation, and entrapped hydrogen. Preheating the base metal can eliminate the first two causes. Preheating reduces the temperature gradient between weld and adjacent base metal, thus decreasing the cooling rate and resulting stresses. Also, if hydrogen is present, preheating allows more time for this gas to escape. Use of low-hydrogen electrodes, with suitable moisture control, is also advantageous in controlling hydrogen content.

High cooling rates occur at arc strikes that do not deposit weld metal. Hence arc strikes outside the area of permanent welds should be avoided. Cracks or blemishes resulting from arc strikes should be ground to a smooth contour and checked for soundness.

To avoid cracks and for other reasons, standard specifications require that under certain conditions, before a weld is made the base metal must be preheated. Table 1.10 lists typical preheat and interpass temperatures. The table recognizes that as plate thickness, carbon content, or alloy content increases, higher preheats are necessary to lower cooling rates and to avoid microcracks or brittle heat-affected zones.

Preheating should bring to the specified preheat temperature the surface of the base metal within a distance equal to the thickness of

TABLE 1.10   Requirements of AWS D1.1 for Minimum Preheat and Interpass Temperatures, °F, for Welds in Buildings*

Thickness of thickest part at point of welding, in	Shielded metal-arc with other than low-hydrogen electrodes ASTM A36,† A53 grade B, A501, A529, A570 all grades	Shielded metal-arc with low-hydrogen electrodes; submerged-arc, gas metal-arc, or flux-cored arc ASTM A36, A53 grade B, A242, A441, A501, A529, A570 all grades A572 grades 42, 50, A588	Shielded metal-arc with low-hydrogen electrodes; submerged-arc, gas metal-arc, or flux-cored arc ASTM A572 grades 60 and 65	Shielded metal-arc with low-hydrogen electrodes; submerged-arc, with carbon or alloy steel wire neutral flux, gas metal-arc, or flux-cored arc ASTM A514
To ¾	0‡	0‡	50	50
Over ¾ to 1½	150	50	150	125
Over 1½ to 2½	225	150	225	175
Over 2½	300	225	300	225

*In joints involving combinations of base metals, preheat as specified for the higher-strength steel being welded.
†Use only low-hydrogen electrodes when welding A36 steel more than 1 in thick for dynamically loaded structures.
‡When the base-metal temperature is below 32°F, the base metal should be preheated to at least 70°F and the minimum temperature maintained during welding.

the part being welded, but not less than 3 in of the point of welding. This temperature should be maintained as a minimum interpass temperature while welding progresses.

Preheat and interpass temperatures should be sufficient to prevent crack formation. Temperatures above the minimums in Table 1.10 may be required for highly restrained welds.

For A514, A517, and A852 steels, the maximum preheat and interpass temperature should not exceed 400°F for thicknesses up to 1½ in, inclusive, and 450°F for greater thicknesses. Heat input during the welding of these quenched and tempered steels should not exceed the steel producer's recommendation. Use of stringer beads to avoid overheating is advisable.

Peening sometimes is used on intermediate weld layers for control of shrinkage stresses in thick welds to prevent cracking. It should be done with a round-nose tool and light blows from a power hammer after the weld has cooled to a temperature warm to the hand. The root or surface layer of the weld or the base metal at the edges of the weld should not be peened. Care should be taken to prevent scaling or flaking of weld and base metal from overpeening.

When required by plans and specifications, welded assemblies should be stress-relieved by heat treating. (See AWS D1.1 for temperatures and holding times required.) Finish machining should be done after stress relieving.

Tack and other temporary welds are subject to the same quality requirements as final welds. For tack welds, however, preheat is not mandatory for single-pass welds that are remelted and incorporated into continuous submerged-arc welds. Also, defects such as undercut, unfilled craters, and porosity need not be removed before final submerged-arc welding. Welds not incorporated into final welds should be removed after they have served their purpose, and the surface should be made flush with the original surface.

Before a weld is made over previously deposited weld metal, all slag should be removed, and the weld and adjacent material should be brushed clean.

Groove welds should be terminated at the ends of a joint in a manner that will ensure sound welds. Where possible, this should be done with the aid of weld tabs or runoff plates. AWS D1.1 does not require removal of weld tabs for statically loaded structures but does require it for dynamically loaded structures. The ends of the welds then should be made smooth and flush with the edges of the abutting parts.

After welds have been completed, slag should be removed from them. The metal should not be painted until all welded joints have been completed, inspected, and accepted. Before paint is applied, spatter, rust, loose scale, oil, and dirt should be removed.

AWS D1.1 presents details of techniques acceptable for welding buildings. These techniques include handling of electrodes and fluxes and maximum welding currents.

1.3.6 Weld quality

A basic requirement of all welds is thorough fusion of weld and base metal and of successive layers of weld metal. In addition, welds should not be handicapped by craters, undercutting, overlap, porosity, or cracks. (AWS D1.1 gives acceptable tolerances for these defects.) If craters, excessive concavity, or undersized welds occur in the effective length of a weld, they should be cleaned and filled to the full cross section of the weld. Generally, all undercutting (removal of base metal at the toe of a weld) should be repaired by depositing weld metal to restore the original surface. Overlap (a rolling over of the weld surface with lack of fusion at an edge), which may cause stress concentrations, and excessive convexity should be reduced by grinding away of excess material (see Figs. 1.13 and 1.14). If excessive porosity, excessive slag inclusions, or incomplete fusion occur, the defective portions should be removed and rewelded. If cracks are present, their extent should be determined by acid etching, magnetic-particle inspection, or other equally positive means. Not only the cracks but also sound metal 2 in beyond their ends should be removed and replaced with the weld metal. Use of a small electrode for this purpose reduces the chances of further defects due to shrinkage. An electrode not more than 5/32 in in diameter is desirable for depositing weld metal to compensate for size deficiencies.

AWS D1.1 limits convexity C to the values in Table 1.11.

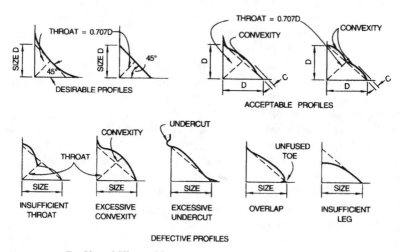

Figure 1.13 Profiles of fillet welds.

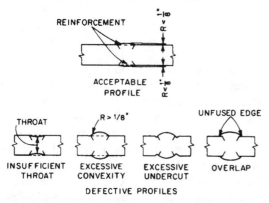

Figure 1.14 Profiles as groove welds.

TABLE 1.11   AWS D1.1 Limits on Convexity of Fillet Welds

Measured leg size or width of surface bead, in	Maximum convexity, in
$\frac{5}{16}$ or less	$\frac{1}{16}$
Over $\frac{5}{16}$ but less than 1	$\frac{1}{8}$
1 or more	$\frac{3}{16}$

Weld-quality requirements should depend on the job the welds are to do. Excessive requirements are uneconomical. Size, length, and penetration are always important for a stress-carrying weld and should completely meet design requirements. Undercutting, on the other hand, should not be permitted in main connections, such as those in trusses and bracing, but small amounts might be permitted in less important connections, such as those in platform framing for an industrial building. Type of electrode, similarly, is important for stress-carrying welds but not so critical for many miscellaneous welds. Again, poor appearance of a weld is objectionable if it indicates a bad weld or if the weld will be exposed where aesthetics is a design consideration, but for many types of structures, such as factories, warehouses, and incinerators, the appearance of a good weld is not critical. A sound weld is important, but a weld entirely free of porosity or small slag inclusions should be required only when the type of loading actually requires this perfection.

Welds may be inspected by one or more methods: visual inspection; nondestructive tests, such as ultrasonic, x-ray, dye penetration, and magnetic particle; and cutting of samples from finished welds. Designers should specify which welds are to be examined, extent of the examination, and methods to be used.

1.4   References

1. American Institute of Steel Construction, *Manual of Steel Construction,* 2d ed., vol. I, LRFD, Chicago, IL, 1994.
2. American Welding Society, *Structural Welding Code,* D1.1, Miami, FL, 1998.
3. Research Council on Structural Connections, "Specification for Structural Joints Using ASTM A325 or A490 Bolts," American Institute of Steel Construction, Chicago, IL, 1988.
4. Steel Structures Technology Center, *Structural Bolting Handbook,* Novi, MI, 1995.

Chapter

2

Designs of Connections for Axial, Moment, and Shear Forces

William A. Thornton
Cives Steel Company
Roswell, GA

Thomas Kane
Cives Steel Company
Roswell, GA

(Courtesy of The Steel Institute of New York.)

2.1 Introduction

Connection design is an interesting subject because it requires a great deal of rational analysis in arriving at a solution. There are literally an infinite number of possible connection configurations, and only a very small number of these have been subjected to physical testing. Even within the small group that has been tested, changes in load directions, geometry, material types, fastener type, and arrangement very quickly result in configurations which have not been tested and therefore require judgment and rational analysis on the part of the designer. This chapter provides design approaches to connections based on test data, when available, supplemented by rational design or art and science in the form of equilibrium (admissible force states), limit states, and ductility considerations. The limit states are those of the American Institute of Steel Construction [AISC LRFD *Manual of Steel Construction,* 2d ed. (1994)].

2.1.1 Philosophy

Connection design is both an art and a science. The science involves equilibrium, limit states, load paths, and the lower-bound theorem of limit analysis. The art involves the determination of the most efficient load paths for the connection and this is necessary because most connections are statically indeterminate.

The lower-bound theorem of limit analysis states: If a distribution of forces within a structure (or connection, which is a localized structure) can be found which is in equilibrium with the external load and which satisfies the limit states, then the externally applied load is less than or at most equal to the load which would cause connection failure. In other words, any solution for a connection which satisfies equilibrium and the limit states yields a safe connection. This is the science of *connection design.* The art involves finding the internal force distribution (or load paths) which maximizes the external load at which a connection fails. This maximized external load is also the true failure load when the internal force distribution results in satisfaction of compatibility (no gaps and tears) within the connection in addition to satisfying equilibrium and the limit states.

It should be noted that, strictly speaking, the lower-bound theorem applies only to yield limit states in structures which are ductile. Therefore, in applying it to connections, limit states involving stability and fracture (lack of ductility) must be considered to preclude these modes of failure.

2.1.2 General procedures

Determine the external (applied) loads, also called *required strengths,* and their lines of action. Make a preliminary layout, preferably to scale. The connection should be as compact as possible to conserve material and to minimize interferences with utilities, equipment, and access. Decide on where bolts and welds will be used and select bolt type and size. Decide on a load path through the connection. For a statically determinate connection, there is only one, but for indeterminate connections there are many possibilities. Use judgment, experience, and published information to arrive at the best load path. Now provide sufficient strength, stiffness, and ductility, using the limit states identified for each part of the load path, to give the connection sufficient design strength, that is, to make the connection adequate to carry the given loads. Complete the preliminary layout, check specification required spacings, and finally check to ensure that the connection can be fabricated and erected. The examples of this chapter will demonstrate this procedure.

2.1.3 Economic considerations

For any given connection situation, it is usually possible to arrive at more than one satisfactory solution. Where there is a possibility of using bolts or welds, let the economics of fabrication and erection play a role in the choice. Different fabricators and erectors in different parts of the country have their preferred ways of working, and as long as the principles of connection design are followed to achieve a safe connection, local preferences should be accepted. Some additional considerations which will result in more economical connections (Thornton, 1995*b*) are

1. For shear connections, provide the actual loads and allow the use of single-plate and single-angle shear connections. Do not specify full depth connections or rely on the AISC uniform load tables.

2. For moment connections, provide the actual moments and the actual shears. Also, provide a "breakdown" of the total moment, that is, give the gravity moment and lateral moment due to wind or seismic loads separately. This is needed to do a proper check for column web-doubler plates. If stiffeners are required, allow the use of fillet welds in place of complete-joint-penetration welds. To avoid the use of stiffeners, consider redesigning with a heavier column to eliminate them.

3. For bracing connections, in addition to providing the brace force, also provide the beam shear and axial transfer force. The transfer force is the axial force that must be transferred to the opposite side of the column. The transfer force is not necessarily the beam axial force that is obtained from a computer analysis of the structure. See Thornton (1995*b*) for a discussion of this. A misunderstanding of transfer forces can lead to both uneconomic and unsafe connections.

2.1.4   Types of connections

There are three basic forces to which connections are subjected. These are axial force, shear force, and moment. Many connections are subject to two or more of these simultaneously. Connections are usually classified according to the major load type to be carried, such as shear connections, which carry primarily shear, moment connections, which carry primarily moment, and axial force connections, such as splices, bracing and truss connections, hangers, etc., which carry primarily axial force. Subsequent sections of this chapter will deal with these three basic types of connections.

2.1.5   Organization

This chapter will cover axial force connections first, then moment connections, and lastly, shear connections. This is done to emphasize the ideas of load paths, limit states, and the lower-bound theorem, which (except for limit states) are less obviously necessary to consider for the simpler connections.

This chapter is based on the limit states of the AISC LRFD Specification (AISC, 1994). The determination of loads, that is, required strengths, is dependent upon the specific building code required for the project, based on location, local laws, etc. At present (1998), much transition is taking place in the determination of seismic loads and connection requirements. Wherever examples involving seismic loads are presented in this chapter, the solutions presented are indicative of the authors' experience in current practice with many structural engineers, and may need to be supplemented with additional requirements from local seismic codes. Chapter 5 deals with connections in high seismic regions and covers these additional requirements.

2.2 Axial Force Connections

2.2.1 Bracing connections

2.2.1.1 Introduction. The lateral force-resisting system in buildings may consist of a vertical truss. This is referred to as a *braced frame* and the connections of the diagonal braces to the beams and columns are the *bracing connections*. Figure 2.1 shows various bracing arrangements. For the bracing system to be a true truss, the bracing connections should be concentric, that is, the gravity axes of all members at any joint should intersect at a single point. If the gravity axes are not concentric, the resulting couples must be considered in the design of the members. The examples of this section will be of concentric type, but the nonconcentric type can also be handled as will be shown.

2.2.1.2 Example 1. Consider the bracing connection of Fig. 2.2. The brace load is 855 kips, the beam shear is 10 kips, and the beam axial force is 411 kips. The horizontal component of the brace force is 627 kips, which means that $627 - 411 = 216$ kips is transferred to the opposite side of the column from the brace side. There must be a connection on this side to "pick up" this load, that is, provide a load path.

The design of this connection involves the design of four separate connections. These are: (1) the brace-to-gusset connection, (2) the gusset-to-column connection, (3) the gusset-to-beam connection, and (4) the beam-to-column connection. A fifth connection is the connection on the other side of the column, which will not be considered here.

1. *Brace-to-gusset:* This part of the connection is designed first because it provides a minimum size for the gusset plate which is then used to design the gusset-to-column and gusset-to-beam connections. Providing an adequate load path involves the following limit states:
   a. *Bolts (A325SC-B-N 1⅛-in-diameter standard holes):* The preceding notation indicates that the bolts are slip-critical, the surface class is B, and threads are not excluded from the shear planes. The slip-critical design strength per bolt is

   $$\phi r_{str} = 1 \times 1.13 \times 0.5 \times 56 = 31.6 \text{ kips}$$

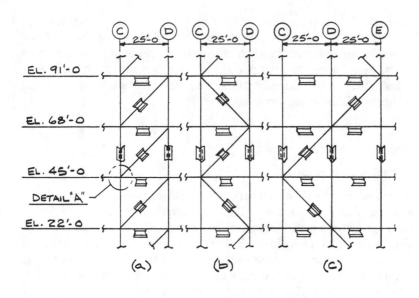

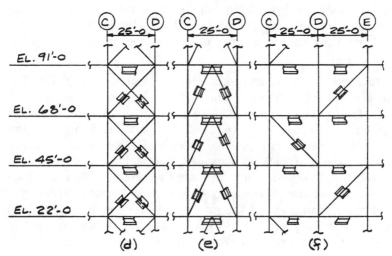

Figure 2.1 Various vertical bracing arrangements.

The bearing design strength is

$$\phi r_v = 0.75 \times \frac{\pi}{4} \times 1.125^2 \times 48 = 35.8 \text{ kips}$$

Since $31.6 < 35.8$, use 31.6 kips as the design strength. The number of bolts required is $855/(31.6 \times 2) = 13.5$. Therefore, use 14 bolts on each side of the connection.

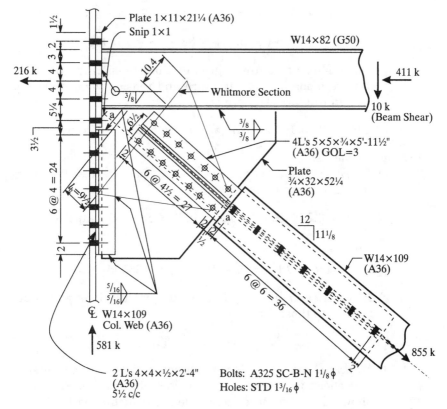

Figure 2.2 Example 1, bracing connection design.

b. *Brace checks:*

   (1) *Bearing:* Because the edge distance $L_e = 2 > 1.5 \times 1.125 = 1.6875$ and the spacing $s = 4.5 > 3 \times 1.125 = 3.375$, the bearing strength per bolt is

   $$\phi r_p = 0.75 \times 2.4 \times 1.125 \times 0.525 \times 58 = 61.7 \text{ kips}$$

   Thus, the design strength of 14 bolts is

   $$\phi r_p = 14 \times 61.7 = 864 \text{ kips} > 855, \text{ ok}$$

   (2) *Block shear rupture:*

   $$A_{gv} = (2 + 6 \times 6) \times 0.525 \times 2 = 39.9 \text{ in}^2$$
   $$A_{gt} = 6.75 \times 0.525 = 3.54 \text{ in}^2$$
   $$A_{nv} = 39.9 - 6.5 \times 1.25 \times 0.525 \times 2 = 31.4 \text{ in}^2$$
   $$A_{nt} = 3.54 - 1 \times 1.25 \times 0.525 = 2.88 \text{ in}^2$$

Shear fracture = $31.4 \times 0.6 \times 58 = 1093$ kips

Tension fracture = $2.88 \times 58 = 167$ kips

Since shear fracture is greater than tension fracture, the failure mode is shear fracture and tension yield; thus, the design fracture strength is

$\phi R_{bs} = 0.75(1093 + 3.54 \times 36) = 915$ kips > 855 kips, ok

c. *Gusset checks:*
(1) *Bearing:*   Again, edge distance and spacing exceed the minimums required, so

$\phi r_p = 0.75 \times 2.4 \times 1.125 \times 0.75 \times 58 = 88.1$ kips/bolt

and

$\phi R_p = 14 \times 88.1 = 1233$ kips > 855 kips, ok

(2) *Block shear rupture:*   Performing calculations which are similar to those for the brace

$\phi R_{bs} = 1109 > 855$ kips, ok

(3) *Whitmore section:*   Since the brace load can be compression, this check is used to check for gusset buckling. Figure 2.2 shows the "Whitmore section" length which is normally $l_w = (27 \tan 30) \times 2 + 6.5 = 37.7$ in, but the section passes out of the gusset and into the beam web at its upper side. Because of the fillet weld of the gusset to the beam flange, this part of the Whitmore section is not ineffective, that is, load can be passed through the weld to be carried on this part of the Whitmore section. The effective length of the Whitmore section is thus:

$$l_{we} = (37.7 - 10.4) + 10.4 \times \frac{0.510}{0.75} \times \frac{50}{36} = 27.3 + 10.1 = 37.1 \text{ in}$$

The gusset buckling length is, from Fig. 2.1, $l_b = 9.5$ in, and the slenderness ratio is

$$\frac{Kl_b}{r} = \frac{0.5 \times 9.5 \times \sqrt{12}}{0.75} = 21.9$$

In this formula the theoretical fixed-fixed factor of 0.5 is used rather than the usually recommended value of 0.65 for columns, because of the conservatism of this buckling check as determined by Gross (1990) from full-scale tests. From the AISC LRFD Specification, Table 3-36,

$$\frac{Kl_b}{r} = 21.9$$

the design buckling strength is

$$\phi F_{cr} = 29.84 \text{ ksi}$$

and the Whitmore section buckling strength is thus

$$\phi R_{wb} = 29.84 \times 37.1 \times 0.75 = 831 \text{ kips} < 855 \text{ kips}$$

The design buckling load of 831 kips is slightly less (2.9%) than the required strength, but this is deemed acceptable because of the very conservative nature of the buckling check.

d. *Brace-to-gusset connection angles:*
   (1) *Gross and net area:* The gross area required is

$$\frac{8.55}{0.9 \times 36} = 26.4 \text{ in}^2$$

Try $4Ls$ $5 \times 5 \times \frac{3}{4}$, $A_{gt} = 6.94 \times 4 = 27.8$ in^2, ok
The net area is

$$A_{nt} = 27.8 - 4 \times 0.75 \times 1.25 = 24.1 \text{ in}^2$$

The effective net area is the lesser of $0.85A_{gt}$ or UA_{nt}, where $U = 1 - 1.52/27 = 0.94$—use $U = 0.9$. Thus $0.85A_{gt} = 0.85 \times 27.8 = 23.6$ and $UA_{nt} = 0.9 \times 24.1 = 21.7$ and then $A_e = 21.7$. Therefore, the net tensile design strength is $\phi R_t = 0.75 \times 58 \times 21.7 = 944$ kips > 855 kips, ok.
   (2) *Bearing:*

$$\phi R_p = 0.75 \times 2.4 \times 58 \times 0.75 \times 1.125 \times 14$$
$$= 1233k > 855 \text{ kips, ok}$$

   (3) *Block shear rupture (tearout):* The length of the connection on the gusset side is the shorter of the two and is, therefore, the more critical. Per angle,

$$A_{nv} = (29 - 6.5 \times 1.25) \times 0.75 = 15.66 \text{ in}^2$$
$$A_{nt} = (2 - 0.5 \times 1.25) \times 0.75 = 1.03 \text{ in}^2$$

Since $0.6F_u A_{nv} > F_u A_{nt}$

$$\phi R_{bs} = 0.75(0.6 \times 58 \times 15.66 + 36 \times 2 \times 0.75) 4$$
$$= 1800 \text{ kips} > 855 \text{ kips, ok}$$

This completes the design checks for the brace-to-gusset connection. All elements of the load path, which consists of the bolts, the brace web, the gusset, and the connection angles, have been checked. The remaining connection interfaces require a method to determine the forces on them. Research (Thornton, 1991, 1995b) has shown that the best method for doing this is the uniform force method (UFM). The force distributions for this method are shown in Fig. 2.3.

From the design of the brace-to-gusset connection, a certain minimum size of gusset is required. This is the gusset shown in Fig. 2.2. Usually, this gusset size, which is a preliminary size, is sufficient for the final design. From Figs. 2.2 and 2.3, the basic data are

$$\tan \theta = \frac{12}{11.125} = 1.079$$

$$e_B = \frac{14.31}{2} = 7.155$$

$$e_c = 0$$

The quantities α and β locate the centroids of the gusset edge connections, and in order for no couples to exist on these connections, α and β must satisfy the following relationship given in Fig. 2.3b,

$$\alpha - \beta \tan \theta = e_B \tan \theta - e_C$$

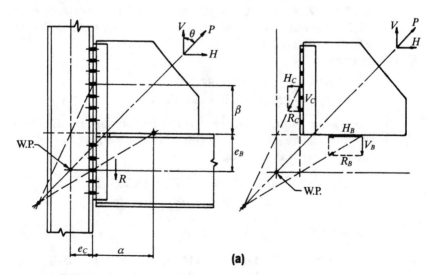

(a)

Figure 2.3a The uniform force method.

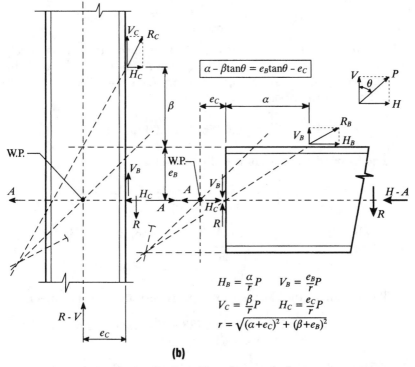

(b)

Figure 2.3b   Force distribution for the uniform force method.

Thus, $\alpha - 1.079\beta = 7.155 \times 1.079 - 0 = 7.720$.

From the geometry given in Fig. 2.2, a seven-row connection at 4-in pitch will give $\beta = 17.5$ in. Then $\alpha = 7.720 + 1.079 \times 17.5 = 26.6$ in and the horizontal length of the gusset is $(26.6 - 1) \times 2 + 1 = 52.2$ in. Choose a gusset length of 52¼ in. With $\alpha = 26.6$ and $\beta = 17.5$,

$$r = \sqrt{(\alpha + e_C)^2 + (\beta + e_B)^2}$$

$$= \sqrt{(26.6 + 0)^2 + (17.5 + 7.155)^2} = 36.27$$

and

$$V_C = \frac{\beta}{r}\,P = \frac{17.5}{36.27} \times 855 = 413 \text{ kips}$$

$$H_C = \frac{e_C}{r}\,P = 0 \text{ kips}$$

$$H_B = \frac{\alpha}{r}\,P = \frac{26.6}{36.27} \times 855 = 627 \text{ kips}$$

$$V_B = \frac{e_B}{r}\,P = \frac{7.155}{36.27} \times 855 = 169 \text{ kips}$$

2. *Gusset-to-column:* The loads are 413 kips shear and 0 kips axial.
   a. *Bolts and clip angles:* The bolts are A325SC-B-N 1⅛ φ, stan-
      dard holes and the clip angles are: try Ls 4 × 4 × ½.
      The shear per bolt is

$$V = \frac{413}{14} = 29.5 \text{ kips} < 31.6 \text{ kips, ok}$$

The bearing strength of the clip angles is

$$\phi R_p = 0.75 \times 2.4 \times 58 \times 0.5 \times 1.125 \times 14$$
$$= 822 \text{ kips} > 413 \text{ kips, ok}$$

The bearing strength of the W 14 × 109 column web is

$$\phi R_p = 822 \times \frac{0.525}{0.5} = 863 \text{ kips} > 413 \text{ kips, ok}$$

The net shear (or block shear rupture) strength of the clips is

$$\phi R_{bs} = 0.75 \times 0.6 \times 58\,(28 - 7 \times 1.25) \times 0.5 \times 2$$
$$= 502 \text{ kips} > 413 \text{ kips, ok}$$

   b. *Fillet weld of clip angles to gusset:* The length of this clip angle
      weld is 28 in. From the AISC LRFD Manual, vol. II, Table 8-42, $l =$
      28, $kl = 3.0$, $k = 0.107$, $al = 4 - xl = 4 - 0.009 \times 28 = 3.75$ and $a =$
      0.134. By interpolation, $c = 1.72$, and the required fillet weld size is
      $D = 413/(1.72 \times 28 \times 2) = 4.29$, so the required fillet weld size is ⁵⁄₁₆,
      and no proration is required because of the ¾-in-thick gusset.
3. *Gusset-to-beam:* The loads are 627 kips shear and 169 kips axial.
   a. *Gusset stresses:*

$$f_v = \frac{627}{0.75 \times 51.25} = 16.3 \text{ ksi} < 0.9 \times 0.6 \times 36 = 19.4 \text{ ksi, ok}$$

$$f_a = \frac{169}{0.75 \times 51.25} = 4.40 \text{ ksi} < 0.9 \times 36 = 32.4 \text{ ksi, ok}$$

   b. *Weld of gusset to beam bottom flange:* The resultant force per
      inch of weld is

$$f_r = \sqrt{16.3^2 + 4.40^2} \times \frac{0.75}{2} = 6.33 \text{ kips/in}$$

The required weld size is

$$D = \frac{6.33}{1.392} \times 1.4 = 6.37$$

which indicates that a $\frac{7}{16}$-in fillet weld is required. The factor 1.4 is a ductility factor from the work of Richard (1986). Even though the stress in this weld is calculated as being uniform, it is well known that there will be local peak stresses, especially in the area where the brace-to-gusset connection comes close to the gusset-to-beam weld. An indication of high stress in this area is also indicated by the Whitmore section cutting into the beam web. Also, as discussed later, frame action will give rise to distortion forces which modify the force distribution given by the UFM.

As an alternate method to determine the size of this weld, LRFD, Table 8.38, can be used. The resultant force $P_u = \sqrt{627^2 + 169^2} = 649$ kips and the angle between the resultant and the weld axis is $\tan^{-1}(169/627) = 15.1°$. From Table 8.38, for $15°$ inclination of load, and $k = a = 0.0$, $c = 2.97$. Thus $D = (649/2.97 \times 51.25)1.4 = 5.97$, or a $\frac{3}{8}$-in fillet weld is required. Figure 2.2 shows the $\frac{3}{8}$-in fillet weld.

c. *Checks on the beam web:* The 627-kip shear is passed into the beam through the gusset-to-beam weld. All of this load is ultimately distributed over the full cross section of the W 14×82, 411 kips pass to the right and 216 kips are transferred across the column. The length of web required to transmit 627 kips of shear is l_{web}, where $627 = 0.9 \times 0.6 \times 50 \times 0.510 x l_{web}$. Thus

$$l_{web} = \frac{627}{0.9 \times 0.6 \times 50 \times 0.510} = 45.5 \text{ in}$$

which is reasonable. Note that this length can be longer than the gusset-to-beam weld, but probably should not exceed about half the beam span.

The vertical component can cause beam web yielding and crippling.

(1) *Web yielding:* The web yield design strength is

$$\phi R_{wy} = 1 \times 0.66 \times 50(51.25 + 2.5 \times 1.625)$$
$$= 1825 \text{ kips} > 169 \text{ kips, ok}$$

(2) *Web crippling:* The web crippling design strength is

$$\phi R_{wcp} = 0.75 \times 135 t_w^2 \left[1 + 3\left(\frac{N}{d}\right)\left(\frac{t_w}{t_f}\right)^{1.5} \right] \sqrt{\frac{F_y t_f}{t_w}}$$

$$= 0.75 \times 135 \times 0.510^2 \left[1 + 3\left(\frac{51.25}{14.31}\right)\left(\frac{0.510}{0.855}\right)^{1.5} \right]$$

$$\times \sqrt{\frac{50 \times 0.855}{0.510}}$$

$$= 1435 \text{ kips} > 169 \text{ kips, ok}$$

The preceding two checks on the beam web seldom control but should be checked "just in case." The web crippling
formula used is that for locations not near the beam end
because the beam-to-column connection will effectively prevent crippling near the beam end. The physical situation is
closer to that at some distance from the beam end rather
than that at the beam end.

4. *Beam-to-column:*   The loads are 216 kips axial, the specified
transfer force and a shear which is the sum of the nominal minimum beam shear of 10 kips and the vertical force from the gusset-
to-beam connection of 169 kips. Thus, the total shear is 10 + 169 =
179 kips.

a. *Bolts and end plate:*   As established earlier in this example, the
bolt design strength in shear is $\phi r_{str} = 31.6$ kips. In this connection, since the bolts also see a tensile load, there is an interaction between tension and shear which must be satisfied. If V is
the factored shear per bolt, the design tensile strength is

$$\phi r_t' = 1.13 T_b\left(1 - \frac{V}{\phi r_{str}}\right) \le 0.75 \times 90 A_b$$

where T_b is the bolt pretension of 56 kips for A325 1⅛-in-diame-
ter bolts and A_b is the bolt nominal area = $\pi/4 \times 1.125^2 = 0.994$
in^2.

For $V = 179/10 = 17.9$ kips < 31.6 kips (ok),

$$\phi r_t' = 1.13 \times 56\left(1 - \frac{17.9}{31.6}\right) = 27.4 \text{ kips}$$

and $0.75 \times 90 \times 0.994 = 67.1$ kips

Thus $\phi r_t' = 27.4$ kips and $\phi R_t' = 10 \times 27.4 = 274$ kips > 216
kips, ok. Checking the interaction for an N-type bearing connection

$$\phi r_t' = \phi(117 A_b - 1.9V) \le \phi(90) A_b$$

Thus $\phi r_t' = 0.75(117 \times 0.994 - 1.9 \times 17.9) = 61.7$ kips and
$0.75 \times 90 \times 0.994 = 67.1$ kips, so $\phi r_t' = 61.7$ kips > 27.4 kips
means that bearing does not control.

To determine the end-plate thickness required, the critical
dimension is the distance, b, from the face of the beam web to
the center of the bolts. For 5½-in cross centers $b = (5.5 -
0.510)/2 = 2.5$ in. To make the bolts above and below the flanges
approximately equally critical, they should be placed no more
than 2½ in above and below the flanges. Figure 2.2 shows them

placed at 2 in. Let the end plate be 11 in wide. Then $a = (11 - 5.5)/2 = 2.75 < 1.25 \times 2.5 = 3.125$, ok. The edge distance at the top and bottom of the end plate is 1.5 in, which is more critical than 2.75 in, and will be used in the following calculations. The notation for a and b follows that of the AISC LRFD *Manual of Steel Construction* (1994) as does the remainder of this procedure.

$$b' = b - \frac{d}{2} = 2.5 - \frac{1.125}{2} = 1.9375$$

$$a' = a + \frac{d}{2} = 1.5 + \frac{1.125}{2} = 2.0625$$

$$\rho = \frac{b'}{a'} = 0.94$$

$$\beta = \frac{1}{\rho}\left(\frac{\phi r_t'}{T} - 1\right)$$

where T = factored tension per bolt = 21.6 kips

$$\beta = \frac{1}{0.94}\left(\frac{27.4}{21.6} - 1\right) = 0.286$$

$$\alpha' = \min\left[\frac{1}{\delta}\left(\frac{\beta}{1 - \beta}\right), 1\right]$$

where $\delta = 1 - d'/p = 1 - 1.1875/4 = 0.70$.

In the preceding expression p is the tributary length of end plate per bolt. For the bolts adjacent to the beam web, this is obviously 4 in. For the bolts adjacent to the flanges, it is also approximately 4 in for p since at $b = 2.0$ in, a $45°$ spread from the center of the bolt gives $p = 4$ in. Note also that p cannot exceed one-half of the width of the end plate. Thus

$$\alpha' = \min\left[\frac{1}{0.70}\left(\frac{0.286}{1 - 0.286}\right), 1\right] = 0.572$$

The required end plate thickness is

$$t_{\text{req'd}} = \sqrt{\frac{4.44Tb'}{pF_y(1 + \delta a')}} = \sqrt{\frac{4.44 \times 21.6 \times 1.9375}{4.0 \times 36 \times (1 + 0.70 \times 0.572)}}$$
$$= 0.960 \text{ in}$$

Use a 1-in end plate, 11 in wide and $14\frac{1}{4} + 2 + 2 + 1\frac{1}{2} + 1\frac{1}{2} = 21.25$ in long.

b. Weld of beam to end plate: All of the shear of 179 kips exists in the beam web before it is transferred to the end plate by the weld of the beam to the end plate. The shear capacity of the beam web is

$$\phi R_v = 0.9 \times 0.6 \times 50 \times 0.510 \times 14.31 = 197 \text{ kips} > 179 \text{ kips, ok}$$

The weld to the end plate that carries this shear is the weld to the beam web plus the weld around to about the k_1 distance inside the beam profile and $2k_1$ on the outside of the flanges. This length is thus

$$2(d - 2t_f) + 4\left(k_1 - \frac{t_w}{2}\right) + 4k_1 = 2(14.31 - 2 \times 0.855)$$
$$+ 4\left(1 - \frac{0.510}{2}\right) + 4 \times 1 = 32.2 \text{ in}$$

The force in this weld per inch due to shear is

$$f_v = \frac{179}{32.2} = 5.56 \text{ kips/in}$$

The length of weld that carries the axial force of 216 kips is the entire profile weld whose length is $4 \times 10.13 - 2 \times 0.510 + 2 \times 14.31 = 68.1$ in. The force in this weld per inch due to axial force is

$$f_a = \frac{216}{68.1} = 3.17 \text{ kips/in}$$

Also, where the bolts are close together, a "hot spot" stress should be checked. The most critical bolt in this regard is the one at the center of the W 14×82. The axial force in the weld local to these bolts is

$$f_a' = \frac{2 \times 21.6}{8} = 5.4 \text{ kips/in}$$

The controlling resultant force in the weld is thus

$$f_R = \sqrt{5.56^2 + 5.4^2} = 7.75 \text{ kips/in}$$

and the required weld size is

$$D = \frac{7.75}{1.392} = 5.57 \qquad \text{or} \qquad \frac{3}{8} \text{ fillet weld}$$

As a final check, make sure that the beam web can deliver the axial force to the bolts. The tensile load for 2 bolts is 2 × 21.6 = 43.2 kips, and 4 in of the beam web must be capable of delivering this load, that is, providing a load path. The tensile capacity of 4 in of beam web is 4 × 0.510 × 0.90 × 50 = 91.8 kips > 43.2 kips, ok.

2.2.1.3 Some observations on the design of gusset plates. It is a tenet of all gusset-plate design that it must be able to be shown that the stresses on any cut section of the gusset do not exceed the yield stresses on this section. Now, once the resultant forces on the gusset horizontal and vertical sections are calculated by the UFM, the resultant forces on any other cut section, such as section a-a of Fig. 2.2, are easy to calculate (see the appropriate free body diagram incorporating this section, as shown in Fig. 2.4, where the resultant forces on section a-a are shown), but the determination of the stresses is not. The traditional approach to the determination of stresses, as mentioned in many books (Blodget, 1966; Gaylord and Gaylord, 1972; and Kulak, Fisher, and Struik, 1987), and papers (Whitmore, 1952, and Vasarhelyi, 1971) is to use the formulas intended for long slender members, that is, $f_a = P/A$ for axial stress, $f_b = Mc/I$ for bending stress, and $f_v = V/A$ for shear stress. It is well known that these are not correct for gusset plates (Timoshenko, 1970). They are recommended only because there is seemingly no alternative. Actually, the uniform force method, coupled with the Whitmore section

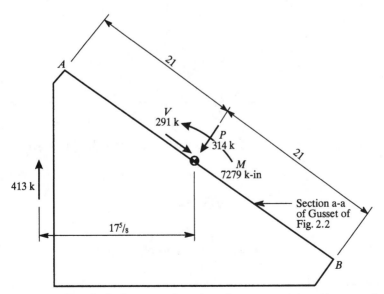

Figure 2.4 Free-body diagram of portion of gusset cut at section a-a of Fig. 2.2.

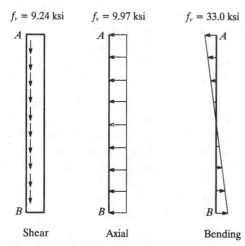

$f_v = 9.24$ ksi $f_v = 9.97$ ksi $f_v = 33.0$ ksi

A   A   A

B   B   B

Shear   Axial   Bending

Figure 2.5 Traditional cut section stresses.

and the block shear fracture (tearout) limit state, is an alternative as will be shown subsequently.

Applying the slender member formulas to the section and forces of Fig. 2.4, the stresses and stress distribution of Fig. 2.5 result. The stresses are calculated as
shear:

$$f_v = \frac{291}{0.75 \times 42} = 9.24 \text{ ksi}$$

axial:

$$f_a = \frac{314}{0.75 \times 42} = 9.97 \text{ ksi}$$

bending:

$$f_b = \frac{7279 \times 6}{0.75 \times 42^2} = 33.0 \text{ ksi}$$

These are the basic "elastic"* stress distributions. The peak stress occurs at point A and is
shear:

$$f_v = 9.24 \text{ ksi}$$

normal:

$$f_a + f_b = 9.97 + 33.0 = 43.0 \text{ ksi}$$

---

*Actually the shear stress is not elastic because it is assumed uniform. The slender beam theory elastic shear stress would have a parabolic distribution with a peak stress of $9.24 \times 1.5 = 13.9$ ksi at the center of the section.

The shear yield stress (design strength) is $\phi F_v = \phi(0.6F_y) = 0.9(0.6 \times 36) = 19.4$ ksi. Since $9.24 < 19.4$, the section has not yielded in shear. The normal yield stress (design strength) is $\phi F_n = \phi F_y = 0.9(36) = 32.4$ ksi. Since $43.0 > 32.4$, the yield strength has been exceeded at point A. At this point, it appears that the design is unsatisfactory (that is, not meeting AISC requirements). But consider that the normal stress exceeds yield over only about 11 in of the 42-in-long section starting from point A. The remaining $42 - 11 = 31$ in have not yet yielded. This means that failure has not occurred because the elastic portion of the section will constrain unbounded yield deformations, that is, the deformation is "self-limited." Also, the stress of 43.0 ksi is totally artificial! It cannot be achieved in an elastic–perfectly plastic material with a design yield point of 32.4 ksi. What *will* happen is that when the design yield point of 32.4 ksi is reached, the stresses on the section will redistribute until the design yield point is reached at *every* point of the cross section. At this time, the plate will fail by unrestrained yielding if the applied loads are such that higher stresses are required for equilibrium.

To conclude, on the basis of 43.0 ksi at point A, that the plate has failed is thus false. What must be done is to see if a redistributed stress state on the section can be achieved which nowhere exceeds the design yield stress. Note that if this can be achieved, all AISC requirements will have been satisfied. AISC specifies that the design yield stress shall not be exceeded, but does *not* specify the formulas used to determine this.

The shear stress, f_v, and the axial stress, f_a, are already assumed uniform. Only the bending stress, f_b, is nonuniform. To achieve simultaneous yield over the entire section, the bending stress must be adjusted so that when combined with the axial stress, a uniform normal stress is achieved. To this end, consider Fig. 2.6. Here the bend-

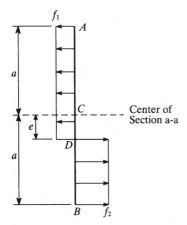

Figure 2.6 Admissible bending stress distribution of section a-a.

ing stress is assumed uniform but of different magnitudes over the upper and lower parts of the section. Note that this can be done because **M** in Fig. 2.4, although shown at the centroid of the section, is actually a free vector which can be applied anywhere on the section or indeed anywhere on the free body diagram. This being the case, there is no reason to assume that the bending stress distribution is symmetrical about the center of the section. Considering the distribution shown in Fig. 2.6, because the stress from A to the center is too high, the zero point of the distribution can be allowed to move down the amount e toward B. Equating the couple **M** of Fig. 2.4 to the statically equivalent stress distribution of Fig. 2.6 and taking moments about point D,

$$\mathbf{M} = \frac{t}{2}\left[f_1(a + e)^2 + f_2(a - e)^2\right]$$

where t is the gusset thickness. Also, from equilibrium

$$f_1\,(a + e)t = f_2(a - e)t$$

The preceding two equations permit a solution for f_1 and f_2 as

$$f_1 = \frac{\mathbf{M}}{at(a + e)}$$

$$f_2 = \frac{\mathbf{M}}{at(a - e)}$$

For a uniform distribution of normal stress,

$$f_1 + f_a = f_2 - f_a$$

from which e can be obtained as

$$e = \frac{1}{2}\left[\sqrt{\left(\frac{\mathbf{M}}{at\,f_a}\right)^2 + 4a^2} - \frac{\mathbf{M}}{at\,f_a}\right]$$

Substituting numerical values,

$$e = \frac{1}{2}\left[\sqrt{\left(\frac{7279}{(21)(0.75)(9.97)}\right)^2 + 4(21)^2} - \frac{7279}{(21)(0.75)(9.97)}\right] = 8.10 \text{ in}$$

Thus,

$$f_1 = \frac{7279}{(21)(0.75)(21 + 8.10)} = 15.9 \text{ ksi}$$

$$f_2 = \frac{7279}{(21)(0.75)(21 - 8.10)} = 35.8 \text{ ksi}$$

and the normal stress at point A is

$$f_{n_A} = f_1 + f_a = 15.9 + 9.97 = 25.9 \text{ ksi}$$

and at point B

$$f_{n_B} = f_2 - f_a = 35.8 - 9.97 = 25.9 \text{ ksi}$$

Now the entire section is uniformly stressed. Since

$$f_v = 9.24 \text{ ksi} < 19.4 \text{ ksi}$$
$$f_n = 25.9 \text{ ksi} < 32.4 \text{ ksi}$$

at all points of the section, the design yield stress is nowhere exceeded and the connection is satisfactory.

It was stated previously that there is an alternative to the use of the inappropriate slender-beam formulas for the analysis and design of gusset plates. The preceding analysis of the special section a-a demonstrates the alternative which results in a true limit state (failure mode or mechanism) rather than the fictitious calculation of "hot spot" point stresses which, since their associated deformation is totally limited by the remaining elastic portions of the section, cannot correspond to a true failure mode or limit state. The uniform force method performs exactly the same analysis on the gusset horizontal and vertical edges and on the associated beam-to-column connection. It is capable of producing forces on all interfaces which give rise to uniform stresses. Each interface is designed to just fail under these uniform stresses. Therefore, true limit states are achieved at every interface. For this reason, the uniform force method achieves a good approximation to the greatest lower-bound solution (closest to the true collapse solution) in accordance with the lower-bound theorem of limit analysis.

The uniform force method is a complete departure from the so-called traditional approach to gusset analysis using slender-beam theory formulas. It has been validated against all known full-scale gusseted-bracing connection tests (Thornton, 1991, 1995b). It does not require the checking of gusset sections such as that studied in this section (section a-a of Fig. 2.4). The analysis at this section was done to prove a point. But the uniform force method does include a check in the brace-to-gusset part of the calculation which is closely related to the special section a-a of Fig. 2.4. This is the shear rupture or

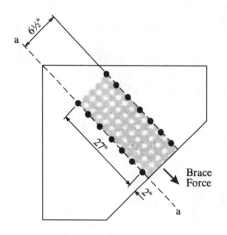

Brace
Force

Figure 2.7 Block shear rupture and its relation to gussed section a-a.

"tearout" section of Fig. 2.7 (Hardash and Bjorhovde, 1985, and Richard et al., 1983) which is included in section J4 of the AISC LRFD Specification (AISC, 1994). Following the notation of the AISC LRFD commentary, the tearout capacity is calculated as follows:

The gross shear area is

$$A_{vg} = (27 + 2.0) \times 0.75 \times 2 = 43.5 \text{ in}^2$$

The net shear area is

$$A_{vn} = 43.5 - 6.5 \times 1.25 \times 0.75 \times 2 = 31.3 \text{ in}^2$$

The gross tension area is

$$A_{tg} = 6.5 \times 0.75 = 4.875 \text{ in}^2$$

The net tension area is

$$A_{tn} = 4.875 - 1 \times 1.25 \times 0.75 = 3.9375 \text{ in}^2$$

The shear fracture design strength is

$$\phi F_{vf} = \phi(0.6 F_u \times A_{vn}) = 0.75 \times 0.6 \times 58 \times 31.3 = 817 \text{ kips}$$

The tension fracture design strength is

$$\phi F_{tf} = \phi(F_u \times A_{tn}) = 0.75 \times 58 \times 3.9375 = 171 \text{ kips}$$

The controlling limit state is the one with the larger fracture component.

Therefore, associated with the shear fracture, calculate tension yield as

$$\phi F_{ty} = \phi(F_y \times A_{tg}) = 0.75 \times 36 \times 4.875 = 132 \text{ kips}$$

The shear fracture (tearout) design strength is thus

$$\phi F_{to} = 817 + 132 = 949 \text{ kips}$$

Since 949 kips > 855 kips, the gusset plate is satisfactory for tearout.

Comparing the tearout limit state to the special section a-a limit state, a reserve capacity in tearout = $(949 - 855)/949 \times 100 = 9.9\%$ is found, and the reserve capacity of the special section = $(32.4 - 25.9)/32.4 \times 100 = 20\%$, which shows that tearout gives a conservative prediction of the capacity of the closely related special section.

A second check on the gusset performed as part of the UFM is the Whitmore section check. From Fig. 2.2, the Whitmore section area is

$$A_w = (37.7 - 10.4) \times 0.75 + 10.4 \times 0.510 \times \frac{50}{36} = 27.8 \text{ in}^2$$

The Whitmore section design strength in tension is

$$\phi F_w = \phi(F_y \times A_w) = 0.9(36 \times 27.8) = 901 \text{ kips}$$

The reserve capacity of the Whitmore section in tension is $(901 - 855)/901 \times 100 = 5\%$ which again gives a conservative prediction of capacity when compared to the special section a-a.

With these two limit states, block shear rupture (tearout) and Whitmore, the special section limit state is closely bounded and rendered unnecessary. The routine calculations associated with tearout and Whitmore are sufficient in practice to eliminate the consideration of any sections other than the gusset-to-column and gusset-to-beam sections.

2.2.1.4 Example 2. This connection, shown in Fig. 2.8, occurs at point A of Fig. 2.1. The member on the right of the joint is a "collector" that adds load to the bracing truss. The design in this example is for seismic loads. The brace consists of 2MC12 × 45s with toes 1½ in apart. The gusset thickness is thus chosen to be 1½ in and is then checked. The completed design is shown in Fig. 2.8. In this case, because of the high specified beam shear of 170 kips, it is proposed to use a special case of the uniform force method which sets the vertical component of the load between the gusset and the beam, V_B, to zero. Figure 2.9 shows the resultant force distribution. This method is called "special case 2" of the uniform force method and is discussed in the AISC books (AISC, 1992, and AISC, 1994).

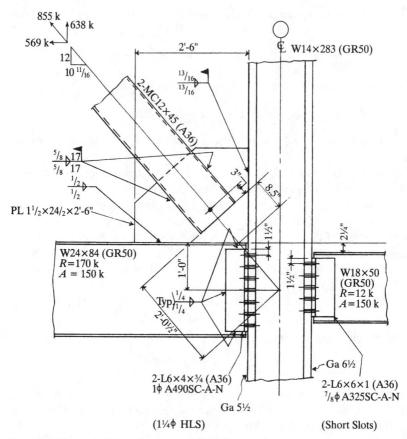

Figure 2.8 Example 2, bracing connection design.

1. *Brace-to-gusset connection:*
   a. *Weld:* The brace is field-welded to the gusset with fillet welds. Because of architectural constraints, the gusset size is to be kept to 30 in horizontally and 24½ in vertically. From the geometry of the gusset and brace, about 17 in of fillet weld can be accommodated. The weld size is

$$D = \frac{855}{4 \times 17 \times 1.392} = 9.03$$

   A ⅝-in fillet weld is indicated, but the flange of the $MC12 \times 45$ must be checked to see if an adequate load path exists. The average thickness of 0.700 in occurs at the center of the flange which is 4.012 in wide. The thickness at the toe of the flange, because of the usual inside flange slope of 2/12 or 16⅔%, is 0.700 −

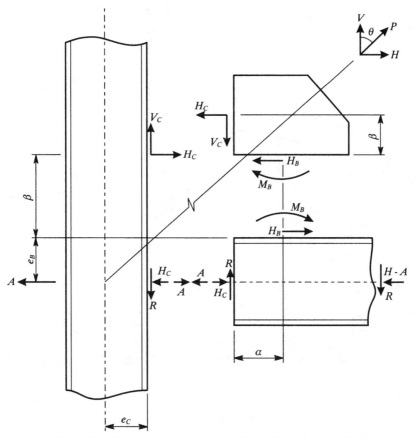

Figure 2.9 Force distribution for special case 2 of the uniform force method.

$\frac{2}{12} \times 2.006 = 0.366$ in (see Fig. 2.10). The thickness at the toe of the fillet is $0.366 + \frac{2}{12} \times 0.625 = 0.470$ in. The design shear rupture strength of the $MC12$ flange at the toe of the fillet is

$$\phi R_v = 0.75 \times 0.6 \times 58 \times 0.470 \times 17 \times 4 = 834 \text{ kips}$$

The design tensile rupture strength of the toe of the MC flange under the fillet is

$$\phi R_t = 0.75 \times 36 \left(\frac{0.366 + 0.470}{2} \right) 0.625 \times 4 = 28 \text{ kips}$$

Thus the total strength of the load path in the channel flange is $834 + 28 = 862$ kips > 855 kips, ok.

b. Gusset-to-brace shear rupture (tearout):

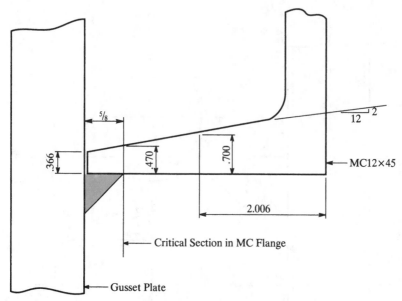

Figure 2.10 Critical section at toe of fillet weld.

shear fracture:

$$\phi R_v = 0.75 \times 0.6 \times 58 \times 1.5 \times 17 \times 2 = 1331 \text{ kips}$$

tension fracture

$$\phi R_t = 0.75 \times 58 \times 1.5 \times 12 = 783 \text{ kips}$$

$$\phi R_{bs} = 1331 + 0.75 \times 36 \times 1.5 \times 12 = 1817 \text{ kips} > 855 \text{ kips, ok}$$

c. *Whitmore section:* The theoretical length of the Whitmore section is $(17 \tan 30)2 + 12 = 31.6$ in. The Whitmore section extends into the column by 5.40 in. The column web is stronger than the gusset since $1.29 \times 50/36 = 1.79 > 1.5$ in. The Whitmore also extends into the beam web by 6.80 in, but since $0.470 \times 50/36 = 0.653 < 1.5$ in, the beam web is not as strong as the gusset. The effective Whitmore section length is

$$l_{w \text{ eff}} = (31.6 - 6.80) + 6.80 \times \frac{0.470}{1.5} \times \frac{50}{36} = 27.8 \text{ in}$$

The effective length is based on $F_y = 36$ and the gusset thickness of 1.5 in.

Since the brace force can be tension or compression, compression will control. The slenderness ratio of the unsupported length of gusset is

$$\frac{Kl}{r} = \frac{0.5 \times 8.5 \sqrt{12}}{1.5} = 9.8$$

From LRFD, Table 3-36, the buckling strength is

$$\phi F_a = 30.4 \text{ ksi}$$

and the buckling strength of the gusset is

$$\phi R_{wb} = 27.8 \times 1.5 \times 30.4 = 1268 > 855 \text{ kips, ok}$$

This completes the brace-to-gusset part of the design. Before proceeding, the distribution of forces to the gusset edges must be determined. From Fig. 2.8,

$$e_B = \frac{24.10}{2} = 12.05 \qquad e_c = 8.37 \qquad \beta = 12.25 \qquad \overline{\alpha} = 15.0$$

$$\theta = \tan^{-1} \frac{10.6875}{12} = 41.6°$$

$$V_C = P \cos \theta = 855 \times 0.747 = 638 \text{ kips}$$

$$H_C = \frac{V_C e_C}{e_B + \beta} = \frac{638 \times 8.37}{12.05 + 12.25} = 220 \text{ kips}$$

$$H_B = P \sin \theta - H_C = 855 \times 0.665 - 220 = 349 \text{ kips}$$

$$M_B = H_B e_B = 349 \times 12.05 = 4205 \text{ kip-in}$$

Note that, in this special case 2, the calculations can be simplified as shown here. The same results can be obtained formally with the UFM by setting $\beta = \overline{\beta} = 12.25$ and proceeding as follows. With $\tan \theta = 0.8906$

$$\alpha - 0.8906 \, \beta = 12.05 \times 0.8906 - 8.37 = 2.362$$

Setting $\beta = \overline{\beta} = 12.25$, $\alpha = 13.27$. Since $\overline{\alpha} = 15.0$, there will be a couple, M_B, on the gusset-to-beam edge. Continuing

$$r = [(13.27 + 8.37)^2 + (12.25 + 12.05)^2]^{1/2} = 32.54$$

$$\frac{P}{r} = 26.27$$

$$H_B = \frac{\alpha}{r} P = 349 \text{ kips}$$

$$H_C = \frac{e_C}{r} P = 220 \text{ kips}$$

$$V_B = \frac{e_B}{r} P = 317 \text{ kips}$$

$$V_C = \frac{\beta}{r} P = 322 \text{ kips}$$

$$M_B = | \ V_B(\alpha - \overline{\alpha}) \ | = 548 \text{ kip-in}$$

This couple is clockwise on the gusset edge. Now, introducing special case 2, in the notation of the AISC LRFD *Manual of Steel Construction* (1994), vol. II, pp. 11–22, set $\Delta V_B = V_B = 317$ kips. This reduces the vertical force between the gusset and beam to zero, and increases the gusset-to-column shear, V_C, to $317 + 322 = 639$ kips and creates a counterclockwise couple on the gusset-to-beam edge of $\Delta V_B \overline{\alpha} = 317 \times 15.0 = 4755$ kip-in. The total couple on the gusset-to-beam edge is thus $M_B = 4755 - 548 = 4207$ kip-in. It can be seen that these gusset interface forces are the same as those obtained from the simpler method.

2. *Gusset-to-column connection:*   The loads are 638 kips shear and 220 kips axial.
   a. *Gusset stresses:*

$$f_v = \frac{638}{1.5 \times 24.5} = 17.4 \text{ ksi} < 0.9 \times 0.6 \times 36 = 19.4 \text{ ksi, ok}$$

$$f_a = \frac{220}{1.5 \times 24.5} = 5.98 \text{ ksi} < 0.9 \times 36 = 32.4 \text{ ksi, ok}$$

   b. *Weld of gusset to column flange:*   Using AISC LRFD, Table 8-38, $P_u = \sqrt{638^2 + 220^2} = 675$ kips and the angle from the longitudinal weld axis is $\tan^{-1} (220/638) = 19°$, so using the table for 15° with $k = a = 0.0, c = 2.97$. Thus

$$D = \frac{675}{2.97 \times 24.5} \ 1.4 = 12.98$$

   which indicates that a ¹³⁄₁₆ fillet is required. The ductility factor 1.4 is used because the weld is assumed to be uniformly loaded, but research shows that the ratio of peak-to-average stresses is about 1.4 (Richard, 1986).

   c. *Checks on column web:*
      (1) *Web yielding (under normal load H_C):*

$$\phi R_{wy} = 1.0 \times 0.66 \times 50 \times 1.290(24.5 + 5 \times 2\tfrac{3}{4}) = 1628 \text{ kips}$$
$$> 220 \text{ kips, ok}$$

      (2) *Web crippling (under normal load H_C):*

$$\phi R_{wcp} = 0.75 \times 135 \times 1.290^2 \left[1 + 3\left(\frac{24.5}{16.74}\right)\left(\frac{1.290}{2.070}\right)^{1.5} \right]$$

$$\times \sqrt{\frac{50 \times 2.070}{1.290}} = 4769 \text{ kips} > 220 \text{ kips, ok}$$

(3) *Web shear:* The horizontal force, H_C, is transferred to the column by the gusset-to-column connection and back into the beam by the beam-to-column connection. The situation is similar to that shown in Fig. 2.6, with $f_1\,(a + e)t = f_2(a - e)t = H_C$. Thus, the column web sees $H_C = 220$ kips as a shear. The column shear capacity is

$$\phi R_v = 0.9 \times 0.6 \times 50 \times 1.290 \times 16.74 = 583 \text{ kips} > 220 \text{ kips, ok}$$

3. *Gusset-to-beam connection:* The loads are 349 kips shear and a 4205-kip-in couple.
   a. *Gusset stresses:*

   $$f_v = \frac{349}{1.5 \times 30} = 7.76 \text{ ksi} < 19.4 \text{ ksi, ok}$$

   $$f_a = 0$$

   $$f_b = \frac{4205 \times 4}{1.5 \times 30^2} = 12.5 \text{ ksi} < 32.4 \text{ ksi, ok}$$

   b. *Weld of gusset-to-beam flange:*

   $$f_R = \sqrt{7.76^2 + 12.5^2} \times \frac{1.5}{2} = 11.0 \text{ kips/in}$$

   $$f_{ave} = \sqrt{7.76^2 + 6.25^2} \times \frac{1.5}{2} = 7.47 \text{ kips/in}$$

Since $11.0/7.47 = 1.47 > 1.4$, the weld size based on the peak force in the weld, f_R, effectively includes a ductility factor, therefore

$$D = \frac{11.0}{1.392} = 7.9$$

A ½ fillet weld is indicated.

An alternate method for calculating the weld size required is to use Table 8-38 of the AISC LRFD, vol. II, p. 8-163, special case $k = 0$, $P_u = 349$, and $al = 4205/349 = 12.05$ in, thus $a = 12.05/30 = 0.40$ and $c = 2.00$ and the required weld size is

$$D = \frac{349}{2.0 \times 30} = 5.8$$

A ⅜ fillet is indicated. This method does not give an indication of peak and average stresses, but it will be safe to use the ductility factor. Thus, the required weld size would be

$$D = 5.8 \times 1.4 = 8.12$$

Thus, a ⁹⁄₁₆ (8.12) fillet is indicated which is about the same size (within 3%) as the ½ (7.9) required by the classical method. The ½ fillet weld will be ok in this case.

c. *Checks on beam web:*

(1) *Web yield:* Although there is no axial component, the couple $M_B = 4205$ kip-in is statically equivalent to equal and opposite vertical shears at a lever arm of one-half the gusset length or 15 in. The shear is thus

$$V_s = \frac{4205}{15} = 280 \text{ kips}$$

This shear is applied to the flange as a transverse load over 15 in of flange. It is convenient for analysis purposes to imagine this load doubled and applied over the contact length $N = 30$ in. The design web yielding strength is

$$\phi R_{wy} = 1.0 \times 0.66 \times 50 \times 0.470(30 + 5 \times 1\tfrac{9}{16})$$

$$= 586 \text{ kips} > 280 \times 2 = 560 \text{ kips, ok}$$

(2) *Web crippling:*

$$\phi R_{wcp} = 0.75 \times 135 \times 0.470^2\left[1 + 3\left(\frac{30}{24.10}\right)\left(\frac{0.470}{0.770}\right)^{1.5}\right]$$

$$\times \sqrt{\frac{50 \times 0.770}{0.470}} = 563 \text{ kips} > 560 \text{ kips, ok}$$

(3) *Web shear:*

$$\phi P_v = 0.9 \times 0.6 \times 50 \times 0.470 \times 24.10 = 306 \text{ kips} > 280 \text{ kips, ok}$$

The maximum shear due to the couple is centered on the gusset 15 in from the beam end. It does not reach the beam-to-column connection where the beam shear is 170 kips. Because of the total vertical shear capacity of the beam and the gusset acting together, there is no need to check the beam web for a combined shear of V_s and R of $280 + 170 = 450$ kips.

4. *Beam-to-column connection:* The shear load is 170 kips and the axial force is $H_C \pm A = 220 \pm 150$ kips. Since the W18 × 50 is a collector, it adds load to the bracing system. Thus, the axial load is $220 + 150 = 370$ kips. However, the AISC book on connections

(AISC, 1992) addresses this situation and states that because of frame action (distortion), which will always tend to reduce H_C, it is reasonable to use the larger of H_C and A as the axial force. Thus the axial load would be 220 kips in this case. The connection will be designed in this way and then the method will be justified.

a. *Bolts and clips:*   The bolts are A490 SC-A-N 1-in diameter in oversize 1¼-in-diameter holes. Thus, for shear

$$\phi r_{str} = 0.85 \times 1.13 \times 0.33 \times 64 = 20.3 \text{ kips/bolt}$$

and for tension

$$\phi r_t = 0.75 \times 113 \times 0.7854 = 66.6 \text{ kips/bolt}$$

The clips are Ls $4 \times 4 \times ¾$ with seven rows of bolts. For shear

$$\phi R_v = 20.3 \times 14 = 284 \text{ kips} > 170 \text{ kips, ok}$$

For tension, the bolts and clips are checked together for prying action.

Since all of the bolts are subjected to tension simultaneously, there is interaction between tension and shear. The reduced tensile capacity is

$$\phi r_t' = 1.13 \times 64\left(1 - \frac{170/14}{20.3}\right) = 29.1 \text{ kips/bolt}$$

Since 29.1 kips > 220/14 = 15.7 kips, the bolts are ok for tension. The bearing type interaction expression should also be checked but it will not control. Prying action is now checked using the method and notation of the AISC LRFD *Manual of Steel Construction:*

$$b = \frac{5.5 - 0.470}{2} - 0.75 = 1.765$$

$$a = \frac{8 + 0.470 - 5.5}{2} = 1.485$$

Check $1.25b = 1.25 \times 1.765 = 2.206$. Since $2.206 > 1.485$, use $a = 1.485$

$$b' = 1.765 - \frac{1.0}{2} = 1.265$$

$$a' = 1.485 + \frac{1.0}{2} = 1.985$$

$$\rho = 0.637$$

$$\delta = 1 - \frac{1.25}{3} = 0.583$$

$$t_c = \sqrt{\frac{4.44 \times 29.1 \times 1.265}{3 \times 36}} = 1.23$$

$$\alpha' = \frac{1}{0.583 \times 1.637}\left[\left(\frac{1.23}{0.75}\right)^2 - 1\right] = 1.77$$

Since $\alpha' > 1$, use $\alpha' = 1.0$

The design strength per bolt including prying is

$$T_d = 29.1\left(\frac{0.75}{1.23}\right)^2 (1 + 0.583 \times 1.0) = 17.1 \text{ kips} > 15.70 \text{ kips, ok}$$

In addition to the prying check, the clips should also be checked for gross and net shear. These will not control in this case.

b. *Weld of clips to beam web:* The weld is a C-shaped weld with length $l = 21$ in, $kl = 3.5$ in, $k = 3.5/21 = 0.167$. From the AISC LRFD *Manual of Steel Construction,* Table 8-42, $xl = 0.0220 \times 21 = 0.462$, so $al = 6 - 0.462 = 5.538$, and $a = 5.538/21 = 0.264$. Since $\tan^{-1} 220/170 = 52.3°$, use the chart on p.8-190 for 45°. By interpolation, $C = 1.92$. A ¼ fillet weld has a capacity of $\phi R_w = 1.92 \times 4 \times 2 \times 21 = 323$ kips. In order to support this weld, the web thickness required is $0.9 \times 0.6 \times 50 \times t_w \geq 1.392 \times 4 \times 2$. Thus $t_{w,\text{req'd}} \geq 0.41$ in. Since the actual web thickness is 0.470 in, the weld is fully effective and has the calculated capacity. Thus, since 323 kips $> \sqrt{220^2 + 170^2} = 278$ kips, the ¼ fillet weld is ok.

c. *Bending of the column flange:* Because of the axial force, the column flange can bend just as the clip angles. A yield-line analysis derived from Mann and Morris (1979) can be used to determine an effective tributary length of column flange per bolt. The yield lines are shown in Fig. 2.11. From Fig. 2.11,

$$p_{\text{eff}} = \frac{(n-1)p + \pi\overline{b} + 2\overline{a}}{n}$$

where $\overline{b} = (5.5 - 1.290)/2 = 2.105$
$\overline{a} = (16.110 - 5.5)/2 = 5.305$
$p = 3$
$n = 7$

Thus,

$$p_{\text{eff}} = \frac{6 \times 3 + \pi \times 2.105 + 2 \times 5.305}{7} = 5.032$$

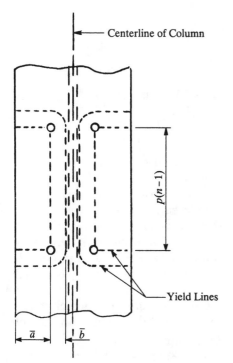

— Centerline of Column

$p(n-1)$

Yield Lines

$\overline{a}$ $\overline{b}$

Figure 2.11 Yield lines for flange bending.

Using p_{eff} in place of p, and following the AISC procedure,

$$b = \overline{b} = 2.105$$

$$b' = 2.105 - \frac{1.0}{2} = 1.605$$

$$a = \min\left(\frac{4 + 4 + 0.470 - 5.5}{2}, \, 5.305, \, 1.25 \times 2.105\right)$$

$$= \min\,(1.485, \, 5.305, \, 2.63) = 1.485$$

$$a' = 1.485 + 0.5 = 1.985$$

$$\rho = \frac{b'}{a'} = 0.81$$

$$\delta = 1 - \frac{1.0625}{5.032} = 0.79$$

Note that standard holes are used in the column flange.

$$t_c = \sqrt{\frac{4.44 \times 29.1 \times 1.605}{5.032 \times 50}} = 0.91$$

$$\alpha' = \frac{1}{0.79 \times 1.81}\left[\left(\frac{0.91}{2.070}\right)^2 - 1\right] = -0.564$$

Since $\alpha' < 0$, use $\alpha' = 0$

$$T_d = 29.1 \text{ kips/bolt} > 15.7 \text{ kips/bolt, ok}$$

When $\alpha' < 1$, the bolts, and not the flange, control the strength of the connection.

2.2.1.5 Frame action. The method of bracing connection design presented here, the uniform force method, is an equilibrium-based method. Every proper method of design for bracing connections, and in fact for every type of connection, must satisfy equilibrium. The set of forces derived from the uniform force method, as shown in Fig. 2.3, satisfy equilibrium of the gusset, the column, and the beam with axial forces only. Such a set of forces is said to be "admissible." But equilibrium is not the only requirement that must be satisfied to establish the true distribution of forces in a structure or connection. Two additional requirements are (1) the constitutive equations which relate forces to deformations and (2) the compatibility equations which relate deformations to displacements.

If it is assumed that the structure and connection behave elastically (an assumption as to constitutive equations) and that the beam and column remain perpendicular to each other (an assumption as to deformation—displacement equations), then an estimate of the moment in the beam due to distortion of the frame (frame action) (Thornton, 1991) is given by

$$M_D = 6\,\frac{P}{Abc}\,\frac{I_b I_c}{[(I_b/b) + (2I_c/c)]}\,\frac{(b^2 + c^2)}{bc}$$

in which the subscript, D = distortion
I_b = moment of inertia of beam = 2370 in⁴
I_c = moment of inertia of column = 3840 in⁴
P = brace force = 855 kips
A = brace area = 26.4 in²
b = length of beam to inflection point (assumed at beam midpoint) = 175 in
c = length of columns to inflection points (assumed at column midlengths) = 96 in

With

$$\frac{2I_c}{c} = 80 \quad \text{and} \quad \frac{I_b}{b} = 13.5$$

$$M_D = \frac{6 \times 855 \times 2370 \times 3840}{26.4 \times 175 \times 96(13.5 + 80)} \frac{175^2 + 96^2}{(175 \times 96)} = 2670 \text{ kip-in}$$

This moment M_D is only an estimate of the actual moment that will exist between the beam and column. The actual moment will depend on the strength of the beam-to-column connection. The strength of the beam-to-column connection can be assessed by considering the forces induced in the connection by the moment, M_D, as shown in Fig. 2.12. The distortion force, F_D, is assumed to act as shown through the gusset edge connection centroids. If the brace force, P, is a tension, the angle between the beam and column tends to decrease, compressing the gusset between them, so F_D is a compression. If the brace force, P, is a compression, the angle between the beam and column tends to increase and F_D is a tension. Figure 2.12 shows how the distortion

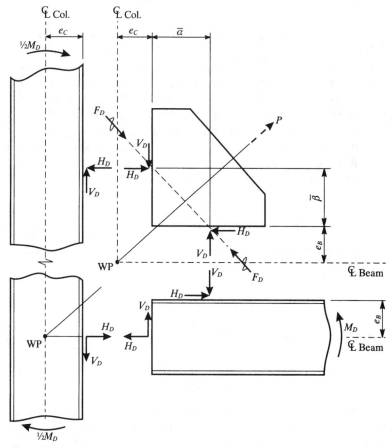

Figure 2.12 Distribution of distortion forces.

force, F_D, is distributed throughout the connection. From Fig. 2.12, the following relationships exist between F_D, its components (H_D and V_D), and M_D:

$$F_D = \sqrt{H_D{}^2 + V_D{}^2}$$

$$\overline{\beta}H_D = \overline{\alpha}V_D$$

$$H_D = \frac{M_D}{\overline{\beta} + e_B}$$

For the elastic case with no angular distortion

$$H_D = \frac{2670}{(12.25 + 12.05)} = 110 \text{ kips}$$

$$V_D = \frac{\overline{\beta}}{\overline{\alpha}}\, H_D = \frac{12.25}{15} \times 110 = 89.8 \text{ kips}$$

It should be remembered that these are just estimates of the distortion forces. The actual distortion forces will be dependent also upon the strength of the connection. But it can be seen that these estimated distortion forces are not insignificant. Compare, for instance, H_D to H_C. H_C is 220 kips tension when H_D is 110 kips compression. The net axial design force would then be $220 - 110 = 110$ kips rather than 220 kips.

The strength of the connection can be determined by considering the strength of each interface, including the effects of the distortion forces. The following interface forces can be determined from Figs. 2.3 and 2.12. For the gusset-to-beam interface:

$$T_B(\text{tangential force}) = H_B + H_D$$
$$N_B(\text{normal force}) = V_B - V_D$$

For the gusset-to-column interface:

$$T_C = V_C + V_D$$
$$N_C = H_C - H_D$$

For the beam-to-column interface:

$$T_{BC} = |V_B - V_D| + R$$
$$N_{BC} = |H_C - H_D| \pm A$$

The only departure from a simple equilibrium solution to the bracing connection design problem was in the assumption that frame action would allow the beam-to-column connection to be designed for an axial force equal to the maximum of H_c and A, or max (220, 150) =

220 kips. Thus, the design shown in Fig. 2.8 has its beam-to-column connection designed for $N_{BC} = 220$ kips and $T_{BC} = 170$ kips. Hence

$$N_{BC} = |220 - H_D| + 150 = 220$$

means that $H_D = 150$ kips
and

$$V_D = \frac{12.25}{15}\, 150 = 122.5 \text{ kips}$$

From

$$T_{BC} = |V_B - 122.5| + 170 = 170$$

$$V_B = 122.5 \text{ kips}$$

Note that in order to maintain the beam-to-column loads of 170 kips shear and 220 kips tension, the gusset-to-beam shear, V_B, must increase from 0 to 122.5 kips. Figure 2.13 shows the transition from the original load distribution to the final distribution as given in Fig. 2.13d. Note also that N_{BC} could have been set as $17.1 \times 14 = 239$ kips, rather than 220 kips, because this is the axial capacity of the connection at 170 kips shear. The N_{BC} value of 220 kips is used to

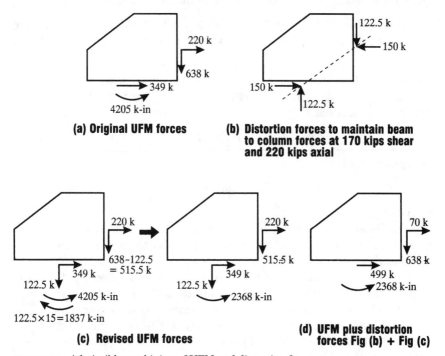

(a) Original UFM forces

(b) Distortion forces to maintain beam to column forces at 170 kips shear and 220 kips axial

(c) Revised UFM forces

(d) UFM plus distortion forces Fig (b) + Fig (c)

Figure 2.13 Admissible combining of UFM and distortion forces.

cover the case when there is no excess capacity in the beam-to-column connection. Now, the gusset-to-beam and gusset-to-column interfaces will be checked for the redistributed loads of Fig. 2.13d.

Gusset-to-beam

1. *Gusset stresses:*

$$f_v = \frac{499}{1.5 \times 30} = 11.1 \text{ ksi} < 19.4 \text{ ksi, ok}$$

$$f_b = \frac{2368 \times 4}{1.5 \times 30^2} = 7.02 \text{ ksi} < 32.4 \text{ ksi, ok}$$

2. *Weld of gusset to beam flange:*

$$f_R = \sqrt{11.1^2 + 7.02^2}\,\frac{1.5}{2} = 9.85$$

$$D = \frac{9.85}{1.392} = 7.08$$

A ½-in fillet weld is indicated, which is what was provided. No ductility factor is used here because the loads include a redistribution.

Gusset-to-column. This connection is ok without calculations because the loads of Fig. 2.13d are no greater than the original loads of Fig. 2.13a.

Discussion. From the foregoing analysis, it can be seen that the AISC suggested procedure for the beam-to-column connection, where the actual normal force

$$N_{BC} = |H_C - H_D| \pm A$$

is replaced by

$$N_{BC} = \max(H_c, A)$$

is justified.

It has been shown that the connection is strong enough to carry the distortion forces of Fig. 2.13b, which are larger than the elastic distortion forces.

In general, the entire connection could be designed for the combined uniform force method forces and distortion forces, as shown in Fig. 2.13d for this example. This set of forces is also admissible. The UFM forces are admissible because they are in equilibrium with the applied forces. The distortion forces are in equilibrium with zero external forces. Under each set of forces, the parts of the connection are also in equilibrium. Therefore, the sum of the two loadings is admissible because each individual loading is admissible. A safe design is thus guaranteed by the lower-bound theorem of limit analysis. The difficulty is in determining the distortion forces. The elastic distortion forces could be used, but they are only an estimate of the true distortion forces. The distortion forces depend as much on the properties of the connection, which are inherently inelastic and affect the maintenance of the angle between the members, as on the properties and lengths of the members of the frame. For this example, the distortion forces are $[(150 - 110)/110]100 = 36\%$ greater than the elastic distortion forces. In full-scale tests by Gross (1990) as reported by Thornton (1991), the distortion forces were about 2½ times the elastic distortion forces while the overall frame remained elastic. Because of the difficulty in establishing values for the distortion forces, and because the uniform force method has been shown to be conservative when they are ignored (Thornton, 1991, 1995b), they are not included in bracing connection design, except implicitly as noted here to justify replacing $| H_C - H_D | \pm A$ with max (H_C, A).

2.2.1.6 An exposition on load paths. The uniform force method produces a load path which is consistent with the gusset-plate boundaries. For instance, if the gusset-to-column connection is to a column web, no horizontal force is directed perpendicular to the column web because, unless it is stiffened, the web will not be able to sustain this force. This is clearly shown in the physical test results of Gross (1990) where it was reported that bracing connections to column webs were unable to mobilize the column weak axis stiffness because of web flexibility.

A mistake that is often made in connection design is to assume a load path for a part of the connection, and then to fail to follow through to make the assumed load path capable of carrying the loads (satisfying the limit states). Note that load paths include not just connection elements, but also the members to which they are attached. As an example, consider the connection of Fig. 2.14a. This is a configuration similar to that of Fig. 2.1a, detail A, with minimal transfer force into and out of the braced bay. It is proposed to consider the welds of the gusset to the beam flange and to the ½-in end plate as a single L-shaped weld. This will be called the L weld method, and is similar to model 4, the parallel force method, which is discussed by Thornton (1991). This is an apparently perfectly acceptable proposal and will result in very small welds because the centroid of the weld group will lie on or near the line of action of the brace. In the example of Fig. 2.14a, the geometry is arranged to cause the weld centroid to lie exactly on the line of action to simplify the calculation. This makes the weld uniformly loaded, and the force per inch is $f = 300/(33 + 20)$

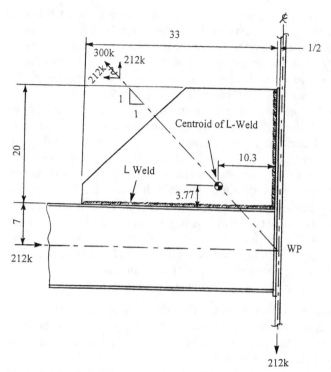

Figure 2.14a Bracing connection to demonstrate the consequences of an assumed load path.

= 5.66 kips/in in a direction parallel to the brace line of action which has horizontal and vertical components of 5.66 × 0.7071 = 4.00 kips/in. This results in free body diagrams for the gusset, beam, and column as shown in Fig. 2.14b. Imagine how difficult it would be to obtain the forces on the free body diagram of the gusset and other members if the weld were not uniformly loaded! Every inch of weld would have a force of different magnitude and direction. Note that while the gusset is in equilibrium under the parallel forces alone, the beam and the column require the moments as shown to provide equilibrium. For comparison, the free body diagrams for the uniform force method are given in Fig. 2.14c. These forces are always easy to obtain and no moments are required in beam or column to satisfy equilibrium.

From the unit force f = 5.66 kip/in, the gusset-to-beam and gusset-to-end plate weld sizes are D = 5.66/(2 × 1.392) = 2.03 sixteenths, actual required size. For comparison, the gusset-to-beam weld for the uniform force method would be

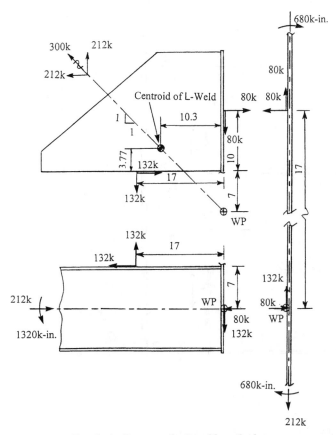

Figure 2.14b Free body diagrams for L weld method.

$$D = \frac{\sqrt{87^2 + 212^2}}{2 \times 33 \times 1.392} \times 1.4 = 3.49$$

actual required size, a 72% increase over the L weld method weld of $D = 2.03$. While the L weld method weld is very small, as expected with this method, now consider the load paths through the rest of the connection.

Gusset-to-column

Bolts. The bolts are A325N-⅞-in diameter with $\phi r_v = 21.6$ kips and $\phi r_t = 40.6$ kips. The shear per bolt is $80/12 = 6.67$ kips < 21.6 kips, ok. The tension per bolt is $80/12 = 6.67$ kips, but ϕr_t must be reduced due to interaction. Thus $\phi r_t' = 0.75\,(117 \times 0.6013 - 1.9 \times 6.67) = 43.3$ kips > 40.6 kips so use $\phi r_t' = 40.6$ kips. Since $40.6 > 6.67$, the bolts are ok for shear and tension.

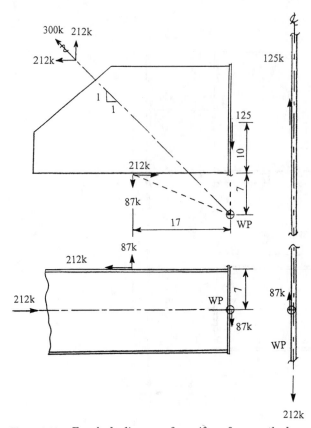

Figure 2.14c Free body diagrams for uniform force method.

End plate.   This involves the standard prying action calculations as follows:

$$b = \frac{5.5 - 0.375}{2} = 2.56 \qquad a = \frac{8 - 5.5}{2} = 1.25 < 1.25b$$

so use

$$a = 1.25 \qquad b' = 2.56 - \frac{0.875}{2} = 2.12 \qquad a' = 1.25 + \frac{0.875}{2} = 1.69$$

$$\rho = \frac{b'}{a'} = 1.25 \qquad \delta = 1 - \frac{0.9375}{3} = 0.69 \qquad p = 3$$

$$t_c = \sqrt{\frac{4.44 \times 40.6 \times 2.12}{3 \times 36}} = 1.88$$

try an end plate ½ in thick.

Calculate

$$\alpha' = \frac{1}{0.698(1 + 1.25)} \left[\left(\frac{1.88}{0.5} \right)^2 - 1 \right] = 8.46$$

Since $\alpha' > 1$, use $\alpha' = 1$, and the design tension strength is

$$T_d = 40.6 \left(\frac{0.5}{1.88} \right)^2 1.69 = 4.85 \text{ kips} < 6.67 \text{ kips, no good}$$

Try an end plate ¾ in thick;

$$\alpha' = \frac{1}{0.69 \times 2.25} \left[\left(\frac{1.88}{0.75} \right)^2 - 1 \right] = 3.40$$

Again, since $\alpha' > 1$, use $\alpha' = 1$, and the design tension strength is

$$T_d = 40.6 \left(\frac{0.75}{1.88} \right)^2 1.69 = 10.9 \text{ kips} > 6.67 \text{ kips, ok}$$

The ¾-in end plate is ok for this connection.

Column web.   The column web sees a transverse force of 80 kips. Figure 2.14d shows a yield-line analysis (Anand and Bertz, 1981) of the column web. The normal force ultimate strength of the yield pattern shown is

$$P_u = 8m_p \left[\sqrt{\frac{2T}{T - g}} + \frac{l}{2(T - g)} \right]$$

where $m_p = \frac{1}{4} F_y t_w^2$. For the present problem, $m_p = 0.25 \times 50 \times (0.44)^2 = 2.42$ kips-in/in, $T = 11.25$ in, $g = 5.5$ in, $l = 15$ in, so

$$P_u = 8 \times 2.42 \left(\sqrt{\frac{2 \times 11.25}{(11.25 - 5.5)}} + \frac{15}{2(11.25 - 5.5)} \right) = 63.5 \text{ kips}$$

Thus $\phi P_u = 0.9 \times 63.5 = 57.2$ kips < 80 kips, no good, and the column web is unable to sustain the horizontal force from the gusset without stiffening or a column web-doubler plate. Figure 2.15 shows a possible stiffening arrangement.

It should be noted that the yield line pattern of Fig. 2.14d compromises the foregoing end plate/prying action calculation. That analysis assumed double curvature with a prying force at the toes of the end plate a distance, *a*, from the bolt lines. But the column web will bend

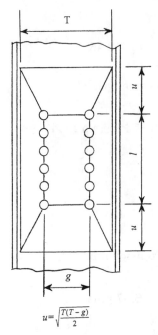

$$u=\sqrt{\frac{T(T-g)}{2}}$$

Figure 2.14d Deformation
method for yield-line analysis
of column web.

away as shown in Fig. 2.14d and the prying force will not develop.
Thus, single curvature bending in the end plate must be assumed,
and the required end-plate thickness is given by (AISC, 1994, vol. II,
p. 11-10)

$$t_{\text{req}} = \sqrt{\frac{4.44r_{ut}\,b'}{pF_y}} = \sqrt{\frac{4.44 \times 6.67 \times 2.12}{3 \times 36}} = 0.76 \text{ in}$$

and a ⅞-in-thick end plate is required.

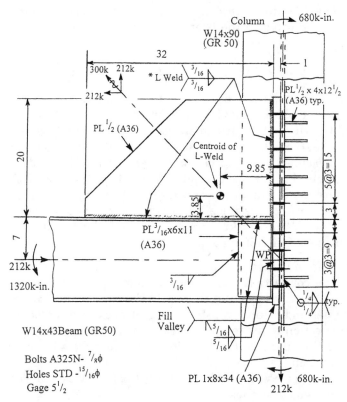

Figure 2.15 Design by L weld method.

Gusset-to-beam. The weld is already designed. The beam must be checked for web yield and crippling, and web shear.

Web yield

$$\phi R_{wcp} = 0.75 \times 135 \times 0.305^2 \left[1 + \left(\frac{34}{13.66} \right) \left(\frac{0.305}{0.530} \right)^{1.5} \right]$$

$$\times \sqrt{\frac{50 \times 0.530}{0.305}} = 183 \text{ kips} > 132 \text{ kips, ok}$$

Web shear. The 132-kip vertical load between the gusset and the beam flange is transmitted to the beam-to-column connection by the beam web. The shear design strength is

$$\phi R_{vw} = 0.9 \times 0.6 \times 0.305 \times 13.66 \times 50 = 112 \text{ kips} < 132 \text{ kips, no good}$$

To carry this much shear, a web-doubler plate is required. Starting at the toe of the gusset plate, $132/33 = 4.00$ kips of shear is added per inch. The doubler must start at a distance, x, from the toe where $4.00x = 112$, $x = 28.0$ in. Therefore, a doubler of length $34 - 28.0 = 6$ in is required, measured from the face of the end plate. The doubler thickness, t_d, required is $0.9 \times 0.6 \times 50 \times (t_d + 0.305) \times 13.4 = 132$, $t_d = 0.06$ in, so use a minimum thickness $\frac{3}{16}$-in plate of grade 50 steel. If some yielding before ultimate load is reached is acceptable, grade 36 plate can be used. The thickness required would be $t_d = 0.06 \times 50/36 = 0.083$ in, so a $\frac{3}{16}$-in A36 plate is ok also.

Beam-to-column. The fourth connection interface (the first interface is the brace-to-gusset connection, not considered here), the beam-to-column, is the most heavily loaded of them all. The 80 kips horizontal between the gusset and column must be brought back into the beam through this connection to make up the beam (strut) load of 212 kips axial. This connection also sees the 132-kip vertical load from the gusset-to-beam connection.

Bolts. The shear per bolt is $132/8 = 16.5$ kips < 21.6 kips, ok. The reduced tension design strength is

$$\phi r_t' = 0.75 (117 \times 0.6013 - 1.9 \times 16.5) = 29.3 \text{ kips} < 40.6 \text{ kips}$$

so use $\phi r_t' = 29.3$ kips. Since 29.3 kips $> 80/8 = 10.0$ kips, the bolts are ok for tension and shear.

End plate. As discussed for the gusset-to-column connection, there will be no prying action and hence double curvature in the end plate, so the required end plate thickness is

$$t_{req} = \sqrt{\frac{4.44 \times 10.0 \times 2.12}{3 \times 36}} = 0.93 \text{ in}$$

A 1-in end plate is required. This plate will be run up to form the gusset-to-column connection, so the entire end plate is a 1-in plate (A36).

Column web. Using the yield-line analysis for the gusset-to-column connection, $T = 11.25, g = 5.5, l = 9$

$$\phi P_u = 0.9 \times 8 \times 2.42 \left[\sqrt{\frac{2 \times 11.25}{5.75}} + \frac{9}{2 \times 5.75} \right]$$
$$= 48 \text{ kips} < 80 \text{ kips, no good}$$

Again, the column web must be stiffened as shown in Fig. 2.15, or a doubler must be used.

Stiffener. If stiffeners are used, the most highly loaded one will carry the equivalent tension load of three bolts or 30.0 kips to the column flanges. The stiffener is treated as a simple supported beam 12½ in long loaded at the gage lines. Figure 2.15 shows the arrangement. The shear in the stiffener is 30.0/2 = 15.0 kips, and the moment is 15.0(12.5 − 5.5)/2 = 52.5 kip-in. Try a stiffener of A36 steel ½ × 4:

$$f_v = \frac{15.0}{0.5 \times 4} = 7.50 \text{ ksi} < 19.4 \text{ ksi, ok}$$

$$f_b = \frac{52.5 \times 4}{0.5 \times 4^2} = 26.3 \text{ ksi} < 32.4 \text{ ksi, ok}$$

The ½ × 4 stiffener is ok. Check buckling, $b/t = 4/0.5 = 8 < 95/\sqrt{36} = 15.8$, ok.

Weld of stiffener to column web. Assume that about 3 in of weld at each gage line is effective, that is, $1.5 \times 1 \times 2 = 3$. Then

$$D = \frac{10.0 + 0.5 \times 10.0}{2 \times 3 \times 1.5 \times 1.392} = 1.19 \qquad \text{Use } \frac{3}{16} \text{ fillet welds}$$

Weld of stiffener to column flange

$$D = \frac{15.0}{2(4 - 0.75)1.392} = 1.66 \qquad \text{Use } \frac{3}{16} \text{ fillet welds}$$

Weld of end plate to beam web and doubler plate. The doubler is ³⁄₁₆ in thick and the web is 0.305 in thick, so 0.1875/0.4925 = 0.38 or 38% of the load goes to the doubler and 62% goes to the web. The load is $\sqrt{132^2 + 80^2} = 154$ kips. The length of the weld is $13.66 - 2 \times 0.530 = 12.6$ in. The weld size to the doubler is $D = 0.38 \times 154/(2 \times 12.6 \times 1.392) = 1.67$ and that to the web is $D = 0.62 \times 154/(2 \times 12.6 \times 1.392) = 2.72$, so ³⁄₁₆-in minimum fillets are indicated.

Additional discussion. The 80-kip horizontal force between the gusset and the column must be transferred to the beam-to-column connections. Therefore, the column section must be capable of making this transfer. The weak axis shear capacity (design strength) of the column is

$$\phi R_v = 0.9 \times 0.6 \times 50 \times 0.710 \times 14.520 \times 2 = 557 \text{ kips} > 80 \text{ kips, ok}$$

It was noted earlier that the column and beam require couples to be in equilibrium. These couples could act on the gusset-to-column and gusset-to-beam interfaces, since they are free vectors, but this would totally change these connections. Figure 2.14b shows them acting in the members instead, because this is consistent with the L weld method. For the column, the moment is $80 \times 17 = 1360$ kip-in and is shown with half above and half below the connection. The bending strength of the column is $\phi M_{py} = 0.9 \times 50 \times 133 = 5985$ kip-in, so the $1360/2 = 680$ kip-in is 11% of the capacity, which probably does not seriously reduce the column's weak axis bending strength. For the beam, the moment is $132 \times 17 - 132 \times 7 = 1320$ kip-in (should be equal and opposite to the column moment since the connection is concentric—the slight difference is due to numerical roundoff). The bending strength of the beam is $\phi M_{px} = 0.9 \times 50 \times 69.6 = 3132$ kip-in, so the 1320-kip-in couple uses up 42% of the beam's bending strength. This will greatly reduce its capacity to carry 212 kips in compression and is probably not acceptable.

This completes the design of the connection by the L weld method. The reader can clearly see how the loads filter through the connection, that is, the load paths involved. The final connection as shown in Fig. 2.15 has small welds of the gusset to the beam and the end plate, but the rest of the connection is very expensive. The column stiffeners are expensive, and also compromise any connections to the opposite side of the column web. The 1-in end plate must be flame cut because it is generally too thick for most shops to shear. The web-doubler plate is an expensive detail and involves welding in the beam k-line area which may be prone to cracking (AISC, 1997). Finally, although the connection is satisfactory, its internal admissible force distribution which satisfies equilibrium requires generally unacceptable couples in the members framed by the connection.

As a comparison, consider the design that is achieved by the uniform force method. The statically admissible force distribution for this connection is given in Fig. 2.14c. Note that all elements (gusset, beam, and column) are in equilibrium with no couples. Note also how easily these internal forces are computed. The final design for this method, which can be verified by the reader, is shown in Fig. 2.16. There is no question that this connection is less expensive than its L weld counterpart in Fig. 2.15, and it does not compromise the strength of the column and strut. To summarize, the L weld method seems a good idea at the outset, but a complete "trip" through the load paths ultimately exposes it as a fraud, that is, it produces expensive and unacceptable connections. As a final comment, a load path assumed for part of a connection affects every other part of the connection, including the members that frame to the connection.

2.2.1.7 Bracing connections utilizing shear plates. All of the bracing connection examples presented here have involved connections to the

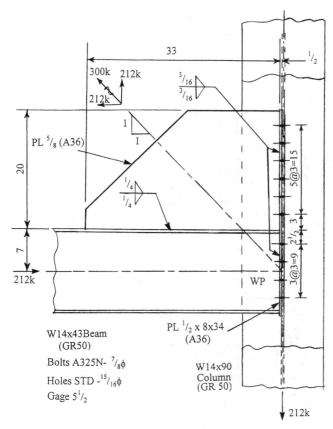

Figure 2.16 Design by uniform force method.

column using end plates or double clips, or are direct welded. The uniform force method is not limited to these attachment methods. Figures 2.17 and 2.18 show connections to a column flange and web, respectively, using shear plates. These connections are much easier to erect than the double angle or shear plate type because the beams can be brought into place laterally and easily pinned. For the column web connection of Fig. 2.18, there are no common bolts, which enhances erection safety. The connections shown were used on an actual job and were designed for the tensile strength of the brace to resist seismic loads in a ductile manner.

2.2.1.8 Connections with nonconcentric workpoints. The uniform force method can be easily generalized to this case as shown in Fig. 2.19a, where x and y locate the specified nonconcentric work point (WP) from the intersection of the beam and column flanges. All of the forces on the connection interfaces are the same as for the concentric uni-

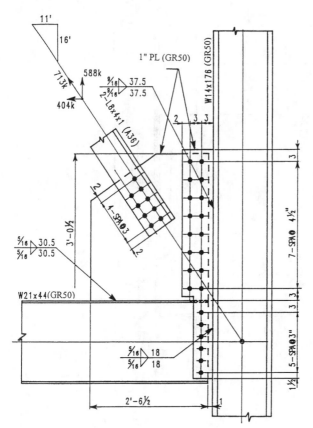

Figure 2.17 Bracing connection to a column flange utilizing a shear plate.

form force method, except that there is an extra moment on the gusset plate $M = Pe$, which can be applied to the stiffer gusset edge. It should be noted that this nonconcentric force distribution is consistent with the findings of Richard (1986) who found very little effect on the force distribution in the connection when the work point is moved from concentric to nonconcentric locations. It should also be noted that a nonconcentric work point location induces a moment in the structure of $M = Pe$, and this may need to be considered in the design of the frame members. In the case of Fig. 2.19a, since the moment $M = Pe$ is assumed to act on the gusset-to-beam interface, it also must be assumed to act on the beam outside of the connection, as shown. In the case of a connection to a column web this will be the actual distribution (Gross, 1990), unless the connection to the column mobilizes the flanges, as for instance is done in Fig. 2.15 by means of stiffeners.

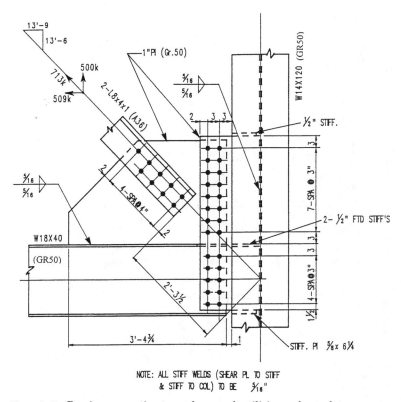

Figure 2.18 Bracing connection to a column web utilizing a shear plate.

An alternate analysis, where the joint is considered rigid, that is, a connection to a column flange, the moment, M, is distributed to the beam and column in accordance with their stiffnesses (the brace is usually assumed to remain an axial force member and so is not included in the moment distribution) can be performed. If η denotes the fraction of the moment that is distributed to the beam, then horizontal and vertical forces H' and V', respectively, acting at the gusset-to-beam, gusset-to-column, and beam-to-column connection centroids due to the distribution of M are

$$H' = \frac{(1 - \eta)M}{\overline{\beta} + e_B}$$

$$V' = \frac{M - H'\overline{\beta}}{\overline{\alpha}}$$

These forces, shown in Fig. 2.19b, are to be added algebraically to the concentric uniform force method forces acting at the three connection

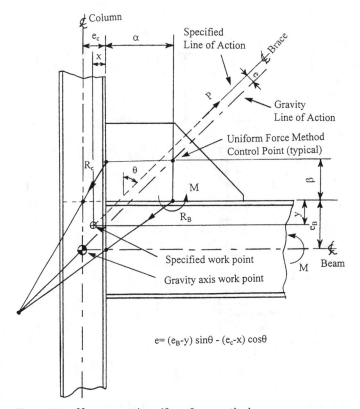

Figure 2.19a   Nonconcentric uniform force method

interfaces. Note that for connections to column webs, $\eta = 1, H' = 0$, and V' $= M/\bar{\alpha}$, unless the gusset-to-column web and beam-to-column web connections positively engage the column flanges, as for instance, in Fig. 2.15.

Example   Consider the connection of Sec. 2.2.1.4 as shown in Fig. 2.8, but consider that the brace line of action passes through the corner of the gusset rather than to the gravity axis intersection of the beam and column. Using the data of Fig. 2.8, $e_c = 8.37$, $e_B = 12.05$, $\bar{\alpha} = 15.0$, $\bar{\beta} = 12.25$, and

$$\theta = \tan^{-1} \frac{10^{11}\!/_{16}}{12} = 41.6°$$

Since the specified work point is at the gusset corner, $x = y = 0$, and $e = 12.05$ sin $41.6° - 8.37$ cos $41.6° = 1.77$ in. Thus, $M = Pe = 855 \times 1.77 = 1513$ kip-in, and using the frame data of Sec. 2.2.1.5,

$$\eta = \frac{13.5}{13.5 + 80} = 0.144$$

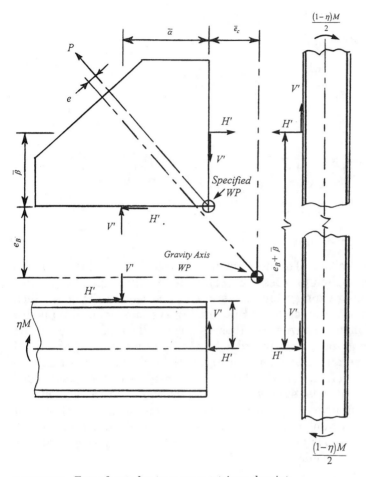

Figure 2.19b Extra forces due to nonconcentric work point.

$$H' = \frac{(1 - 0.144)1513}{(12.25 + 12.05)} = 53 \text{ kips}$$

$$V' = \frac{1513 - 53 \times 12.25}{15} = 58 \text{ kips}$$

These forces are shown on the gusset in Fig. 2.19c. This figure also shows the original uniform force method forces of Fig. 2.13a. The design of this connection will proceed in the same manner as shown in Sec. 2.2.1.4, but the algebraic sum of the original forces and the additional forces due to the nonconcentric work point are used on each interface.

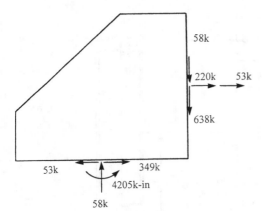

58k

220k   53k

638k

53k   349k

4205k-in

58k

Figure 2.19c Uniform force method and nonconcentric forces combined.

2.2.2 Truss connections

2.2.2.1 Introduction. The uniform force method as originally formulated can be applied to trusses as well as to bracing connections. After all, a vertical bracing system is just a truss as seen in Fig. 2.1, which shows various arrangements. But bracing systems generally involve orthogonal members whereas trusses, especially roof trusses, often have a sloping top chord. In order to handle this situation, the UFM has been generalized as shown in Fig. 2.20 to include non-orthogonal members. As before, α and β locate the centroids of the gusset-edge connections and must satisfy the constraint shown in the box on Fig. 2.20. This can always be arranged when designing a connection, but in checking a given connection designed by some other method, the constraint may not be satisfied. The result is gusset-edge couples which must be considered in the design.

2.2.2.2 A numerical example. As an application of the UFM to a truss, consider the situation of Fig. 2.21. This is a top chord connection in a large aircraft hangar structure. The truss is cantilevered from a core support area. Thus, the top chord is in tension. The design shown in Fig. 2.21 was obtained by generalizing the KISS ("keep it simple, stupid") method (Thornton, 1995b) shown in Fig. 2.22 for orthogonal members to the nonorthogonal case. The KISS method is the simplest admissible design method for truss and bracing connections. On the negative side, however, it generates large, expensive, and unsightly connections. The problem with the KISS method is the couples required on the gusset edges to satisfy equilibrium of all parts. In the Fig. 2.21 version of the KISS method, the truss diagonal horizontal and vertical components are placed at the gusset-edge centroids as shown. The couples 15,860 kip-in on the top edge and 3825 kip-in on the vertical edge are necessary for equilibrium of the gusset, top chord, and truss vertical,

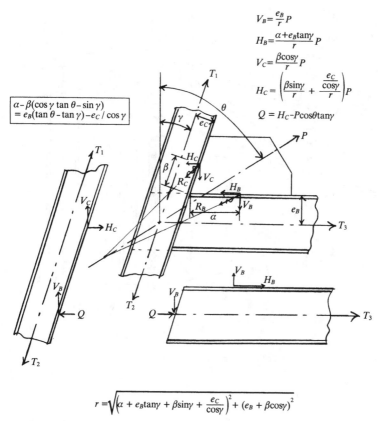

$$V_B = \frac{e_B}{r} P$$

$$H_B = \frac{\alpha + e_B \tan\gamma}{r} P$$

$$V_C = \frac{\beta\cos\gamma}{r} P$$

$$H_C = \left(\frac{\beta\sin\gamma}{r} + \frac{\frac{e_C}{\cos\gamma}}{r} \right) P$$

$$Q = H_C - P\cos\theta\tan\gamma$$

$$\boxed{\begin{array}{l} \alpha - \beta(\cos\gamma\tan\theta - \sin\gamma) \\ = e_B(\tan\theta - \tan\gamma) - e_C / \cos\gamma \end{array}}$$

$$r = \sqrt{\left(\alpha + e_B\tan\gamma + \beta\sin\gamma + \frac{e_C}{\cos\gamma} \right)^2 + (e_B + \beta\cos\gamma)^2}$$

Figure 2.20 Generalized uniform force method.

with the latter two experiencing only axial forces away from the connection. It is these couples which require the ¾-in chord doubler plate, the ⁷⁄₁₆-in fillets between the gusset and chord, and the 38-bolt, ⅞-in thick end plate on the vertical edge.

The design shown in Fig. 2.23 is also obtained by the KISS method with the brace force resolved into tangential components on the gusset edges. Couples still result, but they are much smaller than in Fig. 2.21. The resulting connection requires no chord doubler plate, ⁵⁄₁₆-in fillets of the gusset to the chord, and a 32-bolt, ¾-in thick end plate on the vertical edge. This design is much improved over that of Fig. 2.21.

When the uniform force method of Fig. 2.20 is applied to this problem, the resulting design is as shown in Fig. 2.24. The vertical connection has been reduced to only 14 bolts and a ⅝-in end plate.

The designs of Figs. 2.21, 2.23, and 2.24 are all satisfactory designs for some admissible force system. For instance, the design of Fig. 2.21 will be satisfactory for the force systems of Figs. 2.23 and 2.24, and

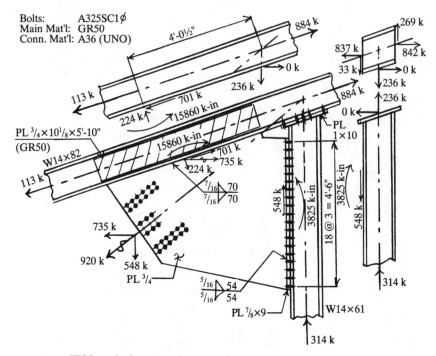

Figure 2.21 KISS method—gusset forces arc brace components.

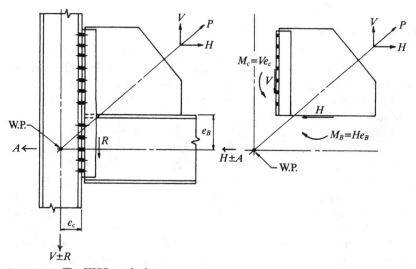

Figure 2.22 The KISS method.

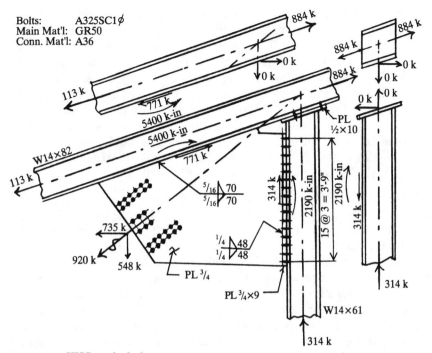

Figure 2.23 KISS method—brace components are tangent to gusset edges.

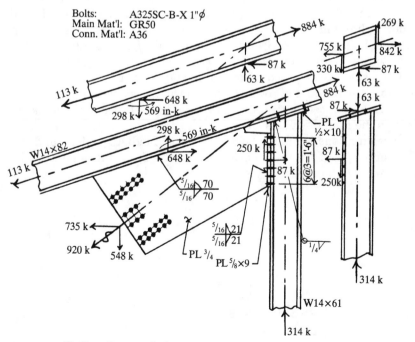

Figure 2.24 Uniform force method.

the design of Fig. 2.23 will be satisfactory for the force system of Fig. 2.24. How can it be determined which is the "right" or "best" admissible force system to use? The lower-bound theorem of limit analysis provides an answer. This theorem basically says that, for a given connection configuration, that is, Fig. 2.21, 2.23, or 2.24, the statically admissible force distribution which maximizes the capacity of the connection is closest to the true force distribution. Conversely, for a given load, the smallest connection satisfying the limit states is closest to the true required connection. Of the three admissible force distributions given in Figs. 2.21, 2.23, and 2.24, the distribution of Fig. 2.24, based on the UFM, is the "best" or "right" distribution.

2.2.2.3   A numerical example.   To demonstrate the calculations required to design the connections of Figs. 2.21, 2.23, and 2.24, for the statically admissible forces of these figures, consider, for instance, the UFM forces and the resulting connection of Fig. 2.24.

The geometry of Fig. 2.24 is arrived at by trial and error. First, the brace-to-gusset connection is designed and this establishes the minimum size of the gusset. For calculations for this part of the connection, see Sec. 2.2.1.2. Normally, the gusset is squared off as shown in Fig. 2.23, which gives 16 rows of bolts in the gusset-to-truss vertical connection. The gusset-to-top chord connection is pretty well constrained by geometry to be about 70 in long plus about 13½ in for the cutout. Starting from the configuration of Fig. 2.23, the UFM forces are calculated from the formulas of Fig. 2.20 and the design is checked. It will be found that Fig. 2.23 is a satisfactory design via the UFM, even though it fails via the KISS method forces of Fig. 2.21. Although the gusset-to-top chord connection cannot be reduced in length because of geometry, the gusset-to-truss vertical is subject to no such constraint. Therefore, the number of rows of bolts in the gusset-to-truss vertical is sequentially reduced until failure occurs. The last achieved successful design is the final design as shown in Fig. 2.24.

The calculations for Fig. 2.24 and the intermediate designs and initial design of Fig. 2.23 are performed in the following manner. The given data for all cases is

$$P = 920 \text{ kips}$$

$$e_B = 7 \text{ in}$$

$$e_c = 7 \text{ in}$$

$$\gamma = 17.7°$$

$$\theta = 36.7°$$

The relationship between α and β is

$$\alpha - \beta(0.9527 \times 0.7454 - 0.3040) = 7(0.7454 - 0.3191) - \frac{7}{0.9527}$$

$$\alpha - 0.4061\beta = -4.363$$

This relationship must be satisfied for there to be no couples on the gusset edges. For the configuration of Fig. 2.24 with seven rows of bolts in the gusset-to-truss vertical connection (which is considered the gusset-to-beam connection of Fig. 2.20), $\overline{\alpha} = 18.0$ in. Then,

$$\beta = \frac{18 + 4.363}{0.4061} = 55.07 \text{ in}$$

From Fig. 2.24, the centroid of the gusset-to-top chord (which is the gusset-to-column connection of Fig. 2.20) is $\overline{\beta} = 13.5 + 70/2 = 48.5$ in. Since $\overline{\beta} \neq \beta$, there will be a couple on this edge unless the gusset geometry is adjusted to make $\overline{\beta} = \beta = 55.07$. In this case, we will leave the gusset geometry unchanged and work with the couple on gusset-to-top chord interface.

Rather than choosing $\overline{\alpha} = 18.0$ in, we could have chosen $\overline{\beta} = 48.5$ and solved for $\alpha \neq \overline{\alpha}$. In this case, a couple will be required on the gusset-to-truss vertical interface unless gusset geometry is changed to make $\overline{\alpha} = \alpha$.

Of the two possible choices, the first is the better one because the rigidity of the gusset-to-top chord interface is much greater than that of the gusset-to-truss vertical interface. This is so because the gusset is direct welded to the center of the top chord flange and is backed up by the chord web, whereas the gusset-to-truss vertical involves a flexible end plate and the bending flexibility of the flange of the truss vertical. Thus, any couple required to put the gusset in equilibrium will tend to migrate to the stiffer gusset-to-top chord interface.

With $\alpha = 18.0$, $\beta = 55.07$

$$r = \left[\left(18.0 + 7 \times 0.3191 + 55.07 \times 0.3040 + \frac{7}{0.9527} \right)^2 + (7 + 55.07 \times 0.9527)^2 \right]^{1/2}$$

$$= 74.16 \text{ in}$$

and from the equations of Fig. 2.20

$$V_C = 648 \text{ kips}$$

$$H_C = 298 \text{ kips}$$

$$V_B = 87 \text{ kips}$$

$$H_B = 250 \text{ kips}$$

For subsequent calculations, it is necessary to convert the gusset-to-top chord forces to normal and tangential forces as follows: the tangential or shearing component is

$$T_C = V_C \cos \gamma + H_C \sin \gamma = (\beta + e_C \tan \gamma) \, P/r$$

The normal or axial component is

$$N_C = H_C \cos \gamma - V_C \sin \gamma = \frac{e_C P}{r}$$

The couple on the gusset-to-top chord interface is then

$$M_C = |N_C(\beta - \bar{\beta})|$$

Thus

$$T_C = (55.07 + 7 \times 0.3191) \, \frac{920}{74.16} = 711 \text{ kips}$$

$$N_C = 7 \, \frac{920}{74.16} = 86.6 \text{ kips}$$

$$M_C = 86.6(55.07 - 48.5) = 569 \text{ kip-in}$$

Each of the connection interfaces will now be designed.

1. *Gusset-to-top chord:*
   a. *Weld:*   Weld length is 70 in.

$$f_v = \frac{711}{2 \times 70} = 5.08 \text{ kips/in}$$

$$f_a = \frac{86.6}{2 \times 70} = 0.61 \text{ kip/in}$$

$$f_b = \frac{569 \times 2}{70^2} = 0.23 \text{ kip/in}$$

$$f_R = \sqrt{(5.08)^2 + (0.61 + 0.23)^2} = 5.1 \text{ kip/in}$$

$$D = \frac{5.1}{1.392} = 3.7$$

Check ductility

$$f_{\text{ave}} = \sqrt{5.08^2 + \left(0.61 + \frac{0.23}{2}\right)^2} = 5.1 \text{ kips/in}$$

Since $1.4 \times f_{ave} = 7.1 > 5.1$, the size weld for ductility requirement

$$D = \frac{7.1}{1.392} = 5.1$$

Use $\frac{5}{16}$ fillet weld (2% violation of ductility requirement ok).
b. *Gusset stress:*

$$f_v = \frac{711}{0.75 \times 70} = 13.5 \text{ ksi} < 19.4 \text{ ksi, ok}$$

$$f_a + f_b = \frac{86.6}{0.75 \times 70} + \frac{569 \times 4}{0.75 \times 70^2} = 2.27 \text{ ksi} < 32.4 \text{ ksi, ok}$$

c. *Top chord web yield:* The normal force between the gusset and the top chord is $T_c = 86.6$ kips and the couple is $M_c = 569$ kips-in. The contact length, N, is 70 in. The couple, M_c, is statically equivalent to equal and opposite normal forces $V_s = M_c/(N/2) = 569/35 = 16.2$ kips. The normal force, V_s, acts over a contact length of $N/2 = 35$ in. For convenience, an equivalent normal force acting over the contact length, N, can be defined as

$$N_{C \text{ equiv.}} = N_C + 2V_s = 86.6 + 2 \times 16.2 = 119 \text{ kips}$$

Now, for web yielding

$$\phi R_{wy} = 1.0(5k + N) F_{yw} t_w = 1.0(5 \times 1.625 + 70) \times 50 \times 0.510$$
$$= 1992 \text{ kips} > 119 \text{ kips, ok}$$

d. *Top chord web crippling:*

$$\phi R_{wcp} = 0.75 \times 135 t_w^2 \left[1 + 3\left(\frac{N}{d}\right)\left(\frac{t_w}{t_f}\right)^{1.5} \right] \sqrt{\frac{F_{yw} t_f}{t_w}}$$

$$= 0.75 \times 135 \times 0.510^2 \left\{ 1 + 3\left[\frac{70}{14.31}\left(\frac{0.510}{0.855}\right)^{1.5} \right] \right\} \sqrt{\frac{50 \times 0.855}{0.510}}$$
$$\times 1871 \text{ kips} > 119 \text{ kips, ok}$$

In the web-crippling check, the formula used is that for a location greater than $d/2$ from the chord end because $\bar{\beta} = 13.5 + 70/2 = 48.5$ in $> 14.31/2 = 7.2$ in. $\bar{\beta}$ is the position of the equivalent normal force.

The checks for web yield and crippling could have been dismissed by inspection in this case, but were completed to illustrate the method. Another check that should be made when there is a couple acting on a gusset edge is to ensure that the transverse shear induced on the supporting member, in this case the top chord W14 × 82, can be sustained. In this case, the induced transverse shear is V_s = 16.2 kips. The shear capacity of the W14 × 82 is 0.510 × 14.3 × 0.9 × 0.6 × 50 = 197 kips > 16.2 kips, ok. Now consider for contrast, the couple of 15,860 kips-in shown in Fig. 2.21. For this couple, V_s = 15,860/35 = 453 kips > 197 kips, so a ¾-in doubler plate of GR50 steel is required as shown in Fig. 2.21.

2. *Gusset-to-truss vertical:*
   a. *Weld:*

$$f_v = \frac{250}{2 \times 21} = 5.95 \text{ kips/in}$$

$$f_a = \frac{87}{2 \times 21} = 2.07 \text{ kips/in}$$

$$f_R = \sqrt{5.95^2 + 2.07^2} = 6.30 \text{ kips/in}$$

$$\text{Fillet weld size required} = \frac{6.30}{1.392} = 4.5$$

Because of the flexibility of end-plate and truss vertical flange, there is no need to size the weld to provide ductility. Therefore, use a 5⁄16 fillet weld.

   b. *Bolts and end plate:* The bolts are A325SC-B-X, 1-in-diameter in standard holes. The end plate is 9 in wide and the gage of the bolts is 5½ in. Thus, using the prying action formulation notation of the AISC LRFD *Manual of Steel Construction,* 2d ed., vol. II (1994),

$$b = \frac{5.5 - 0.75}{2} = 2.375$$

$$a = \frac{9 - 5.5}{2} = 1.75 < 1.25 \times 2.375, \text{ ok}$$

$$b' = 2.375 - 0.5 = 1.875$$

$$a' = 1.75 + 0.5 = 2.25$$

$$\rho = \frac{1.875}{2.25} = 0.833$$

$$\delta = 1 - \frac{1.0625}{3} = 0.646$$

$$\text{Shear per bolt} = V = \frac{250}{14} = 17.9 < 28.8 \text{ kips, ok}$$

$$\text{Tension per bolt} = T = \frac{87}{14} = 6.21 \text{ kips}$$

$$\phi r_n' = 1.13 \times 51\left(1 - \frac{17.9}{28.8}\right) = 21.8 \text{ kips} > 6.21 \text{ kips, ok}$$

Try ½ plate

$$t_c = \sqrt{\frac{4.44 \times 21.8 \times 1.875}{3 \times 36}} = 1.30$$

$$\alpha' = \frac{1}{0.646 \times 1.833}\left[\left(\frac{1.30}{0.5}\right)^2 - 1\right] = 4.86$$

Use $\alpha' = 1$

$$T_d = 21.8\left(\frac{0.5}{1.30}\right)^2 1.646 = 5.31 \text{ kips} < 6.21 \text{ kips, no good}$$

Try ⅝ plate

$$\alpha' = \frac{1}{0.646 \times 1.833}\left[\left(\frac{1.30}{0.625}\right)^2 - 1\right] = 2.81$$

$$T_d = 21.8\left(\frac{0.625}{1.30}\right)^2 1.646 = 8.29 \text{ kips} > 6.21 \text{ kips, ok}$$

Use ⅝ plate for the end plate.

c. *Truss vertical flange:* The flange thickness of the W 14 × 61 is 0.645 in, which exceeds the end-plate thickness as well as being grade 50 steel. The truss vertical flange is, therefore, ok by inspection, but a calculation will be performed to demonstrate how the flange can be checked. A formula (Mann and Morris, 1979) for an effective bolt pitch can be derived from yield-line analysis as

$$p_{\text{eff}} = \frac{p(n - 1) + \pi \bar{b} + 2\bar{a}}{n}$$

where the terms are as previously defined in Fig. 2.11. For the present case

$$\bar{b} = \frac{5.5 - 0.375}{2} = 2.5625$$

$$\bar{a} = \frac{10 - 5.5}{2} = 2.25$$

$$n = 7$$

$$p = 3$$

$$p_{\text{eff}} = \frac{3(7-1) + \pi\, 2.5625 + 2 \times 2.25}{7} = 4.36$$

Once p_{eff} is determined, the prying action theory of the AISC *Manual of Steel Construction,* 2nd edition, is applied.

$$b = \bar{b} = 2.5625$$
$$b' = 2.5625 - 0.5 = 2.0625$$

a = smaller of $\bar{a}$ and a for the end plate = $1.75 < 1.25 \times 2.5625$, ok

$$a' = 1.75 + 0.5 = 2.25$$

$$\rho = \frac{b'}{a'} = 0.917$$

$$\delta = 1 - \frac{1.0625}{4.36} = 0.756$$

$$t_c = \sqrt{\frac{4.44 \times 21.8 \times 2.0625}{4.36 \times 50}} = 0.957 \text{ in}$$

$$\alpha' = \frac{1}{0.756 \times 1.917}\left[\left(\frac{0.957}{0.645}\right)^2 - 1\right] = 0.829$$

$$T_d = 21.8\left(\frac{0.645}{0.957}\right)^2 (1 + 0.829 \times 0.756) = 16.1 \text{ kips} > \text{kips, ok}$$

3. *Truss vertical to top chord connection:* The forces on this connection, from Figs. 2.20 and 2.24, are

Vertical = $Q = 298 - 920 \cos (36.7) \tan (17.7) = 63$ kips

Horizontal = 87 kips

Converting these into normal and tangential components

$$T_{BC} = 87 \cos \gamma - 63 \sin \gamma = 64 \text{ kips}$$
$$N_{BC} = 87 \sin \gamma + 63 \cos \gamma = 86 \text{ kips (compression)}$$

a. *Bolts:* Since the normal force is always compression, the bolts see only the tangential or shear force, thus, the number of bolts required is

$$\frac{64}{28.8} = 2.2 \qquad \text{Use four bolts}$$

b. *Weld:*   Use a profile fillet weld of the cap plate to the truss vertical, but only the weld to the web of the vertical is effective because there are no stiffeners between the flanges of the top chord. Thus, the effective length of weld is

$$\frac{13.89 - 2 \times 0.645}{\cos \gamma} = 13.23 \text{ in}$$

$$f_v = \frac{64}{2 \times 13.23} = 2.42 \text{ kips/in}$$

$$f_a = \frac{86}{2 \times 13.23} = 3.25 \text{ kips/in}$$

$$f_R = \sqrt{2.42^2 + 3.25^2} = 4.05 \text{ kips/in}$$

The weld size required is $4.05/1.392 = 2.91$. Use ¼ FW (AISC minimum size). Check the W 14×61 web to support required 2.91/16 FW. For welds of size W on both sides of a web of thickness t_w

$$0.9 \times 0.6 \times F_y x t_w \geq 0.75 \times 0.60 \times 70 \times 0.7071 \times W \times 2$$

or

$$t_w \geq 1.65 \ W \qquad \text{for grade 50 steel}$$

Thus for W = 2.91/16 = 0.182

$$t_{w_{\min}} = 1.65 \times 0.182 = 0.300 \text{ in}$$

Since the web thickness of a W 14×61 is 0.375 in, the web can support the welds.

c. *Cap plate:*   The cap-plate thickness will be governed by bearing. The bearing design strength per bolt is

$$\phi r_p = 0.75 \times 2.4 \times 58 \times t_p \times 1$$

The load per bolt is $64/4 = 16.0$ kips. The required cap-plate thickness is thus

$$t_p = \frac{16.0}{0.75 \times 2.4 \times 58 \times 1} = 0.153$$

Use a ½-in cap plate.

This completes the calculations required to produce the connection of Fig. 2.24.

2.2.2.4   Truss rigid frame end connections.   A rigid frame building is usually pictured as consisting of wide-flange columns and a wide-

Chords WT 7x41   A36
Webs   WT 4x15.5 A36

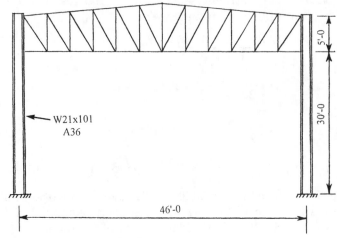

Figure 2.25   Rigid frame with truss rafter.

flange roof rafter, either flat, sloping in one direction (a shed roof) or gabled (ridge at centerline). For longer spans, the roof rafter can be replaced by a truss, such as shown in Fig. 2.25. Compared to the W 21 × 101 columns, this truss is very rigid and essentially induces a fixed condition at the top of the column in the same way that a moment connection between the column and rafter would produce a rigid joint in a wide-flange frame. The moment at the top of the column is produced by the 75 kips between the column and the top and bottom chords. A connection design philosophy similar to that of the uniform force method can be used to design the connections of the top and bottom chords to the column flange.

Top chord connection.   Figure 2.26 shows the completed connection. The procedure to achieve this connection involves understanding the load paths. The 75-kip design axial load (required strength) is a horizontal force. In order for all the bolts to the column to be equally loaded, the centroid of the bolt group between the column and the end plate should coincide with the gravity axis of the top chord because this member carries most of the 75-kip horizontal force. Also, since the top and bottom chord connections work together to produce the moment at the column top, the working point (WP) can be taken at the face of the column flange, and this point should fall at or near the bolt group centroid. The extra connection axial force due to WP location is 60 × 10.68/60 = 10.7 kips, so the total axial force seen by the chord connections will be 75.0 + 10.7 = 85.7 kips. The five-row connection shown

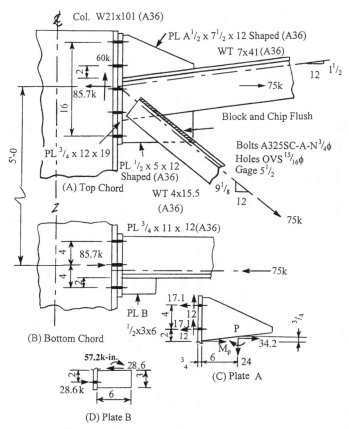

Figure 2.26 Truss-to-column connections.

in Fig. 2.26 is first estimated by "off-to-the-side" calculations to be adequate. The 4-in pitch is chosen to accommodate the truss web member to top chord connection. The detailed calculations for each part of the connection follow.

Bolts. The bolts are A325SC-A-N, ¾-in diameter in oversized ¹⁵/₁₆-in holes for ease of erection. The shear capacity of the bolts is $\phi r_v = 8.88$ kips/bolt and the tensile capacity is $\phi r_t = 29.8$ kips/bolt. The shear per bolt is $V = 60/10 = 6.0$ kips/bolt < 8.88 kips/bolt, ok. The tensile capacity must be reduced due to shear/tension interaction. For an N-type connection

$$\phi F_t' = 0.75(117 - 1.9f_v) \le 0.75 \times 90 = 0.75\left(117 - \frac{1.9 \times 6.0}{0.4418}\right)$$

$$= 68.4 \text{ ksi} > 67.5 \text{ ksi} \qquad \text{Use } 67.5 \text{ ksi}$$

$$\phi r_t' = 67.5 \times 0.4418 = 29.8 \text{ kips}$$

For an SC-type connection

$$\phi r_t' = 1.13 T_b \left(1 - \frac{V}{\phi r_t} \right) < \phi r_t = 1.13 \times 28 \left(1 - \frac{6}{8.88} \right) = 10.26 \text{ kips}$$
$$< 28.8 \text{ kips} \qquad \text{use } \phi r_t' = 10.26 \text{ kips}$$

Since the capacity (design strength) is less for the SC connection than for the N connection, the bolt design tensile strength is $\phi r_t' = 10.26$ kips/bolt. Since the tension load (required strength) per bolt of $T = 85.7/10 = 8.57$ kips is less than 10.26 kips, the bolts are ok for tension and shear.

End plate. Try a ¾-in A36 plate 12 in wide with a column gage of 5½ in. Then

$$b = \frac{5.5 - 0.5}{2} = 2.5$$

$$a = \frac{12 - 5.5}{2} = 3.25$$

check $a \leq 1.25b = 2.5 \times 1.25 = 3.125$ in, so use $a = 3.125$ in

$$b' = 2.5 - \frac{0.75}{2} = 2.125$$

$$a' = 3.125 + \frac{0.75}{2} = 3.50$$

$$\rho = \frac{b'}{a'} = 0.61$$

$$p = 4$$

$$\delta = 1 - \frac{0.9375}{4} = 0.766$$

$$t_c = \sqrt{\frac{4.44 \times 10.26 \times 2.125}{4 \times 36}} = 0.820$$

$$\alpha' = \frac{1}{0.766 \times 1.61} \left[\left(\frac{0.820}{0.75} \right)^2 - 1 \right] = 0.158$$

$$T_d = 10.26 \left(\frac{0.750}{0.820} \right)^2 (1 + 0.766 \times 0.158) = 9.62 \text{ kips} > 8.57 \text{ kips, ok}$$

Column flange. Since the flange thickness of the W 21 × 101 is $t_f = 0.80$ in, which is greater than the end-plate thickness of 0.75 in, and the column is A572-50 material, the column flange is ok "by inspection."

Weld of WT stem and stem extension plate and stiffener plate A to the end plate. The load path assumed for this connection has all the bolts equally loaded. For this reason, the WT flange is not welded to the end plate. If the flange were welded to the end plate, the bolts and weld local to this area would see more load. To prevent a fracture from occurring, it is best not to weld the WT flange to the end plate. With this assumption, each 4 in of weld sees the load from one pair of bolts, so the load per 4 in of weld is 12 kips shear and 17.1 kips tension. Thus, the weld size required is

$$D = \frac{\sqrt{12^2 + 17.1^2}}{2 \times 4 \times 1.392} = 1.88 \qquad \text{use a ¼-in fillet weld}$$

Stress in WT stem, stem extension plate, and stiffener plate A

$$f_v = \frac{12}{0.5 \times 4} = 6.00 \text{ ksi} < 0.9 \times 0.6 \times 36 = 19.4 \text{ ksi, ok}$$

$$f_a = \frac{17.1}{0.5 \times 4} = 8.55 \text{ ksi} < 0.9 \times 36 = 32.4 \text{ ksi, ok}$$

Connection of stiffener plate (plate A) to WT flange. Figure 2.26c shows the free body diagram of plate A. The moment, M_p, is calculated by taking moments about point P

$$\Sigma M_p = 0 = -M_p + 17.1 \times 1.25 + 17.1 \times 5.25 - 12 \times 2$$
$$\times 6.75 \qquad M_p = 50.9 \text{ kip-in}$$

The normal and tangential forces on the plate edge are

$$N = 24 \cos \theta + 34.2 \sin \theta = 28.1 \text{ kips}$$
$$S = 34.2 \cos \theta - 24 \sin \theta = 30.1 \text{ kips}$$

where $\theta = \tan^{-1} (1.5/12) = 7.125$. Stresses on plate A edge

$$f_a = \frac{28.1}{0.5 \times 12} = 4.68 \text{ ksi}$$

$$f_b = \frac{50.9 \times 6}{0.5 \times 12^2} = 4.24 \text{ ksi}$$

$$f_h = 4.68 + 4.24 = 8.92 \text{ ksi} < 32.4 \text{ ksi, ok}$$

$$f_v = \frac{31.0}{0.5 \times 12} = 5.17 \text{ ksi} < 19.4 \text{ ksi, ok}$$

Weld of plate A to WT flange

$$D = \sqrt{8.92^2 + 5.17^2} \, \frac{0.5}{2} \, \frac{1}{1.392} = 1.85 \qquad \text{use 3/16 fillet weld}$$

Weld of WT 4 × 15.5 diagonal to WT 7 stem and stem extension plate. This member has its flange blocked and chipped flush on one side and then is lapped against the WT 7 × 41 stem and the stem extension plate. The weld length required with a ¼ in fillet weld is

$$D = \frac{75}{4 \times 2 \times 1.392} = 6.73 \quad \text{or} \quad 7 \text{ in}$$

as shown in Fig. 2.26. For this type of longitudinal fillet weld, it is required that the length of the welds be at least as long as the distance between them. To eliminate the shear lag effect completely, the length of the welds must be at least twice the distance between them (AISC LRFD *Manual of Steel Construction*, 2d ed., specification Section B3).

Strength of the WT4 × 15.5 diagonal—Fracture. The area of the blocked and chipped portion of the WT 4 × 15.5 is

$$A_n = A_g - \left(\frac{b_f - t_w}{2}\right)t_f = 4.56 - \left(\frac{7.995 - 0.285}{2}\right)0.435 = 2.88$$

and the shear lag factor is $U = 0.75$. The design fracture strength is thus

$$\phi R_n = 0.75 \times 0.75 \times 2.88 \times 58 = 94 > 75 \text{ kips, ok}$$

Yield

$$\phi R_n = 0.9 \times 36 \times 4.56 = 148 > 75 \text{ kips, ok}$$

Support for the ¼-in fillet weld of the WT 4 × 15.5 to the chord stem and extension plate.

This can be checked in two ways. The simplest check is to calculate a thickness required by matching the shear yield stress in the supporting elements (WT 7 stem and stem extension plate A) to the strength of the fillet weld. From the AISC LRFD *Manual of Steel Construction*, Table 9-3, the required thickness of A36 material to support a ¼-in fillet weld is 0.57/2 > 0.5 in, ok. Alternately, if this check fails, a less conservative block shear check can be made as follows:

$$A_{vg} = 0.5 \times 7.0 \times 2 = 7.0 \text{ in}^2$$

$$A_{tg} = 0.5 \times 4.0 = 2.0 \text{ in}^2$$

$$0.6 F_u A_{vg} = 0.6 \times 58 \times 7.0 = 244 \text{ kips}$$

$$F_u A_{tg} = 58 \times 2.0 = 116 \text{ kips}$$

Since $0.6 F_y A_{vg} > F_u A_{tg}$

$$\phi R_{bs} = 0.75(244 + 36 \times 2.0) = 237 \text{ kips} > 75 \text{ kips, ok}$$

Bottom chord connection. The bottom chord connection carries the same 85.7-kip axial force as does the top chord connection. These two forces acting throughout the 5-ft 0-in distance between them form a moment connection at the juncture between the column and the truss. As with the top chord, the bottom chord connection should "straddle" the bottom chord gravity axis. Unlike the top chord connection, the bottom chord sees no shear, so there is no tension/shear interaction to consider, so try a three-row connection as shown in Fig. 2.26b.

Bolts. Per bolt, the tension $T = 85.7/6 = 14.3$ kips < 29.8 kips, ok.

End plate. Try a ¾-in A36 plate 12 in wide with a column gage of 5½ in. Then the geometry is the same as for the top chord, and $b = 2.5$, $a = 3.125$, $b' = 2.125$, $a' = 3.50$, $\rho = 0.61$, $p = 4$, $\delta = 0.766$, and

$$t_c = \sqrt{\frac{4.44 \times 29.8 \times 2.125}{4 \times 36}} = 1.397$$

$$\alpha' = \frac{1}{0.766 \times 1.61}\left[\left(\frac{1.397}{0.75}\right)^2 - 1\right] = 2.00 \qquad \text{use } \alpha' = 1$$

$$T_d = 29.8\left(\frac{0.75}{1.397}\right)^2 1.766 = 15.2 \text{ kips} > 14.3 \text{ kips, ok}$$

Column flange. This is ok by inspection as discussed for the top chord.

Weld of WT stem and stiffener plate to end plate. As discussed for the top chord, there is no weld between the WT flange and the end plate. Therefore,

$$D = \frac{14.3 \times 2}{2 \times 4 \times 1.392 \times 1.5} = 1.71 \qquad \text{Use a ¼-in fillet weld}$$

the AISC minimum fillet weld for the material thicknesses involved. The factor 1.5 in the denominator in the preceding equation is to recognize that transversely loaded fillets are 50% stronger than fillets longitudinally loaded. See AISC (1994), Appendix J.

Connection of stiffener plate to WT flange. This plate, shown in Fig. 2.26d, sees the load from one pair of bolts, $14.3 \times 2 = 28.6$ kips, at a distance from the WT flange of 2 in.

Stresses on plate at WT flange

$$f_v = \frac{28.6}{0.5 \times 6} = 9.53 \text{ ksi} < 19.4 \text{ ksi, ok}$$

$$f_b = \frac{28.6 \times 2 \times 4}{0.5 \times 6^2} = 12.7 \text{ ksi} < 32.4 \text{ ksi, ok}$$

Weld of plate to WT flange

$$D = \sqrt{9.53^2 + 12.7^2}\ \frac{0.5}{2}\ \frac{1}{1.392} = 2.85 \qquad \text{Use a } \frac{3}{16}\text{-in fillet weld}$$

Note that this weld could be sized using the AISC *Manual of Steel Construction,* Table 8-38, p. 8-163, with $k = 0$, $a = 2/6 = 0.333$, $c = 2.21$, $D = 28.6/(2.21 \times 6) = 2.16$. Again, a $\frac{3}{16}$-in filled weld is indicated, although it can be seen that this latter method is less conservative than the vectorial addition of stresses method.

A detailing consideration. If a truss of this type is detailed such that when assembled horizontally on the ground (in the shop or in the field if too large to ship), the top and bottom chord end connection faying surfaces are in the same plane, when the truss is rotated into the vertical position, the weight of the truss will cause deflections which rotate the chord ends out of a common plane. This is commonly called truss "kick" and can cause difficulty in erection. This problem is avoided by detailing the truss with top and bottom chord ends out of plane such that when it is under dead load the top and bottom chord ends come into plane. This is discussed in AISC (1983), p. 8-16.

2.2.3   Hanger connections

The most interesting of the genre is the type that involves prying action, sometimes of both the connection fitting and the supporting member. Figure 2.27 shows a typical example. The calculations to determine the capacity of this connection are as follows: The connection can be broken into three main parts, that is, the angles, the piece W 16 × 57, and the supporting member, the W 18 × 50. The three main parts are joined by two additional parts, the bolts of the angles to the piece W16 and the bolts from the piece W 16 to the W 18. The load path in this connection is unique. The load, P, passes from the angles through the bolts into the piece W 16, thence through bolts again into the supporting W 18. The latter bolt group is arranged to straddle the brace line of action. These bolts then see only direct tension and shear and no additional tension due to moment. Statics is sufficient to establish this. Consider now the determination of the capacity of this connection.

1. *Angles:* The limit states for the angles are gross tension, net tension, block shear rupture, and bearing. The load can be compression as well as tension in this example. Compression will affect the angle design, but tension will control the above limit states.

   a. Gross tension: The gross area, A_{gt}, is $1.94 \times 2 = 3.88$ in^2. The capacity (design strength) is

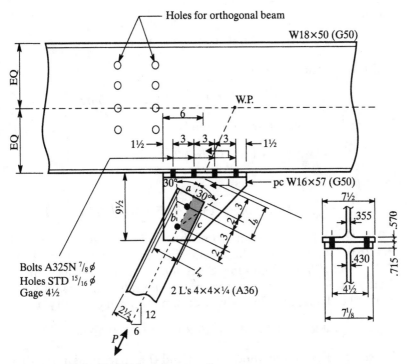

Figure 2.27 Typical bolted hanger connection.

$$\phi R_{gt} = 0.9 \times 36 \times 3.88 = 126 \text{ kips}$$

b. Net tension: The net tension area is $A_{nt} = 3.88 - 0.25 \times 1.0 \times 2 = 3.38 \text{ in}^2$. The effective net tension area, A_e, is less than the net area because of shear lag since only one of the two angle legs is connected. From the AISC LRFD *Manual of Steel Construction* (1994) commentary on Section B3

$$A_e = UA_{nt} = 0.75 \times 3.38 = 2.54 \text{ in}^2$$

The net tension capacity is

$$\phi R_{nt} = 0.75 \times 58 \times 2.54 = 110 \text{ kips}$$

c. Block shear rupture: This failure mode involves the tearing out of the cross-hatched block in Fig. 2.27. The failure is by yield or fracture on the longitudinal line through the bolts (line ab) and a simultaneous yield or fracture failure on the perpendicular line from the bolts longitudinal line to the angle toe (line bc). Since this is a rupture or fracture failure mode, the mode with the larger fracture component is the controlling mode.

For line ab, the gross shear area is

$$A_{gv} = 5 \times 0.25 \times 2 = 2.5 \text{ in}^2$$

and the net shear area is

$$A_{nv} = 2.5 - (1.5 \times 0.25 \times 1.0)2 = 1.75 \text{ in}^2$$

For line bc, the gross tension area is

$$A_{gt} = 1.5 \times 0.25 \times 2 = 0.75 \text{ in}^2$$

and the net tension area is

$$A_{nt} = 0.75 - 0.5 \times 1.0 \times 0.25 \times 2 = 0.5 \text{ in}^2$$

The shear fracture strength of line ab is $0.6F_u A_{nv} = 0.6 \times 58 \times 1.75 = 60.9$ kips, and the tensile fracture strength of line bc is $F_u A_{nt} = 58 \times 0.5 = 29.0$ kips. Since $60.9 > 29.0$, the block shear limit state involves shear fracture on line ab and tension yield on line bc. Thus, the block shear design strength (capacity) is

$$\phi R_{bs} = 0.75(60.9 + 36 \times 0.75) = 65.9 \text{ kips}$$

  d. Bearing: The end distance of 2 in and the spacing of 3 in satisfy the criteria of $1\tfrac{1}{2}d$ and $3d$, respectively, where d is the bolt diameter. Thus, the bearing design strength is

$$\phi R_p = 0.75 \times 2.4 \times 0.875 \times 0.25 \times 58 \times 2 \times 2 = 91.4 \text{ kips}$$

2. *Bolts—angles to piece W 16:* The limit state for the bolts is shear. The shear capacity of one bolt is

$$\phi r_v = 0.75 \times 48 \, \frac{\pi}{4} \, 0.875^2 = 21.6 \text{ kips}$$

This value can also be obtained from AISC LRFD *Manual of Steel Construction,* vol. II, Table 8-11, p. 8-24. The design strength of two bolts in double shear is

$$\phi R_v = 21.6 \times 2 \times 2 = 86.4 \text{ kips}$$

3. *Piece W 16 × 57:* The limit states for this part of the connection are Whitmore section yield and buckling, bearing, and prying action in conjunction with the W 16 flange to W 18 flange bolts. Because there is only one line of bolts, tearout is not a limit state.
  a. Whitmore section: This is the section denoted by l_w on Fig. 2.27. It is formed by 30° lines from the bolt furthest away

from the end of the brace to the intersection of these lines with a line through and perpendicular to the bolt nearest the end of the brace. Whitmore (1952) determined that this 30° spread gave an accurate estimate of the stress in gusset plates at the end of the brace. The length of the Whitmore section $l_w = 3(\tan 30°)2 = 3.46$ in.

(1) *Whitmore yield:*

$$\phi R_{wy} = 0.9 \times 50 \times 3.46 \times 0.430 = 67.0 \text{ kips}$$

where 0.430 is the web thickness of a W 16 × 57.

(2) *Whitmore buckling:* Tests (Gross, 1990) have shown that the Whitmore section can be used as a conservative estimate for gusset buckling. In the present case, the web of the W 16 × 57 is a gusset. If the load, P, is a compression, it is possible for the gusset to buckle laterally in a sidesway mode. For this mode of buckling, the K factor is 1.2. The buckling length is $l_b = 5$ in in Fig. 2.27. Thus the slenderness ratio is

$$\frac{Kl}{r} = \frac{1.2 \times 5 \sqrt{12}}{0.430} = 48.3$$

From the AISC LRFD *Manual of Steel Construction,* vol. I, Table 3-50, p. 6-148, $\phi F_{cr} = 35.8$ ksi, and the Whitmore buckling capacity is

$$\phi R_{wb} = 35.8 \times 3.46 \times 0.430 = 53.3 \text{ kips}$$

b. *Bearing:* Since the end and spacing distance requirements are satisfied

$$\phi R_p = 0.75 \times 2.4 \times 0.875 \times 0.430 \times 65 \times 2 = 88.0 \text{ kips}$$

c. *Prying action:* Prying action explicitly refers to the extra tensile force in bolts which connect flexible plates or flanges subjected to loads normal to the flanges. For this reason, prying action involves not only the bolts but the flange thickness, bolt pitch and gage, and in general, the geometry of the entire connection.

The AISC LRFD *Manual of Steel Construction,* vol. II, p. 11-6, presents a method to calculate the effects of prying. This method was originally developed by Struik (1969) and presented in the book by Kulak, Fisher, and Struik (1987). The form used in the AISC LRFD *Manual of Steel Construction* was developed by Thornton (1985) for ease of calculation and to provide optimum results, that is, maximum capacity for a given connection (analysis) and minimum required thickness for a given load (design). Thornton (1992, 1997) has shown that this method gives a very conservative estimate of ultimate load and shows that very close estimates of ultimate load can be obtained by using the flange ultimate strength, F_u, in place of the yield strength, F_y, in the AISC LRFD *Manual of Steel Construction* formulas. See Sec. 2.5.3.

From the foregoing calculations, the capacity (design strength) of this connection is 53.3 kips. Let us take this as the design load (required strength) and proceed to the prying calculations. The vertical component of 53.3 is 47.7 kips and the horizontal component is 23.8 kips. Thus, the shear per bolt is $V = 23.8/8 = 2.98$ kips and the tension per bolt is $T = 47.7/8 = 5.96$ kips. Since $2.98 < 21.6$, the bolts are ok for shear. The interaction equation for A325N bolts is

$$\phi F'_t = 0.75(117 - 1.9f_v) \le 0.75 \times 90 = 67.5 \text{ ksi}$$

With $V = 2.98$, $f_v = 2.98/0.6013 = 4.96$ ksi and

$$\phi F'_t = 0.75(117 - 1.9 \times 4.96) = 80.7 \text{ ksi} > 67.5 \text{ ksi}$$

so

$$\phi F'_t = 67.5 \text{ ksi} \qquad \text{and} \qquad \phi r'_t = 67.5 \times 0.6013 = 40.6 \text{ kips}$$

Since the design strength per bolt, $\phi r'_t = 40.6$ kips, is greater than the required strength (or load) per bolt, $T = 5.96$ kips, the bolts are ok. Now, to check prying of the W 16 piece, following the notation of the AISC LRFD *Manual of Steel Construction*,

$$b = \frac{4.5 - 0.430}{2} = 2.035$$

$$a = \frac{7.125 - 4.5}{2} = 1.3125$$

Check that $a < 1.25b = 1.25 \times 2.035 = 2.544$. Since $a = 1.3125 < 2.544$, use $a = 1.3125$. If $a > 1.25b$, $a = 1.25b$ would be used:

$$b' = 2.035 - \frac{0.875}{2} = 1.598$$

$$a' = 1.3125 + \frac{0.875}{2} = 1.75$$

$$\rho = \frac{b'}{a'} = 0.91$$

$$p = 3$$

$$\delta = 1 - \frac{d'}{p} = 1 - \frac{0.9375}{3} = 0.6875$$

$$\alpha' = \frac{1}{\delta(1 + \rho)}\left[\left(\frac{t_c}{t}\right)^2 - 1\right]$$

$$t_c = \sqrt{\frac{4.44(\phi r_t')b'}{pF_y}} = \sqrt{\frac{4.44 \times 40.6 \times 1.598}{3 \times 50}} = 1.386$$

$$\alpha' = \frac{1}{0.6875 \times 1.91}\left[\left(\frac{1.386}{0.715}\right)^2 - 1\right] = 2.10$$

Since $\alpha' > 1$, use $\alpha' = 1$ in subsequent calculations. $\alpha' = 2.10$ means that the bending of the W 16×57 flange will be the controlling limit state. The bolts will not be critical. The design tensile strength, T_d, per bolt including the flange strength is

$$T_d = \phi r_t'\left(\frac{t}{t_c}\right)^2(1 + \delta) = 40.6\left(\frac{0.715}{1.386}\right)^2 1.6875$$

$$= 18.2 \text{ kips} > 5.96 \text{ kips, ok}$$

The subsequent d denotes "design" strength.

In addition to the prying check on the piece W 16×57, a check should also be made on the flange of the W 18×50 beam. A method for doing this was presented in Fig. 2.11. Thus,

$$\bar{b} = \frac{4.5 - 0.355}{2} = 2.073$$

$$\bar{a} = \frac{7.5 - 4.5}{2} = 1.50$$

$$n = 4$$

$$p = 3$$

$$p_{\text{eff}} = \frac{3(4 - 1) + \pi\, 2.073 + 2 \times 1.50}{4} = 4.63$$

Now, using the prying formulation from the AISC LRFD *Manual of Steel Construction,*

$$b = \bar{b} = 2.073$$

$$a = 1.3125$$

Note that the prying lever arm is controlled by the narrower of the two flanges:

$$b' = 2.073 - \frac{0.875}{2} = 1.636$$

$$a' = 1.3125 + \frac{0.875}{2} = 1.75$$

$$\rho = 0.93$$

$$p = p_{eff} = 4.63$$

$$\delta = 1 - \frac{0.9375}{4.63} = 0.798$$

$$t_c = \sqrt{\frac{4.44 \times 40.6 \times 1.636}{4.63 \times 50}} = 1.13$$

$$\alpha' = \frac{1}{0.798 \times 1.93}\left[\left(\frac{1.13}{0.570}\right)^2 - 1\right] = 1.90$$

Use $\alpha' = 1$

$$T_d = 40.6\left(\frac{0.570}{1.13}\right)^2 1.798 = 18.6 \text{ kips} > 5.96 \text{ kips, ok}$$

The prying checks in this example were not critical. If they were and a better estimate of the true failure limit state is desired, the ultimate strength rather than the yield strength can be used in the prying formulas (Thornton, 1992, 1997).

The only formula that needs to be modified is

$$t_c = \sqrt{\frac{4.44\phi r_t' b'}{pF_u}}$$

For the piece W 16

$$t_c = \sqrt{\frac{4.44 \times 40.6 \times 1.598}{3 \times 65}} = 1.215$$

$$\alpha' = \frac{1}{0.6875 \times 1.93}\left[\left(\frac{1.215}{0.715}\right)^2 - 1\right] = 1.42$$

Use $\alpha' = 1$

$$T_d = 40.6\left(\frac{0.715}{1.215}\right)^2 1.6875 = 23.7 \text{ kips}$$

Thus, T_d is increased to 23.7 kips from 18.2 kips. Additional checks on the W 18 × 50 beam are for web yielding; since $5k = 5 \times 1.25 = 6.25 > p = 3$, the web tributary to each bolt at the k distance exceeds the bolt spacing and thus $N = 9$.

$\phi R_{wy} = 1.0(9 + 5 \times 1.25)50 \times 0.355 = 271$ kips > 47.7 kips, ok

and for web crippling, web crippling occurs when the load is compression, thus $N = 12$, the length of the piece W16.

$$\phi R_{wcp} = 0.75 \times 135 \times 0.355^2 \left[1 + 3 \left(\frac{12}{17.99} \right) \left(\frac{0.355}{0.570} \right)^{1.5} \right]$$

$$\times \sqrt{\frac{50 \times 0.570}{0.355}} = 227 \text{ kips} > 47.7 \text{ kips, ok}$$

This completes the design calculations for this connection. A load path has been provided through every element of the connection. For this type of connection, the beam designer should make sure that the bottom flange is stabilized if P can be compressive. A transverse beam framing nearby as shown in Fig. 2.27 by the W18 × 50 web hole pattern, or a bottom flange stay, will provide stability.

2.2.4   Column base plates

The geometry of a column base plate is shown in Fig. 2.28. The area of the base plate is $A_1 = B \times N$. The area of the pier which is concentric with A_1 is A_2. If the pier is not concentric with the base plate, only that portion which is concentric can be used for A_2. The design strength of the concrete in bearing is

$$\phi_c F_p = 0.6 \times 0.85 f_c' \sqrt{\frac{A_2}{A_1}}$$

where f_c' is the concrete compressive strength, in ksi, and

$$1 \le \sqrt{\frac{A_2}{A_1}} \le 2$$

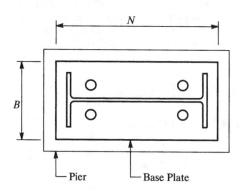

Figure 2.28 Column base plate.

The required bearing strength is

$$f_p = \frac{P}{A_1}$$

where P is the column load (factored), in kips. In terms of these variables, the required base plate thickness is

$$t_p = l \sqrt{\frac{2f_p}{\phi F_y}}$$

where $l = \max \{m, n, \lambda n'\}$

ϕF_y = base plate design strength = $0.9 F_y$

$$m = \frac{N - 0.95d}{2}$$

$$n = \frac{B - 0.8b_f}{2}$$

$$n' = \frac{\sqrt{db_f}}{4}$$

$$\lambda = \frac{2\sqrt{x}}{1 + \sqrt{1 - x}} \leq 1$$

$$x = \frac{4db_f}{(d + b_f)^2} \frac{f_p}{\phi_c F_p}$$

d = depth of column

b_f = flange width of column

For simplicity λ can always be conservatively taken as unity. The formulation given here was developed by Thornton, (1990a, 1990b) based on previous work by Murray (1983), Fling (1970), and Stockwell (1975). It is the method given in the AISC LRFD *Manual of Steel Construction* (1994).

Example   The column of Fig. 2.28 is a W 24 × 84 carrying 600 kips. The concrete has $f'_c = 4.0$ ksi. Try a base plate of A36 steel 4 in bigger than the column in both directions. Since $d = 24\frac{1}{8}$ and $b_f = 9$, $N = 24\frac{1}{8} + 4 = 28\frac{1}{8}$, $B = 9 + 4 = 13$. Try a plate 28 × 13. Assume that 2 in of grout will be used so the minimum pier size is 32 × 17. Thus $A_1 = 28 \times 13 = 364$ in², $A_2 = 32 \times 17 = 544$ in², $\sqrt{A_2/A_1} = 1.22 < 2$ (ok), and

$$\phi_c F_p = 0.6 \times 0.85 \times 4 \times 1.22 = 2.49 \text{ ksi}$$

$$f_p = \frac{600}{364} = 1.65 \text{ ksi} < 2.49 \text{ ksi, ok}$$

$$m = \frac{28 - 0.95 \times 24.125}{2} = 2.54$$

$$n = \frac{13 - 0.8 \times 9}{2} = 2.90$$

$$n' = \sqrt{\frac{24.125 \times 9}{4}} = 3.68$$

$$x = \frac{4 \times 24.125 \times 9.0}{(24.125 + 9.0)^2} \frac{1.65}{2.49} = 0.52$$

$$\lambda = \frac{2.\sqrt{0.52}}{1 + \sqrt{1 - 0.52}} = 0.85$$

$$l = \max\{2.54, 2.90, 0.85 \times 3.68\} = 3.13$$

$$t_p = 3.13 \sqrt{\frac{2 \times 1.65}{0.9 \times 36}} = 0.99 \text{ in}$$

Use a plate $1 \times 13 \times 28$ of A36 steel. If the conservative assumption of $\lambda = 1$ were used, $t_p = 1.17$ in, which indicates a 1¼-in-thick base plate.

Erection considerations. In addition to designing a base plate for the column compression load, loads on base plates and anchor bolts during erection should be considered. A common design load for erection is a 1-kip working load, applied at the top of the column in any horizontal direction. If the column is, say, 40 ft high, this 1-kip force at a lever arm of 40 ft will cause a significant couple at the base plate and anchor bolts. The base plate, anchor bolts, and column-to-base plate weld, should be checked for this construction load condition. The paper by Murray (1983) gives some yield-line methods that can be used for doing this.

2.2.5 Splices—columns and truss chords

Section J1.4 of the AISC LRFD *Manual of Steel Construction,* 2d ed., Specification says that finished to bear compression splices in columns need be designed only to hold the parts "securely in place." For this reason, AISC provides a series of "standard" column splices in the AISC LRFD *Manual of Steel Construction,* vol. II, p. 11-72 to p. 11-91. These splices are nominal in the sense that they are not designed for a particular load. Section J1.4 also requires that splices in trusses be designed for at least 50% of the design load (required strength). The difference between columns and "other compression members," such as compression chords of trusses, is that for columns, splices are usually near lateral support points, such as floors, whereas trusses can have their splices at the midpanel point where there is no lateral support.

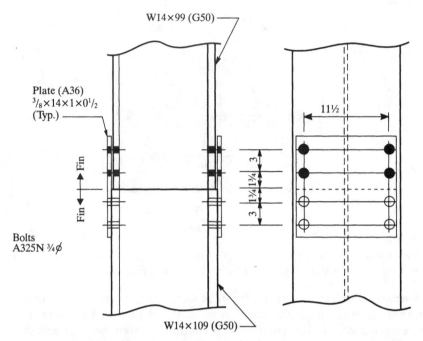

Figure 2.29   An AISC standard column splice.

Column splices.   Figure 2.29 shows a standard AISC column splice for a W 14 × 99 to a W 14 × 109. If the column load remains compression, the strong axis column shear can be carried by friction. The coefficient of static friction of steel to steel is on the order of 0.5 to 0.7, so quite high shears can be carried by friction. Suppose the compression load on this column is 700 kips. How much major axis bending moment can this splice carry? Even though these splices are nominal, they can carry quite significant bending moment. The flange area of the W 14 × 99 is $A_f = 0.780 \times 14.565 = 1.14$ in^2. Thus, the compression load per flange is $700 \times 11.4/29.1 = 274$ kips. In order for a bending moment to cause a tension in the column flange, this load of 274 kips must first be unloaded. Assuming that the flange force acts at the flange centroid, the moment in the column can be represented as:

$$M = T(d - t_f) = T(14.16 - 0.780) = 13.38T$$

If $T = 274$ kips, one flange will be unloaded, and $M = 13.38 \times 274 = 3666$ kip-in = 306 kip-ft. The design strength in bending for this column (assuming sufficient lateral support) is $\phi M_p = 647$ kip-ft. Thus, because of the compression load, the nominal AISC splice, while still seeing no load, can carry almost 50% of the column's bending capacity.

The splice plates and bolts will allow additional moment to be carried. It can be shown that the controlling limit state for the splice material is bolt shear. For one bolt $\phi R_v = 15.9$ kips. Thus for four bolts, $\phi R_v = 15.9 \times 4 = 63.6$ kips. The splice forces are assumed to act at the faying surface of the deeper member. Thus the moment capacity of the splice plates and bolts is $M_s = 63.6 \times 14.32 = 911$ kip-in $= 75.9$ kip-ft. The total moment capacity of this splice with zero compression is thus 75.9 kip-ft and with 700 kips compression, it is $306 + 75.9 = 382$ kip-ft. The role of compression in providing moment capability is often overlooked in column splice design.

Erection stability. As discussed earlier for base plates, the stability of columns during erection must be a consideration for splice design also. The usual nominal erection load for columns is a 1-kip horizontal force at the column top in any direction. In LRFD format, the 1-kip working load is converted to a factored load by multiplying by a load factor of 1.5. This load of $1 \times 1.5 = 1.5$ kips will require connections which will be similar to those obtained in allowable stress design (ASD) with a working load of 1 kip. It has been established that for major axis bending, the splice is good for 75.9 kip-ft. This means that the 1.5-kip load can be applied at the top of a column $75.9/1.5 = 50.6$ ft tall. Most columns will be shorter than 50.6 ft, but if not, a more robust splice should be considered.

Minor axis stability. If the 1.5-kip erection load is applied in the minor or weak axis direction, the forces at the splice will be as shown in Fig. 2.30. The upper shaft will tend to pivot about point O. Taking moments about point O,

$$ PL = T\left(\frac{d}{2} - \frac{g}{2}\right) + T\left(\frac{d}{2} + \frac{g}{2}\right) = Td $$

Thus the erection load, P, that can be carried by the splice is

$$ P = \frac{Td}{L} $$

Note that this erection load capacity (design strength) is independent of the gage g. This is why the AISC splices carry the note, "Gages shown can be modified if necessary to accommodate fittings elsewhere on the column." The standard column gages are 5½ and 7½ in for beams framing to column flanges. Errors can be avoided by making all column gages the same. The gages used for the column splice can also be 5½ or 7½ in without affecting erection stability.

If the upper column of Fig. 2.29 is 40 ft long and T is the shear strength of four (two per splice plate) bolts,

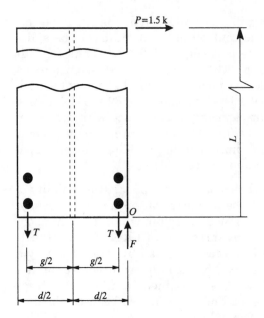

Figure 2.30 Weak-axis stability forces for column splice.

$$P = \frac{4 \times 15.9 \times 14.565}{40 \times 12} = 1.93 \text{ kips}$$

Since $1.93 > 1.5$, this splice is satisfactory for a 40-ft-long column. If it were not, larger or stronger bolts could be used.

2.2.5.1 Column splices for biaxial bending. The simplest method for designing this type of splice is to establish a flange force (required strength) which is statically equivalent to the applied moments and then to design the bolts, welds, plates, and fillers (if required) for this force.

Major axis bending. If M_x is the major axis applied moment and d is the depth of the deeper of the two columns, the flange force (or required strength) is

$$F_{fx} = \frac{M_x}{d}$$

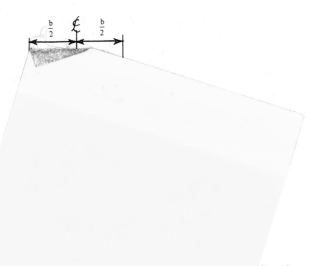

distribution
ling.

Minor axis bending. The force distribution is similar to that shown in Fig. 2.30 for erection stability. The force, F, in the case of actual (factored) design loads can be quite large and will need to be distributed over some finite bearing area as shown in Fig. 2.31. In Fig. 2.31, the bearing area is $2\varepsilon t$, where t is the thickness of the inner flange, ε is the position of the force, F, from the toe of the flange of the smaller column, and T is the force per gage line of bolts. The quantities T and F are for each of the two flanges. If M_y is the weak axis applied moment, $M_f = M_y/2$ is the weak axis applied moment per flange. Taking moments about point 0 gives (per flange)

$$M_f = T\left(\frac{b}{2} - \frac{g}{2} - \varepsilon\right) + T\left(\frac{b}{2} + \frac{g}{2} - \varepsilon\right) = T(b - 2\varepsilon)$$

The bearing area is determined by requiring that the bearing stress reaches its design strength at the load, F. Thus, $0.75\,(1.8F_y)(2\varepsilon)t = F$, and since from vertical equilibrium $F = 2T$, and

$$0.75(1.8F_y)\,t\,\varepsilon = T$$

Thus

$$M_f = 0.75(1.8F_y)t\varepsilon(b - 2\varepsilon)$$

and solving for ε

$$\varepsilon = \frac{1}{4}\,b - \frac{1}{2}\sqrt{\left(\frac{b}{2}\right)^2 - \frac{40}{27}\left(\frac{M_f}{F_{ytt}}\right)} = \frac{1}{4}\,b\left[1 - \sqrt{1 - \frac{8}{3}\frac{M_f}{\phi M_{py}}}\right]$$

where $M_{py} = F_yZ_y = \frac{1}{2}F_ytb^2$. This expression for ε is valid as long as

$$M_f \le \frac{27}{40}\left(\frac{F_ytb^2}{4}\right) = \frac{3}{8}\,\phi M_{py}$$

When $M_f > 3/8\ \phi M_{py}$, the tension, T, on the bolts on the bearing side vanishes and Fig. 2.32 applies. In this case, $F = T = 0.75(1.8F_y)t(2\varepsilon)$,

$$M_f = T\left(\frac{b + g}{2} - \varepsilon\right)$$

and

$$\varepsilon = \frac{1}{4}(b + g) - \frac{1}{2}\sqrt{\left(\frac{b+g}{2}\right)^2 - \frac{40}{27}\left(\frac{M_f}{F_yt}\right)}$$

$$= \frac{1}{4}\,b\gamma\left[1 - \sqrt{1 - \frac{8}{3}\frac{M_f}{\phi M_{py}}\left(\frac{1}{\gamma}\right)^2}\right]$$

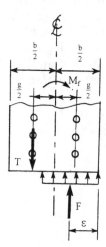

Figure 2.32 Splice force distribution when bolts on bearing side are ineffective.

where $\gamma = 1 + g/b$. This expression for ε is valid as long as

$$M_f \leq \frac{27}{40} \frac{F_y t(b + g)^2}{4} = \frac{3}{8} \gamma^2 \phi M_{py}$$

but T need never exceed M_f/g. The flange force in every case is $F_{fy} = 2T$.

Example   Design a bolted splice for a W 14 × 99 upper shaft to a W 14 × 193 lower shaft. Design the splice for 15% of the axial capacity of the smaller member plus 20% of the smaller member's bending capacity about either the major or minor axis, whichever produces the greater flange force, F_f. The columns are ASTM A572-50, the splice plates are ASTM A36, and the bolts are ASTM A490 1-in-diameter X type. The holes are standard $1\frac{1}{16}$-in diameter. The gage is $7\frac{1}{2}$ in.

The completed splice is shown in Fig. 2.33. The flange force due to tension is

$$F_{f_t} = 0.15\phi \frac{F_y}{2} A_g = \frac{0.15 \times 0.9 \times 50 \times 29.1}{2} = 98.2 \text{ kips}$$

The flange force due to major axis bending is

$$F_{f_x} = \frac{0.20\phi M_{px}}{d} = \frac{0.20 \times 647 \times 12}{14.16} = 110 \text{ kips}$$

The flange force due to minor axis bending is calculated as follows:

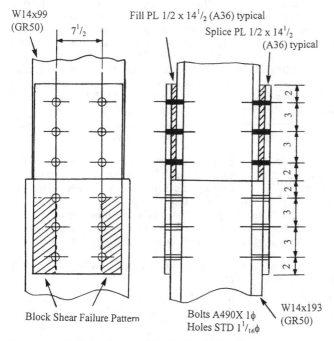

Figure 2.33   Bolted column splice for biaxial bending.

$$M_f = \frac{0.20\phi M_{py}}{2} = \frac{0.20 \times 0.9 \times 50 \times 83.6}{2} = 376 \text{ kip-in}$$

Check that $M_f = 376 \leq 3/8 \times 0.9 \times 50 \times 83.6 = 1410$ kip-in, ok, and calculate

$$\varepsilon = \frac{1}{4} \ 14.565 \left(1 - \sqrt{1 - \frac{376}{1410}} \ \right) = 0.523 \text{ in}$$

Thus, $T = 0.75(1.8 \times 50) \ 0.780 \times 0.523 = 27.5$ kips and $F_{fy} = 2 \times 27.5 = 55.0$ kips. The flange force for design of the splice is thus

$$F_f = F_{f_t} + \max\{F_{f_x}, F_{f_y}\} = 98.2 + \max\{110, 55.0\} = 208 \text{ kips}$$

Suppose that $M_f > 3/8\phi M_{py}$. Let $M_f = 1500$ kip-in, say, $\gamma = 1 + 7.5/14.565 = 1.515$ and check $M_f = 1500$ kip-in $< 3/8 \ \gamma\phi M_{py} = (1.515)^2 \ 1410 = 3236$ kip-in, so proceeding

$$\varepsilon = \frac{1}{4} \ 14.565 \times 1.515 \left(1 - \sqrt{1 - \frac{1500}{3236}} \ \right) = 1.476 \text{ in}$$

$T = 0.75(1.8 \times 50) \ 0.78 \times 2 \times 1.476 = 155$ kips and $F_{fy} = 2T = 311$ kips, which is still less than the maximum possible value of

$$F_{fy} = \frac{1500}{7.5} \times 2 = 400 \text{ kips}$$

Returning to the splice design example, the splice will be designed for a load of 208 kips. Since the columns are of different depths, fill plates will be needed. The theoretical fill thickness is $(15\frac{1}{2} - 14\frac{1}{8})/2 = \frac{11}{16}$ in, but for ease of erection AISC suggests subtracting either $\frac{1}{8}$ in or $\frac{3}{16}$ in, whichever results in $\frac{1}{8}$-in multiples of fill thickness. Thus, use actual fills $\frac{3}{16} - \frac{11}{16} = \frac{1}{2}$ in thick. Since this splice is a bearing splice, the fills either must be developed, or the shear strength of the bolts must be reduced. It is usually more economical to do the latter in accordance with AISC Specification section J6 when the total filler thickness is not more than $\frac{3}{4}$ in. Using J6, the bolt shear design strength is

$$\phi r_v = 44.2[1 - 0.4(0.5 - 0.25)] = 39.8 \text{ kips}$$

The number of bolts required is $208/39.8 = 5.23$ or 6 bolts. By contrast, if the fillers were developed, the number of bolts required would be $208[1 + 0.5/(0.5 + 0.780)]/44.2 = 6.54$ or 8 bolts. By reducing the bolt shear strength instead of developing the fills, $[(8 - 6)/8]100 = 25\%$ fewer bolts are required for this splice. Next, the splice plates are designed. These plates will be approximately as wide as the narrower column flange. Since the W 14 $\times$ 99 has a flange width of $14\frac{5}{8}$ in, use a plate $14\frac{1}{2}$ in wide. The following limit states are checked:

1. *Gross area:* The required plate thickness based on gross area is $t_p = 208/(0.9 \times 36 \times 14.5) = 0.44$ in. Use a $\frac{1}{2}$-in plate so far.

2. *Net area:* The net area is $A_n = (14.5 - 2 \times 1.125)0.5 = 6.125 \text{ in}^2$, but this cannot exceed 0.85 of the gross area or $0.85 \times 14.5 \times 0.5 = 6.16 \text{ in}^2$. Since

6.16 > 6.125, the effective net area $A_e = A_n = 6.125$ in^2. The design strength in gross tension is $\phi R_n = 0.75 \times 58 \times 6.125 = 266$ kips > 208 kips, ok.

3. *Block shear rupture (or tearout)*: Since $b - g < g$, the failure will occur as shown in Fig. 2.33 on the outer parts of the splice plate.

$$A_{gv} = 8 \times 0.5 \times 2 = 8.0 \text{ in}^2$$

$$A_{gt} = (14.5 - 7.5)\,0.5 = 3.5 \text{ in}^2$$

$$A_{nv} = 8.0 - 2.5 \times 1.125 \times 0.5 \times 2 = 5.1875 \text{ in}^2$$

$$A_{nt} = 14.0 - 1 \times 1.25 \times 0.5 = 2.94 \text{ in}^2$$

$$F_u A_{nt} = 58 \times 2.94 = 171 \text{ kips}$$

$$0.6 F_u A_{nv} = 0.6 \times 58 \times 5.1875 = 181 \text{ kips}$$

Since $F_u A_{nt} < 0.6 F_u A_{nv}$, the failure is by shear fracture and tension yielding, and

$$\phi R_n = 0.75(181 + 36 \times 3.5) = 320 \text{ kips} > 208 \text{ kips, ok}$$

4. *Bearing:*

$$\phi R_n = 0.75 \times 2.4 \times 58 \times 1 \times 0.5 \times 6 = 313 \text{ kips} > 208 \text{ kips, ok}$$

5. *Whitmore section:*

$$l_w = (6 \tan 30)2 + 7.5 = 14.43 \text{ in}$$

$$\phi R_n = 0.9 \times 36 \times 14.43 \times 0.5 = 234 \text{ kips} > 208 \text{ kips}$$

Note that if $l_w > 14.5$ in, 14.5 in would have been used in the calculation of design strength.

In addition to the checks for the bolts and splice plates, the column sections should also be checked for bearing and block shear rupture. These are not necessary in this case because $t_f = 0.780 > tp = 0.50$ and the column material is stronger than the plate material.

2.2.5.2 Splices in truss chords. These splices must be designed for 50% of the chord load, even if the load is compression and the members are finished to bear. As discussed earlier, these splices may be positioned in the center of a truss panel, and therefore, must provide some degree of continuity to resist bending. For the tension chord, the splice must be designed to carry the full tensile load.

Example Design the tension chord splice shown in Fig. 2.34. The load is 800 kips (factored). The bolts are A325X, ⅞ in in diameter, $\phi r_v = 27.1$ kips. The load at this location is controlled by the W 14 × 90, so the loads should be apportioned to flanges and web based on this member. Thus, the flange load is

$$P_f = \frac{0.710 \times 14.520}{26.5}\,800 = 311 \text{ kips}$$

and the web load is

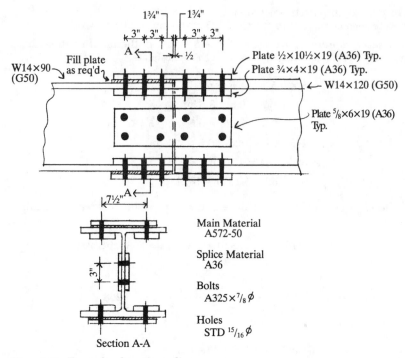

Figure 2.34 Truss chord tension splice.

$$P_w = 800 - 2 \times 311 = 178 \text{ kips}$$

The load path is such that the flange load, P_f, passes from the W 14 × 90 (say) through the bolts into the flange plates and into the W 14 × 120 flanges through a second set of bolts. The web load path is similar:

1. *Flange connection:*
   a. *Bolts:* The number of bolts in double shear is 311/2 × 27.1 = 5.74. Use 6 bolts in 2 rows of 3 as shown in Fig. 2.34.
   b. *Chord net section:* Check to see if the holes in the W 14 × 90 reduce its capacity below 800 kips. Assume that there will be two web holes in alignment with the flange holes.

   $$A_{net} = 26.5 - 4 \times 1 \times 0.710 - 2 \times 1 \times 0.440 = 22.8 \text{ in}^2$$
   $$\phi R_{net} = 22.8 \times 0.75 \times 65 = 1111 \text{ kips} > 800 \text{ kips, ok}$$

   c. *Bearing:*

   $$\phi R_p = 0.75 \times 2.4 \times 65 \times 0.710 \times 0.875 \times 6 = 436 \text{ kips} > 311 \text{ kips, ok}$$

   d. *Block shear rupture (tearout):*

$$A_{nv} = (7.75 - 2.5 \times 1.0)\, 0.710 \times 2 = 7.46 \text{ in}^2$$

$$A_{nt} = \left(\frac{(14.205 - 7.5)}{2} - 0.5 \times 1 \right) 0.710 \times 2 = 4.27 \text{ in}^2$$

$$A_{gv} = 7.75 \times 0.710 \times 2 = 11.0 \text{ in}^2$$

$$A_{gt} = 3.51 \times 0.710 \times 2 = 4.98 \text{ in}^2$$

$$F_u A_{nt} = 65 \times 4.27 = 278 \text{ kips}$$

$$0.6 F_u A_{nv} = 0.6 \times 65 \times 7.46 = 291 \text{ kips}$$

Since $0.6 F_u A_{nv} > F_u A_{nt}$, the tearout capacity (design strength) is

$$\phi R_{to} = 0.75(291 + 50 \times 4.98) = 405 \text{ kips} > 311 \text{ kips, ok}$$

e. *Flange plates:* Since the bolts are assumed to be in double shear, the load path is such that one-half of the flange load goes into the outer plate and one-half goes into the inner plates.
(1) *Outer plate:*
  (a) *Gross and net area:* Since the bolt gage is 7½ in, try a plate 10½ in wide. The gross area in tension required is

$$A_{gt} = \frac{311/2}{0.9 \times 36} = 4.8 \text{ in}^2$$

and the thickness required is $4.8/10.5 = 0.46$ in. Try a plate ½ × 10½

$$A_{gt} = 0.5 \times 10.5 = 5.25 \text{ in}^2$$

$$A_{nt} = (10.5 - 2 \times 1)\, 0.5 = 4.25 \text{ in}^2$$

$$\phi A_{gt} = 0.85 \times 5.25 = 4.46 \text{ in}^2$$

Since $0.85\, A_{gt} > A_{nt}$, use $A_{nt} = 4.25 \text{ in}^2$ as the effective net tension area.

$$\phi R_{nt} = 0.75 \times 58 \times 4.25 = 185 \text{ kips} > 311/2 = 156 \text{ kips,}$$
$$\text{ok}$$

Use a plate ½ × 10½ for the outer flange splice plate for the following limit state checks
(b) *Block shear rupture:*

$$A_{gv} = 7.5 \times 0.5 \times 2 = 7.5 \text{ in}^2$$

$$A_{gt} = 1.5 \times 0.5 \times 2 = 1.5 \text{ in}^2$$

$$A_{nv} = (7.5 - 2.5 \times 1)\, 0.5 \times 2 = 5.0 \text{ in}^2$$

$$A_{nt} = (1.5 - 0.5 \times 1)\, 0.5 \times 2 = 1.0 \text{ in}^2$$

$$F_u A_{nt} = 58 \times 1.0 = 58.0 \text{ kips}$$

$$0.6F_uA_{nv} = 0.6 \times 58 \times 5.0 = 174 \text{ kips}$$

Since $0.6F_uA_{nv} > F_uA_{nt}$,

$$\phi R_{to} = 0.75(174 + 36 \times 1.5) = 171 \text{ kips} > 156 \text{ kips, ok}$$

(c) *Bearing:*

$$\phi R_p = 0.75 \times 2.4 \times 58 \times 0.50 \times 0.875 \times 6$$
$$= 274 \text{ kips} > 156 \text{ kips, ok}$$

Thus, the plate ½ × 10½ (A36) outer splice plate is ok.
(2) *Inner plates:*
  (a) *Gross and net area:* The load to each plate is 156/2 = 78 kips. The gross area in tension required is

$$A_{gt} = \frac{78}{0.9 \times 36} = 2.41 \text{ in}^2$$

Try a plate 4 in wide. Then the required thickness is 2.41/4 = 0.6 in. Try a plate ¾ × 4 (A36).

$$A_{gt} = 0.75 \times 4 = 3 \text{ in}^2$$
$$A_{nt} = (4 - 1.0)\, 0.75 = 2.25 \text{ in}^2$$
$$\phi A_{gt} = 0.85 \times 3 = 2.55 \text{ in}^2$$
$$\phi R_{nt} = 0.75 \times 58 \times 2.25 = 97.9 \text{ kips} > 78 \text{ kips, ok}$$

  (b) *Block shear rupture:* Since there is only one line of bolts, this limit state is not possible.
  (c) *Bearing:*

$$\phi R_p = 0.75 \times 2.4 \times 58 \times 0.75 \times 0.875 \times 3$$
$$= 206 \text{ kips} > 78 \text{ kips, ok}$$

Use the ¾ × 4 (A36) inner splice plates.
2. *Web connection:* The calculations for the web connection involve the same limit states as the flange connection.
  a. Bolts:

$$\text{Number required} = \frac{178}{2 \times 27.1} = 3.28$$

Use four bolts.
  b. Web limit states:
    (1) *Bearing:*

$$\phi R_p = 0.75 \times 2.4 \times 65 \times 0.440 \times 0.875 \times 4$$
$$= 180 \text{ kips} > 178 \text{ kips, ok}$$

(2) *Block shear rupture (tearout)*:   Assume the bolts have a 3-in pitch longitudinally. Then

$$A_{nv} = (4.75 - 1.5 \times 1)\, 0.440 \times 2 = 2.86 \text{ in}^2$$
$$A_{nt} = (3 - 1 \times 1)\, 0.440 = 0.88 \text{ in}^2$$
$$A_{gv} = 4.75 \times 0.440 \times 2 = 4.18 \text{ in}^2$$
$$A_{gt} = 3 \times 0.440 = 1.32 \text{ in}^2$$
$$F_u A_{nt} = 65 \times 0.88 = 57.2 \text{ kips}$$
$$0.6 F_u A_{nv} = 0.6 \times 65 \times 2.86 = 112 \text{ kips}$$

Since $0.6 F_u A_{nv} > F_u A_{nt}$

$$\phi R_{to} = 0.75(112 + 50 \times 1.32) = 134 \text{ kips} < 178 \text{ kips, no good}$$

Since the tearout limit state fails, the bolts can be spaced out to increase the capacity. Increase the bolt pitch from the 3 in assumed previously, to 6 in. Then

$$A_{nv} = (7.75 - 1.5 \times 1)\, 0.440 \times 2 = 5.50 \text{ in}^2$$
$$A_{nt} = 0.88 \text{ in}^2$$
$$A_{gv} = 7.75 \times 0.440 \times 2 = 6.82 \text{ in}^2$$
$$A_{gt} = 1.32 \text{ in}^2$$

As before, $0.6 F_u A_{nv} > F_u A_{nt}$, so

$$\phi R_{to} = 0.75(0.6 \times 65 \times 5.50 + 50 \times 1.32)$$
$$= 210 \text{ kips} > 178 \text{ kips, ok}$$

Use the web bolt pattern shown in Fig. 2.34.
c. *Web plates:*   Try two plates, one each side of web, 6 in wide and ½ in thick.
  (1) *Gross area:*

$$\phi R_{gt} = 0.9 \times 36 \times 0.5 \times 6 \times 2 = 194 \text{ kips} > 178 \text{ kips, ok}$$

  (2) *Net area:*

$$A_{nt} = (6 - 2 \times 1)0.5 \times 2 = 4.0 \text{ in}^2$$
$$0.85 A_{gt} = 0.85 \times 0.5 \times 6 \times 2 = 5.1 \text{ in}^2$$
$$\phi R_{nt} = 0.75 \times 58 \times 4.0 = 174 \text{ kips} < 178 \text{ kips, no good}$$

Increase web plates to ⅝ in thick. Net area will be ok by inspection.
  (3) *Block shear rupture:*   This is checked as shown in previous calculations. It is not critical here.

(4) *Bearing:*

$$\phi R_p = 0.75 \times 2.4 \times 58 \times 0.625 \times 0.875 \times 2 \times 4$$
$$= 457 \text{ kips} > 178 \text{ kips, ok}$$

The completed splice is shown in Fig. 2.34.
If this were a nonbearing compression splice, the splice plates would be checked for buckling. The following paragraph shows the method, which is not required for a tension splice.

(5) *Buckling:*   Because of the 6-in spacing, check the web plate buckling between the rows. The load is $178/(2 \times 2) = 44.5$ kips. The slenderness ratio is $0.65 \times 6\sqrt{12}/0.625 = 22$. From AISC LRFD, Table 3.36, $\phi F_{cv} = 29.83$ ksi, and $29.83 \times 0.625 \times 6 = 112$ kips > 44.5 kips, ok. The plates at the splice line of 4.0-in length can be checked in the same way against a load of $178/2 = 89$ kips/plate. This limit state is ok by inspection, since at a slenderness ratio of 22, each plate is good for 112 kips. This limit state is checked for the flange plates also.

2.3   Moment Connections

2.3.1   Introduction

The most commonly used moment connection is the field-welded moment connection as shown in Fig. 2.35a. This connection was in common use in all regions of the United States. It is still permitted for use in what are referred to as "ordinary moment frames" (AISC, 1997b) with some special requirements for welding and stiffener size. Also, it is still permitted and is in wide use in areas of slow seismicity where the seismic requirements of AISC (1997b) do not apply, and for resistance to wind and gravity forces. The example of this section will deal with this latter case. Chapter 5 will consider the design and detailing of connections specifically for areas of high seismicity.

2.3.2   Example—three-way moment connection

The moment connection of Fig. 2.35a is a three-way moment connection. Additional views are shown in Figs. 2.35b and 2.35c. If the strong axis connection requires stiffeners, there will be an interaction between the flange forces of the strong and weak axis beams. If the primary function of these moment connections is to resist lateral maximum load from wind or seismic sources, the interaction can generally be ignored because the maximum lateral loads will act in only

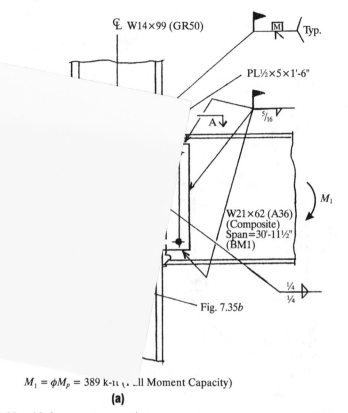

$M_1 = \phi M_p = 389$ k-ı̣ (ı̣ ̣ll Moment Capacity)

(a)

Figure 2.35a Field-welded moment connection.

one direction at any one time. If the moment connections are primarily used to carry gravity loads, such as would be the case when stiff floors with small deflections and high natural frequencies are desired, there will be interaction between the weak and strong beam flange forces. The calculations here will be for a wind or a seismic condition in a region of low to moderate seismicity, but interaction will be included to demonstrate the method.

The load path through this connection that is usually assumed is that the moment is carried entirely by the flanges and the shear entirely by the web. This load path has been verified by testing (Huang et al., 1973) and will be the approach used here. Proceeding to the connection design, the strong axis beam, beam no. 1, will be designed first.

Beam no. 1 W 21 3 62 (A36) composite. The flange connection is a full-penetration weld so no design is required. The column must be checked for stiffeners and doublers.

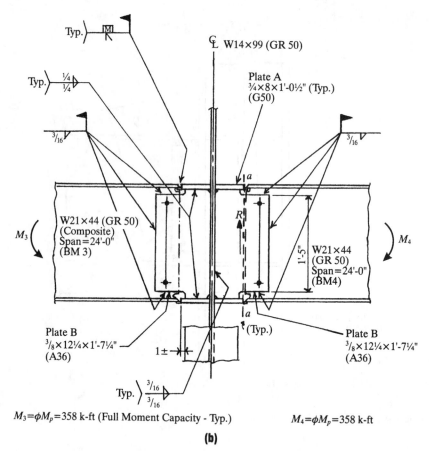

$M_3 = \phi M_p = 358$ k-ft (Full Moment Capacity - Typ.) $M_4 = \phi M_p = 358$ k-ft

(b)

Figure 2.35b Section B-B of Fig. 2.35a.

Stiffeners. The connection is to be designed for the full moment capacity of the beam. Thus, the flange force, F_f, is

$$F_f = \frac{\phi M_p}{d - t_f} = \frac{389 \times 12}{(20.99 - 0.615)} = 229 \text{ kips}$$

From the column load tables of the AISC LRFD *Manual of Steel Construction,* vol. I, p. 3-20:
Web yielding:

$$P_{wy} = P_{wo} + t_b P_{wi} = 174 + 0.615 \times 24.3 = 189 \text{ kips} < 229 \text{ kips}$$

thus stiffeners are required at both flanges.
Web buckling:

$$P_{wb} = 261 \text{ kips} > 229 \text{ kips}$$

no stiffener required at compression flange.

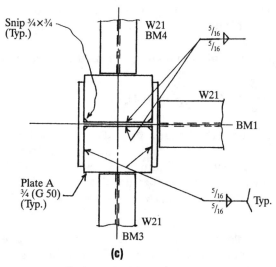

Snip ¾×¾
(Typ.)

W21
BM4

⁵/₁₆
⁵/₁₆

W21

BM1

Plate A
¾ (G 50)
(Typ.)

⁵/₁₆
⁵/₁₆

Typ.

W21

BM3

(c)

Figure 2.35c Section A-A of Fig. 2.35a.

Flange buckling:

$$P_{fb} = 171 \text{ kips} < 229 \text{ kips}$$

stiffener required at tension flange.

From the preceding three checks (limit states), a stiffener is required at both flanges. For the tension flange, the total stiffener force is $229 - 171 = 58$ kips and for the compression flange, the stiffener force is $229 - 189 = 40$ kips. But the loads may reverse, so use the larger of 58 and 40 as the stiffener for both flanges. Then, the force in *each* stiffener is $58/2 = 29$ kips, both top and bottom.

Determination of stiffener size. The minimum stiffener width, w_s, is

$$\frac{b_{fb}}{3} - \frac{t_{wc}}{2} = \frac{8.24}{3} - \frac{0.485}{2} = 2.5 \text{ in}$$

Use a stiffener 6½ in wide to match column.
The minimum stiffener thickness, t_s, is

$$\frac{t_{fb}}{2} = \frac{0.615}{2} = 0.31 \text{ in}$$

Use a stiffener at least ⅜ in thick.
The minimum stiffener length, l_s, is

$$\frac{d_c}{2} - t_{fc} = \frac{14.16}{2} - 0.78 = 6.3 \text{ in}$$

The minimum length is for a "half-depth" stiffener which is not possible in this example because of the weak axis connections. Therefore, use a full-depth stiffener of 12½-in length.

A final stiffener size check is a plate-buckling check which requires that

$$t_s \geq w_s \frac{\sqrt{F_y}}{95} = 6.5 \frac{\sqrt{36}}{95} = 0.41 \text{ in}$$

Therefore, the minimum stiffener thickness is ½ in. The final stiffener size for the strong axis beams is ½ × 6½ × 12½. The contact area of this stiffener against the inside of the column flange is 6.5 − 0.75 = 5.75 due to the snip to clear the column web-to-flange fillet. The stiffener design strength is thus 0.9 × 36 × 5.75 × 0.5 = 93.2 kips > 29 kips, ok.

Welds of stiffeners to column flange and web. Putting aside for the moment that the weak axis moment connections still need to be considered and will affect both the strong axis connection stiffeners and welds, the welds for the ½ × 6½ × 12½ strong axis stiffener are designed as follows. For the weld to the inside of the flange, the *connected* portion of the stiffener must be developed. Thus, the 5¾ contact, which is the connected portion, is designed for 93.2 kips rather than 29 kips, which is the load the stiffener actually "sees." The weld to the flange is thus

$$D_f = \frac{93.2}{2 \times 5.75 \times 1.392 \times 1.5} = 3.9$$

A ¼ fillet weld is indicated, or use the AISC minimum if larger. The factor 1.5 in the denominator of the previous equation comes from the AISC LRFD Specification Appendix J, Section J2.4, for transversely loaded fillets. The weld to the web has a length 12.5 − 0.75 − 0.75 = 11.0, and is designed to transfer the unbalanced force in the stiffener to the web. The unbalanced force in the stiffener is 29 kips in this case. Thus,

$$D_w = \frac{29}{2 \times 11.0 \times 1.392} = 0.95$$

An AISC minimum fillet is indicated.

2.3.2.1 Doublers. The beam flange force (required strength) delivered to the column is $F_f = 229$ kips. The design shear strength of the column $\phi V_v = 0.9 \times 0.6 \times 50 \times 0.485 \times 14.16 = 185$ kips < 229 kips, so a doubler appears to be required. However, if the moment that is causing doublers is $\phi M_p = 389$ kip-ft, then from Fig. 2.36, the column story shear is

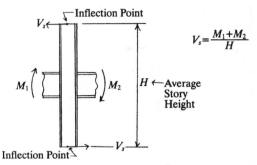

$$V_s = \frac{M_1 + M_2}{H}$$

Figure 2.36 Relationship between column story shear and the moments which induce it.

$$V_s = \frac{\phi M_p}{H}$$

where H is the story height. If $H = 13$ ft

$$V_s = \frac{389}{13} = 30 \text{ kips}$$

and the shear delivered to the column web is $F_f - V_s = 229 - 30 = 199$ kips. Since 199 kips > 185 kips, a doubler (or doublers) is still indicated. If some panel zone deformation is acceptable, the AISC LRFD Specification Section K1.7, Formula K1-11 or K1-12, contains an extra term which increases the panel zone strength. The term is

$$\frac{3b_{fc}t_{fc}^2}{d_b d_c t_{wc}} = \frac{3 \times 14.565 \times 0.780^2}{20.66 \times 14.16 \times 0.485} = 0.187$$

and if the column load is less than $0.75P_y = 0.75A_c F_{yc} = 0.75 \times 29.1 \times 50 = 1091$ kips, which is the usual case,

$$\phi V_v = 185 \times 1.187 = 220 \text{ kips}$$

Since 220 kips > 199 kips, no doubler is required. In a high-rise building where the moment connections are used for drift control, the extra term can still be used, but an analysis which includes inelastic joint shear deformation should be considered.

Placement of doubler plates. If a doubler plate or plates is/are required in this example, the most inexpensive arrangement is to place the doubler plate against the column web between the stiffeners (the panel zone) and to attach the weak axis shear connection plates, plates B, to the face of the doubler. This is permissible provided that the doubler is capable of carrying the entire weak axis shear load,

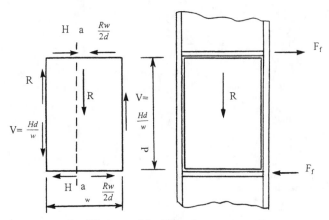

Figure 2.37 Equilibrium of doubler plate with weak axis shear load.

R = 163 kips, on one vertical cross section of the doubler plate. To see this, consider Fig. 2.37. The portion of the shear force induced in the doubler plate by the moment connection flange force, F_f, is H. For the doubler to be in equilibrium under the forces, H, vertical shear forces $V = Hd/w$ must exist. The welds of the doubler at its four edges develop the shear strength of the doubler. Let the shear force, R, from the weak axis connection be applied to the face of the doubler at or near its horizontal center as shown in Fig. 2.37. If it is required that all of the shear, R, can be carried by one vertical section a-a of Fig. 2.37, that is, $0.9 \times 0.6F_y t_d d \geq R$, where t_d is the doubler thickness and F_y is the yield strength of the doubler (and the column), then the free body diagram of Fig. 2.37 is possible. In this figure, all of the shear force, R, is delivered to the side of the doubler where it is opposite in direction to the shear delivered by the moment connection, thereby avoiding overstressing the other side where the two shears would add. Since the doubler and its welds are capable of carrying V or R alone, they are capable of carrying their difference. The same argument applies to the top and bottom edges of the doubler. Also, the same argument holds if the moment and/or weak axis shear reverse(s).

2.3.2.2 Associated shear connections—beam 1. The specified shear for the web connection is R = 163 kips, which is the shear capacity of the W 21 × 62 (A36) beam. The connection is a shear plate with two erection holes for erection bolts. The shear plate is shop-welded to the column flange and field-welded to the beam web. The limit states are plate gross shear, weld strength, and beam web strength.

Plate gross shear. Try a plate ½ × 18

$$\phi R_{gv} = 0.5 \times 18 \times 0.9 \times 0.6 \times 36 = 175 \text{ kips} > 163 \text{ kips, ok}$$

Plate net shear need not be checked here because it is not a valid limit state.

Weld-to-column flange. This weld sees shear only. Thus

$$D = \frac{163}{2 \times 18 \times 1.392} = 3.25 \quad \text{Use ¼ flange width}$$

Weld-to-beam web. This weld sees the shear plus a small couple. Using the AISC *Manual of Steel Construction,* Table 8-42, $l = 18$, $kl = 4.25$, $k = 0.24$, $x = 0.04$, $xl = 0.72$, $al = 4.28$, $a = 0.24$, $c = 2.04$, and

$$D = \frac{163}{2.04 \times 18} = 4.44$$

Thus a ⁵⁄₁₆ fillet weld is satisfactory.

Beam web. To support a ⁵⁄₁₆ fillet weld on both sides of a plate, AISC LRFD *Manual of Steel Construction,* Table 9-3, shows that a 0.72-in web is required. For a ⁵⁄₁₆ fillet on one side, a 0.36-in web is required. Since the W 21 × 62 web is 0.400 in thick, it is ok.

Beams nos. 3 and 4 W 21 3 44 (G50) composite. The flange connection is a full-penetration weld so, again, no design is required. Section A-A of Fig. 2.35a shows the arrangement in plan. See Fig. 2.35c. The connection plates, A, are made ¼ in thicker than the W 21 × 44 beam flange to accommodate under and over rolling and other minor misfits. Also, the plates are extended beyond the toes of the column flanges by ¾ to 1 in to improve ductility. The plates, A, should also be welded to the column web, even if not required to carry load, to provide improved ductility. A good discussion of this is contained in the AISC LRFD *Manual of Steel Construction,* vol. II, pp. 10-61 to 10-65.

The flange force for the W 21 × 44 is based on the full moment capacity as required in this example, so $\phi M_p = 358$ kip-ft and

$$F_f = \frac{358 \times 12}{(20.66 - 0.45)} = 213 \text{ kips}$$

Figure 2.38 shows the distribution of forces on the plates, A, including the forces from the strong axis connection. The weak axis force of 213 kips is distributed one-fourth to each flange and one-half to the web. This is done to cover the case when the beams may not be reacting against each other. In this case, all of the 213 kips must be passed to the flanges. To see this, imagine that beam 4 is removed and the plate, A, for beam 4 remains as a back-up stiffener. One-half of the 213 kips from beam 3 passes into the beam 3 near the side column flanges, while the other half is passed through the column web to the

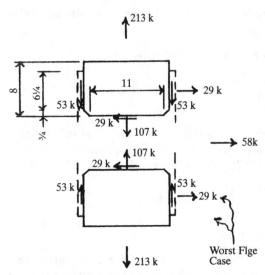

Figure 2.38 Distribution of forces on plates A.

back-up stiffener, and thence into the far side flanges, so that all of the load is passed to the flanges. This is the load path usually assumed, although others are possible.

Merging of stiffeners from strong and weak axis beams. The strong axis beam, beam no. 1, required stiffeners ½ × 6½ × 12½. The weak axis beams no. 3 and no. 4 require plates A ¾ × 8 × 12½. These plates occupy the same space because the beams are all of the same depth. Therefore, the larger of the two plates is used, as shown in Fig. 2.35a.

Since the stiffeners are merged, the welds that were earlier determined for the strong axis beam must be revisited.

Weld-to-flanges. From Fig. 2.38, the worst-case combined flange loads are 53 kips shear and 29 kips axial. The length of weld is 6¼ in. Thus,

$$D_f = \frac{\sqrt{29^2 + 53^2}}{2 \times 6.25 \times 1.392} = 3.5$$

which indicates a ¼ fillet weld, which is also the AISC minimum. However, remember that, for axial load, the contact strength of the stiffener must be developed. The contact strength in this case is 0.9 × 36 × 6.25 × 0.75 = 152 kips because the stiffener has increased in size to accommodate the weak axis beams. But, the delivered load to this stiffener cannot be more than that which can be supplied by the beam, which is 229/2 = 114.5 kips. Thus

$$D_f = \frac{114.5}{2 \times 6.25 \times 1.392 \times 1.5} = 4.39$$

which indicates that a $\frac{5}{16}$ fillet weld is required. This is the fillet weld that should be used and is shown in Fig. 2.35c.

Weld-to-web. From Fig. 2.38,

$$D_w = \frac{\sqrt{29^2 + 107^2}}{2 \times 11.0 \times 1.392} = 3.62$$

Use a $\frac{1}{4}$ fillet weld.

Stresses in stiffeners (plate A). The weak axis beams are G50 steel and are butt-welded to plates A. Therefore, plates A should also be G50 steel. Previous calculations involving this plate assumed it was A36, but changing to G50 will not change the final results in this case because the stiffener contact force is limited by the beam no. 1 delivered force rather than the stiffener strength.

The stiffener stresses for the flange welds are, from Fig. 2.38

$$f_v = \frac{53}{0.75 \times 6.25} = 11.3 \text{ ksi} < 0.9 \times 0.6 \times 50 = 27 \text{ ksi, ok}$$

$$f_a = \frac{29}{0.75 \times 6.25} = 6.19 \text{ ksi} < 0.9 \times 50 = 45 \text{ ksi, ok}$$

and for the web welds

$$f_v = \frac{29}{0.75 \times 11} = 3.5 \text{ ksi} < 27 \text{ ksi, ok}$$

$$f_a = \frac{107}{0.75 \times 11} = 13.0 \text{ ksi} < 45 \text{ ksi, ok}$$

2.3.2.3 Associated shear connections—beams 3 and 4. The specified shear for these beams is $R = 107$ kips.

Weld-to-beam web. As with the strong axis beam web connection, this is a field-welded connection with bolts used for erection only. The design load (required strength) is $R = 107$ kips. The beam web shear, R, is essentially constant in the area of the connection and is assumed to act at the edge of plate A (section A-A of Fig. 2.35b). This being the case, there will be a small eccentricity on the C-shaped field weld. Following the AISC LRFD *Manual of Steel Construction,* Table 8-42, $l = 17$, $kl = 4$, $k = 0.24$, $x = 0.04$, $xl = 0.68$, $al = 4.25 - 0.68 = 3.57$. From Table 8-42 by interpolation, $c = 2.10$, and the weld size required is

$$D = \frac{107}{2.10 \times 17} = 2.99$$

which indicates that a ⁳⁄₁₆ fillet weld is required.

Plate B (shear plate) gross shear. Try a ⅜ in plate of A36 steel. Then

$$\phi R_v = 0.9 \times 0.6 \times 36 \times 0.375 \times 17 = 124 \text{ kips} > 107 \text{ kips, ok}$$

Weld of plate B to column web. This weld carries all of the beam shear, $R = 107$ kips. The length of this weld is 17.75 in. Thus

$$D = \frac{107}{2 \times 17.75 \times 1.392} = 2.17$$

A ⁳⁄₁₆ fillet weld is indicated. Because this weld occurs on both sides of the column web, the column web thickness should satisfy the relationship $0.9 \times 0.6 \times 50t_w \geq 1.392D2$ or $t_w > 0.103 \times 2.17 = 0.22$ in. Since the column web thickness is 0.485 in, the web can support the ⁳⁄₁₆ fillets. The same result can be achieved using AISC LRFD *Manual of Steel Construction,* Table 9-3.

Weld of plate B to plate A. There is a shear flow, $q = VQ/I$, acting on this interface, where $V = R = 107$ kips, Q is the statical moment of plate A with respect to the neutral axis of the I section formed by plates A as flanges and plate B as web. Thus

$$I = \frac{1}{12} 0.375 \times 19.25^3 + 0.75 \times 12.5\left(\frac{19.25 + 0.75}{2}\right)^2 2 = 2100 \text{ in}^4$$

$$Q = 0.75 \times 12.5 \times 10 = 93.8 \text{ in}^3$$

and

$$q = \frac{107 \times 93.8}{2100} = 4.78 \text{ kips/in}$$

Thus,

$$D = \frac{4.78}{2 \times 1.392} = 1.72$$

Since plate A is ¾ in thick, the AISC minimum fillet weld is ¼ in.

Reconsideration of plate A welds. It will be remembered that in the course of this analysis these welds have already been looked at twice. The first time was as stiffener welds for the strong axis beam. The second time was for the combination of forces from the weak and strong axis beam flange connections. Now, additional forces are added to plates A from the weak axis beam web connections. The additional

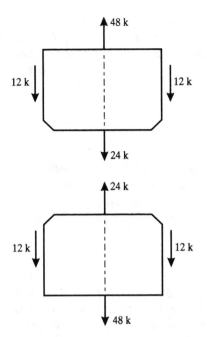

Figure 2.39 Additional forces in plates A due to web connections.

force is $17.8 + 4.78 \times 6.25 = 48$ kips. Figure 2.39 shows this force and its distribution to the plate edges. Rechecking the plate A welds, for the flange weld,

$$D_f = \frac{\sqrt{29^2 + (53 + 12)^2}}{2 \times 6.25 \times 1.392} = 4.09$$

which indicates that a $\frac{5}{16}$ fillet is required. This is the size already determined. For the web weld,

$$D_w = \frac{\sqrt{(107 + 24)^2 + 29^2}}{2 \times 11.0 \times 1.392} = 4.38$$

This weld, which was previously determined to be a $\frac{1}{4}$ fillet weld, must now be increased to a $\frac{5}{16}$ fillet weld.

This completes the calculations required to design the moment connections of Fig. 2.35. Figures 2.35a, b, and c show the completed design. Load paths of sufficient strength to carry all loads from the beams into the column have been provided.

A further consideration. It sometimes happens in the design of this type of connection that the beam is much stronger in bending than the column. In the example just completed, this is not the case. For the strong axis W 21 × 62 beam, $\phi M_p = 389$ kip-ft, while for the column, $\phi M_p = 647$ kip-ft. If the ϕM_p of the column were less than half the ϕM_p of the beam, then the connection should be designed for $2(\phi M_p)$ of the column because this is the maximum moment that can be developed between the beam and the column, that is, that the system can deliver. Similar conclusions can be arrived at for other arrangements.

2.4 Shear Connections

2.4.1 Introduction

These are the most common type of connections on every job. They are generally considered to be "simple" connections in that the beams supported by them are "simple" beams, that is, no bending moment at the beam ends. There are two basic types of shear connections, framed and seated.

2.4.2 Framed connections

These are the familiar double-angle, single-angle, shear plate and shear end-plate connections. They are called *framed connections* because they connect beams, web-to-web, directly. Figure 2.40 shows a typical double-angle connection and Fig. 2.41 shows a shear end-plate connection. These and other types of framed connections can be easily designed using the design aids (charts, tables) contained in the AISC LRFD *Manual of Steel Construction,* vol. II, part 9. A shear end-plate connection will be designed in detail in the next example. The other types are designed in a similar manner.

Example—Shear End Plate Design   The final design is shown in Fig. 2.41. The calculations to achieve it are as follows. The bolts are required to be slip-critical, class A, A325 bolts with threads included in the shear planes. In short-

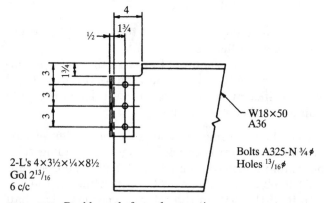

2-L's 4×3½×¼×8½
Gol 2¹³/₁₆
6 c/c

W18×50
A36

Bolts A325-N ¾ ⌀
Holes ¹³/₁₆⌀

Figure 2.40   Double-angle framed connection.

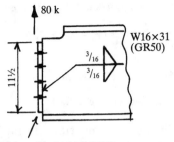

Plate: $^3/_8 \times 8 \times 11^1/_2$ (A36)
Bolts: A325SC-A-N $^7/_8 \phi$ at $5^1/_2$ gage
Holes: HSS

Figure 2.41 Shear end-plate connection.

hand notation, the bolts are A325SC-A-N. The bolt size is $^7/_8$ in diameter. Horizontal short slots are used in the end plate to assist in erection.

Bolts: The slip-critical design strength per bolt is

$$\phi r_v = 0.85 \times 1.13 \times 0.33 \times 39 = 12.4 \text{ kips}$$

For bearing

$$\phi r_v = 0.75 \times 48 \times 0.6013 = 21.6 \text{ kips}$$

Since $12.4 < 21.6$, use $\phi r_v = 12.4$ kips/bolt. The number of bolts required is $80/12.4 = 6.5$. Use 8 bolts, 2 rows of 4.

Weld: The length of the shear plate for four rows of bolts is $11^1/_2$ in with 3-in spacing and $1^1/_4$-in edge distance. Because it is difficult to weld the full length of the plate, a distance equal to the weld size top and bottom is discounted. Thus, the weld length is $11.5 - 2 \times 0.25 = 11.0$ in where a $^1/_4$ fillet has been assumed. The required weld size is

$$D = \frac{80}{2 \times 11 \times 1.392} = 2.6$$

A $^3/_{16}$ fillet weld is satisfactory.

Beam web: The beam web to support $^3/_{16}$ fillets, both sides is 0.31 in thick from AISC Table 9-3. The web of the W16 $\times$ 31 is 0.275 in thick which appears to be too thin. However, the $^3/_{16}$ fillet is the nominal size. The actual size required is 2.6/16. So, $(2.6/3)\, 0.31 = 0.269 < 0.275$ and the web is ok.

End plate: The end plate is A36 steel $^3/_8$ in thick.

Gross shear:

$$\phi R_{gv} = 0.9 \times 0.6 \times 36 \times 0.375 \times 11.5 \times 2 = 168 \text{ kips} > 80 \text{ kips, ok}$$

Net shear:

$$\phi R_{nv} = 0.75 \times 0.6 \times 58\,(11.5 - 4 \times 1.0)\,0.375 \times 2 = 147 \text{ kips} > 80 \text{ kips, ok}$$

Bearing: Because the edge distance of 1.25 < 1.5 × 0.875 = 1.3125, the top bolt has a nominal bearing capacity r_p = 1.25 × 0.375 × 58 = 27.2 kips. For the remaining bolts, the spacing of 3.0 > 3 × 0.875 = 2.625, so r_p = 2.4 × 0.875 × 0.375 × 58 = 45.7 kips. Thus

$$\phi R_p = 0.75 \,(27.2 \times 2 + 45.7 \times 6) = 246 \text{ kips} > 80 \text{ kips, ok}$$

This completes the design calculations for the connection of Fig. 2.41.

One of the principal uses of shear end-plate connections is for skewed connections. Suppose the W 16 beam is skewed 9½° (a 2 on 12 bevel) from the supporting beam or column as shown in Fig. 2.42. The nominal weld size is that determined from the analysis with the plate perpendicular to the beam web (Fig. 2.42a). This is denoted by W', where W' = 2.6/16 = 0.1625. The effective throat for this weld is t_e = 0.7071W' = 0.7071 × 0.1625 = 0.115 in. If the beam web is cut square, the gap on the obtuse side is 0.275 sin 9.5 = 0.0454 < 1/16, so it can be ignored.

The weld size, W, for a skew weld is

$$W = t_e\left(\frac{2\sin\Phi}{2}\right) + g$$

where Φ is the dihedral angle.

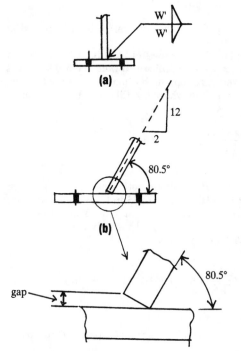

(a)

(b)

gap

Figure 2.42 Geometry of skewed joint.

For the obtuse side, $\Phi = 90 + 9.5 = 99.5$,

$$W = 0.115\left(2 \sin \frac{99.5}{2}\right) + 0 = 0.1755; \ \frac{3}{16} \ Fw$$

For the acute side $\Phi = 90 - 9.5 = 80.5$

$$W = 0.115\left(2 \sin \frac{80.5}{2}\right) + 0 = 0.1486; \ \frac{3}{16} \ Fw$$

In this case, the fillet sizes remain the same as the orthogonal case. In general, the obtuse side weld will increase and the acute side weld will decrease, as will be seen in the next section.

2.4.3 Skewed connections

The shear end plate example of the previous section ended with the calculation of welds for a skewed connection. There are many types of skewed connections. The design recommendations for economy and safety have been reviewed by Kloiber and Thornton (1997). This section is largely taken from this paper.

Skewed connections to beams. The preferred skewed connections for economy and safety are single plates (Fig. 2.43) and end plates (Fig. 2.44). Single bent plates (Fig. 2.45) and eccentric end plates also work well at very acute angles. The old traditional double bent plate connections are difficult to accurately fit and are expensive to fabricate. There are also quality (safety) problems with plate cracking at the bend line as the angle becomes more acute.

Single plates (Fig. 2.43) are the most versatile and economical skewed connection with excellent dimensional control when using

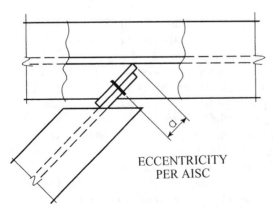

ECCENTRICITY
PER AISC

Figure 2.43 Shear tab (single plate). (*Courtesy of Kloiber and Thornton, with permission from ASCE.*)

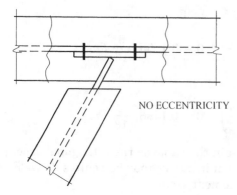

NO ECCENTRICITY

Figure 2.44 Shear end plate. (*Courtesy of Kloiber and Thornton, with permission from ASCE.*)

bend line

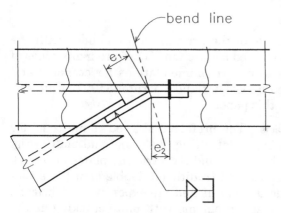

e_1 and e_2 are connection eccentricities
THERE ARE NO MEMBER ECCENTRICITIES

Figure 2.45 Bent plate. (*Courtesy of Kloiber and Thornton, with permission from ASCE.*)

short slotted holes. While capacity is limited, this is usually not a problem because skewed members generally carry a smaller tributary area. Single plates can be utilized for intersection angles of 90° to 30°. Snug-tight bolts are preferred because they are more economical and greatly simplify installation when there are adjacent beams. They also eliminate the "banging bolt" problem which occurs in single plate connections when pretensioned bolts slip into bearing. There are AISC (1994) tables available which can be used to select the required plate size and bolts along with the weld capacity for the required load. This connection has an eccentricity related to the parameter, a, of Fig. 2.43. The actual eccentricity depends on support rigidity, hole type, and bolt installation. The actual weld detail does, however, have to be developed for the joint geometry. Welding details for skewed joints will be discussed presently.

End plates (Fig. 2.44) designed for shear only are able to provide more capacity than single plates and if horizontal slots are utilized with snug-tight bolts in bearing some dimensional adjustment is possible. Hole gages can be adjusted to provide bolt access for more acute skews. A constructability problem can arise when there are opposing beams that limit access to the back side of the connection. These end-plate connections can be sized using the AISC (1994) tables to select plate size, bolts, and weld capacity. Note that there is no eccentricity with this joint. The weld detail, however, has to be adjusted for the actual geometry of the joint in a manner similar to the shear plate.

Single bent plates as in Fig. 2.45 can be sized for either bolted or welded connections using the procedures in the AISC LRFD *Manual of Steel Construction,* 2d ed., for single angle connections. These involve two eccentricities, e_1 and e_2 from the bend line.

Eccentric end plates (Fig. 2.46) can be easily sized for the eccentricity, *e,* using the tables in the AISC LRFD *Manual of Steel Construction,* 2d ed., for eccentric bolt groups.

Skewed connections to columns. Skewed connections to wide-flange columns present special problems. Connections to webs have very limited access and, except for columns where the flange width is less than the depth, or for skews less than 30°, connections to flanges are preferred.

When connecting to column webs, it may be possible to use either a standard end plate or eccentric end plate as shown in Figs. 2.47 and 2.48. Single-plate connections should not be used unless the bolts are positioned outside the column flanges. This will make the connection so eccentric that top and bottom plates, as shown in Fig. 2.49, may be needed to prevent plate buckling and to provide for torsional stiffness of the beam. Extending the single plate increases the connection cost and, unless the connection is designed for the increased eccentricity (*e* of Fig. 2.49), the column design needs to account for it. The eccentricities for these connections are the same as similar connections to beam webs.

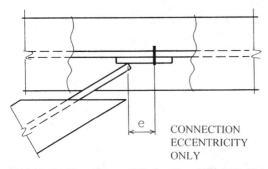

CONNECTION
ECCENTRICITY
ONLY

Figure 2.46 Eccentric end plate. (*Courtesy of Kloiber and Thornton, with permission from ASCE.*)

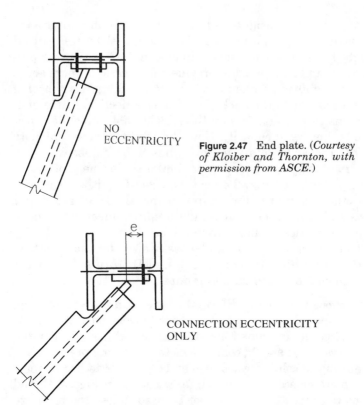

NO
ECCENTRICITY

Figure 2.47 End plate. (*Courtesy of Kloiber and Thornton, with permission from ASCE.*)

CONNECTION ECCENTRICITY
ONLY

Figure 2.48 Eccentric end plate. (*Courtesy of Kloiber and Thornton, with permission from ASCE.*)

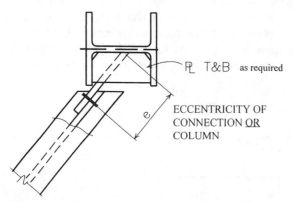

PL T&B as required

ECCENTRICITY OF
CONNECTION OR
COLUMN

Figure 2.49 Single plate (extended shear tab). (*Courtesy of Kloiber and Thornton, with permission from ASCE.*)

Skewed connections to the column flange will also be eccentric when the beam is aligned to the column centerline. However, if the beam alignment is centered on the flange, as shown in Fig. 2.50, the

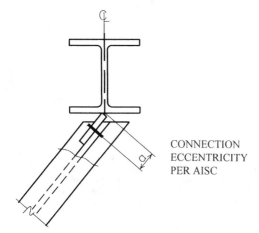

CONNECTION
ECCENTRICITY
PER AISC

Figure 2.50 Single plate (shear tab) centered on column flange. (*Courtesy of Kloiber and Thornton, with permission from ASCE.*)

minor axis eccentricity is eliminated and the major axis eccentricity will not generally govern the column design. The connection eccentricity is related to the parameter, a, here in the same way as was discussed for Fig. 2.43.

When the beam is aligned to the column centerline, either single plates (Fig. 2.51), eccentric end plates (Figs. 2.52 and 2.53), or single bent plates (Fig. 2.54) can be used. The eccentricity for each of these connections is again similar to that for the same connection to a beam web. An additional eccentricity, e_y, which causes a moment about the column weak axis, is present in these connections as shown in Figs. 2.51 through 2.54. The column may need to be designed for this moment.

A special skewed connection is often required when there is another beam framing to the column flange at 90°. If the column flange is not

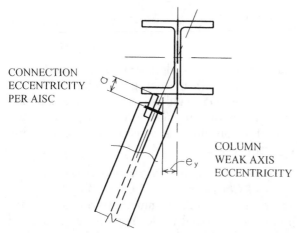

CONNECTION
ECCENTRICITY
PER AISC

COLUMN
WEAK AXIS
ECCENTRICITY

Figure 2.51 Single-plate (shear tab) gravity axis configuration. (*Courtesy of Kloiber and Thornton, with permission from ASCE.*)

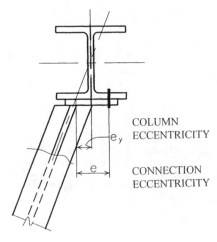

COLUMN
ECCENTRICITY

CONNECTION
ECCENTRICITY

Figure 2.52 Eccentric shear end plate gravity axis configuration. (*Courtesy of Kloiber and Thornton, with permission from ASCE.*)

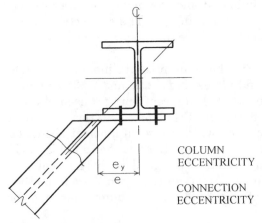

COLUMN
ECCENTRICITY

CONNECTION
ECCENTRICITY

Figure 2.53 Eccentric shear end plate for high skew. (*Courtesy of Kloiber and Thornton, with permission from ASCE.*)

wide enough to accommodate a side-by-side connection, a bent plate can be shop-welded to the column with matching holes for the second beam as shown in Fig. 2.55. The plate weld is sized for the eccentricity, e_2, plus any requirement for development as a fill plate in the orthogonal connection, and the column sees an eccentric moment due to e_y which is equal to e_2 in this case.

2.4.3.1 Methods for determining strength of skewed fillet welds. The AISC LRFD *Manual of Steel Construction,* 2d ed., tables for single plates and end plates are based on using standard AWS equal leg fillet welds. The single-plate weld is sized to develop shear yielding of the plate plus the moment due to eccentricity. However, the weld does

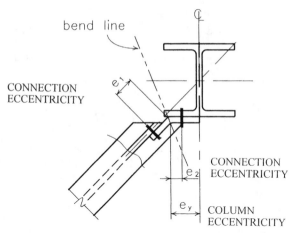

Figure 2.54 Single bent plate—one beam framing to flange. (*Courtesy of Kloiber and Thornton, with permission from ASCE.*)

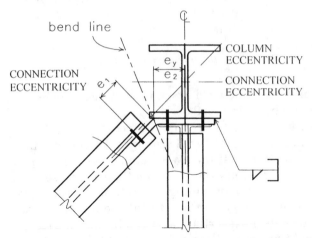

Figure 2.55 Single bent plate—two beams. (*Courtesy of Kloiber and Thornton, with permission from ASCE.*)

not have to be stronger than the plate. This results in a maximum weld size of $0.75t$ for orthogonal connections. The end-plate weld is sized to carry the applied load. These standard orthogonal fillet welds of leg size W (Fig. 2.56b) need to be modified as the skew becomes more acute in order to maintain the required capacity. There are two ways to do this. The AWS D1.1 *Structural Welding Code* provides a method to calculate the effective throat for skewed T joints with varying dihedral angles, which is based on providing equal strength in the obtuse and acute welds. This is shown in Fig. 2.56a and was the method used for the example of the last section. The AISC method is

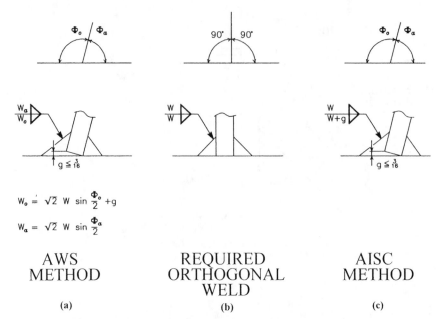

$$W_o \stackrel{'}{=} \sqrt{2} \ W \ \sin \frac{\Phi_o}{2} + g$$

$$W_a = \sqrt{2} \ W \ \sin \frac{\Phi_a}{2}$$

AWS METHOD	REQUIRED ORTHOGONAL WELD	AISC METHOD
(a)	(b)	(c)

Figure 2.56 Skewed fillet weld sizes required to match strength of required orthogonal fillets. (*Courtesy of Kloiber and Thornton, with permission from ASCE.*)

simpler, and simply increases the weld size on the obtuse side by the amount of the gap, as is shown in Fig. 2.56c.

Both methods can be shown to provide a strength equal to or greater than the required orthogonal weld size of W. The main difference with regard to strength is that the AWS method maintains equal strength in both fillets, whereas the AISC method increases the strength on the acute side by maintaining a constant fillet size, W, while the increased size, $W + g$, on the obtuse side actually loses strength because of the gap, g. Nevertheless, it can be shown that the sum of the strengths of these two fillet welds, W and $W + g$, is always greater than the $2W$ of the required orthogonal fillets.

It should be noted that the gap, g, is limited to a maximum value of $\frac{3}{16}$ in for both methods.

The effects of the skew on the effect throat of fillet weld can be very significant as shown in Fig. 2.57. Figure 2.57 also shows how fillet legs W_o and W_a are measured in the skewed configuration. On the acute side of the connection the effective throat for a given fillet weld size gradually increases as the connection intersection angle, ϕ, changes from 90 to 60°. From 60 to 30°, the weld changes from a fillet weld to a partial penetration groove weld (Fig. 2.58) and the effective throat, t_e, decreases due to the allowance, z, for the unwelded portion at the root. While this allowance varies based on the welding process

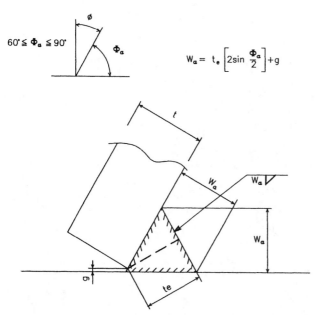

$$60^\circ \leq \Phi_a \leq 90^\circ$$

$$W_a = t_e \left[2\sin \frac{\Phi_a}{2} \right] + g$$

Figure 2.57a Geometry of skewed fillet weld obtuse side. (*Courtesy of Kloiber and Thornton, with permission from ASCE.*)

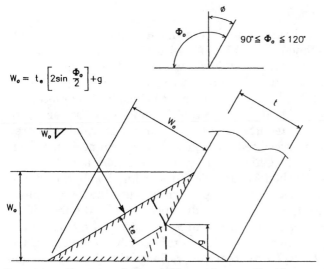

$$90^\circ \leq \Phi_o \leq 120^\circ$$

$$W_o = t_e \left[2\sin \frac{\Phi_o}{2} \right] + g$$

Figure 2.57b Geometry of skewed fillet weld acute side. (*Courtesy of Kloiber and Thornton, with permission from ASCE.*)

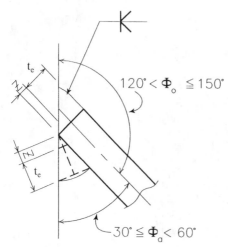

Figure 2.58 Acute angles less than 60° and obtuse angles greater than 120°. (*Courtesy of Kloiber and Thornton, with permission from ASCE.*)

and position, it can conservatively be taken as the throat less ⅛ in for 60 to 45° and less ¼ in for 45 to 30°. Joints less than 30° are not prequalified and generally should not be used.

2.4.4 Seated connections

The second type of shear connection is the seated connection, either unstiffened or stiffened (Fig. 2.59). As with the framed connections, there are tables in the AISC LRFD *Manual of Steel Construction,* part 9, which aid in the design of these connections.

The primary use for this connection is for beams framing-to-column webs. In this case, the seat is inside the column flange toes or nearly so, and is not an architectural problem. Its use also avoids the erection safety problem associated with most framed connections where the same bolts support beams on both sides of the column web.

When a seat is attached to one side of the column web, the column web is subjected to a local bending pattern because the load from the beam is applied to the seat at some distance, *e,* from the face of the web. For stiffened seats, this problem was addressed by Sputo and Ellifrit (1991). The stiffened seat design table (Table 9.9) in the AISC LRFD *Manual of Steel Construction* reflects the results of their research. For unstiffened seats, column web bending also occurs, but no research has been done to determine its effect. This is the case because the loads and eccentricities for unstiffened seats are much smaller than for stiffened seats. Figure 2.60 presents a yield-line

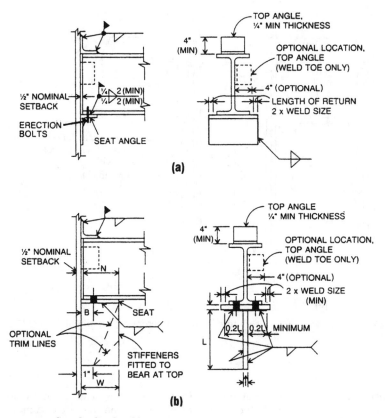

Figure 2.59 Standardized weld seat connections: (*a*) unstiffened seat and (*b*) stiffened seat.

analysis which can be used to assess the strength of the column web. The nominal capacity of the column web is

$$R_w = \frac{2m_p L}{e_f}\left(2\sqrt{\frac{T}{b}} + \frac{T}{L} + \frac{L}{2b}\right)$$

where the terms are defined in Fig. 2.60, and

$$m_p = \frac{1}{4}\, t_w^2 F_y$$

$$b = \frac{T-c}{2}$$

Since this is a yield limit state, $\phi = 0.9$.

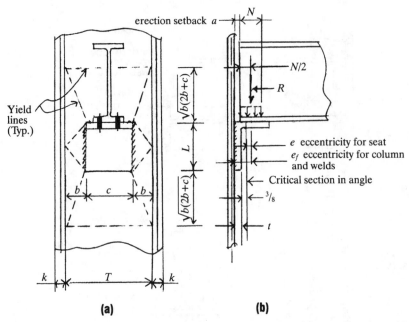

Figure 2.60 Column web yield lines and design parameters for unstiffened seated connection.

Example A W 14 × 22 beam of grade 50 steel is to be supported on an unstiff-ened seat to a W 14 × 90 (G50) column. The given reaction (required strength) is 33 kips. Design the unstiffened seat.

The nominal erection set back $a = \frac{1}{2}$ in. For calculations, to account for underrun, use $a = \frac{3}{4}$. Try a seat 6 in long ($c = 6$). From the AISC LRFD *Manual of Steel Construction*, 2d ed., Table 9-7, with a beam web of approximately ¼ in, a ⅝-in angle gives a capacity of 37.7 kips > 33 kips, ok. Choosing an angle 6 × 4 × ⅝, Table 9-7 indicates that a ⁵⁄₁₆ fillet weld of the seat vertical leg (the 6-in leg) to the column web is satisfactory (41.0 kips). Consider this to be a preliminary design, which needs to be checked.

The first step in the checking of this design is to determine the required bear-ing length, N. Note that N is not the horizontal angle leg length less a, but rather it cannot exceed this value. The bearing length for an unstiffened seat starts at the end of the beam and spreads from this point, because the toe of the angle leg tends to deflect away from the bottom flange of the beam. The bearing length cannot be less than k and can be written in a general way as

$$N = \max\left\{\frac{R - \phi R_1}{\phi R_2}, \ \frac{R - \phi R_3}{\phi R_4} \ \text{or} \ \frac{R - \phi R_5}{\phi R_6}, \ k\right\}$$

where R_1 through R_6 are defined in the AISC LRFD *Manual of Steel Construction*, 2d ed, p. 4-33, and are tabulated in the factored uniform load tables. For the W 14 × 22, $\phi R_1 = 25.2$, $\phi R_2 = 11.5$, $\phi R_3 = 23.0$, $\phi R_4 = 2.86$, $\phi R_5 = 20.4$, and $\phi R_6 = 3.81$. Thus

$$N = \max\left\{\frac{33 - 25.2}{11.5}, \frac{33 - 23.0}{2.86} \text{ or } \frac{33 - 20.4}{3.81}, 0.875\right\}$$

$$= \max\{0.67, 3.50 \text{ or } 3.31, 0.875\}$$

Therefore N is either 3.50 or 3.31, depending on whether $N/d \leq 0.2$ or $N/d > 0.2$, respectively. With $d = 13.74$, $3.50/13.74 = 0.255$, and $3.31/13.74 = 0.241$. Since clearly $N/d > 0.2$, $N = 3.31$ in.

It was stated earlier that $(N + a)$ cannot exceed the horizontal angle leg. Using $a = \frac{1}{2} + \frac{1}{4} = \frac{3}{4}$, $N + a = 3.31 + 0.75 = 4.06$ which is close enough to 4 in to be ok.

The design strength of the seat angle critical section is

$$\phi R_b = \frac{1}{4}\frac{ct^2}{e}\phi F_y$$

where the terms are defined in Fig. 2.60. From Fig. 2.60, $e_f = N/2 + a = 3.31/2 + 0.75 = 2.41$ and $e = e_f - t - 0.375 = 2.41 - 0.625 - 0.375 = 1.41$ Then

$$\phi R_b = \frac{0.25 \times 6 \times 0.625^2 \times 0.9 \times 36}{1.41} = 13.5 \text{ kips}$$

Since 13.5 kips<33 kips, the seat is unsatisfactory. The required seat thickness can be determined from

$$t_{\text{req'd}} = \sqrt{\frac{4Re}{\phi F_y c}} = \sqrt{\frac{4 \times 33 \times 1.41}{0.9 \times 36 \times 6}} = 0.98 \text{ in}$$

Therefore, an angle 1 in thick can be used, although a thinner angle between $\frac{5}{8}$ and 1 in may work since e depends on t. There is no L $6 \times 4 \times 1$ available, so use a $6 \times 6 \times 1$. The extra length of the horizontal leg is irrelevant.

It can be clearly seen from the preceding result that the AISC *Manual of Steel Construction*, 2d ed., Table 9-7, should not be relied upon for final design. The seats and capacities given in Table 9-7 are correct for the seats themselves. It is the beam that is failing. Tables 9-6 and 9-7 were originally derived based on the bearing length required for web yielding (that is, ϕR_1 and ϕR_2). The new requirements of web crippling (ϕR_3 and ϕR_4 or ϕR_5 and ϕR_6) must be considered in addition to the capacities given in Tables 9-6 and 9-7. If web yielding is more critical than web crippling, the tables will give satisfactory capacities. This problem has been addressed in a recent paper (Carter et al., 1997), and the AISC *Manual of Steel Construction*, 2d ed., tables will be revised in the future.

Next, the weld of the seat vertical leg to the column web is checked. Table 9-7 indicated a $\frac{5}{16}$-in fillet was required. This can be checked using AISC *Manual of Steel Construction*, Table 8-38. With $e_x = e_f = 2.41$, $l = 6$, $a = 2.41/6 = 0.40$, $c = 2.00$, and

$$\phi R_{\text{weld}} = 2.00 \times 5 \times 6 = 60.0 \text{ kips} > 33 \text{ kips, ok}$$

The weld sizes given in Tables 9-6 and 9-7 will always be found to be conservative because they are based on using the full horizontal angle leg minus a as the bearing length, N. Finally, checking the column web,

$$m_p = 0.25 \times 0.440^2 \times 50 = 2.42 \frac{\text{kip-in}}{\text{in}}$$

$$T = 11.25$$

$$c = 6$$

$$L = 6$$

$$b = \frac{11.25 - 6}{2} = 2.625$$

$$\phi R_{web} = \frac{0.9 \times 2 \times 2.42 \times 6}{2.41}\left(2\sqrt{\frac{11.25}{2.625}} + \frac{11.25}{6} + \frac{6}{2 \times 2.625}\right)$$

$$= 77.6 \text{ kips} > 33 \text{ kips, ok}$$

This completes the calculations for this example. The final design is shown in Fig. 2.61.

2.4.5 Beam shear splices

If a beam splice takes moment as well as shear, it is designed with flange plates in a manner similar to the truss chord splice treated in Sec. 2.2.5.2. The flange force is simply the moment divided by the cen-

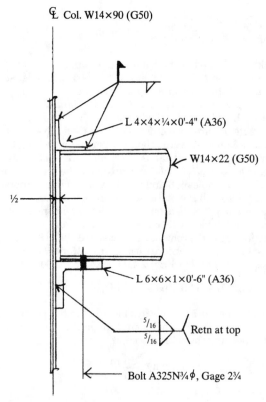

Figure 2.61 Unstiffened seat design.

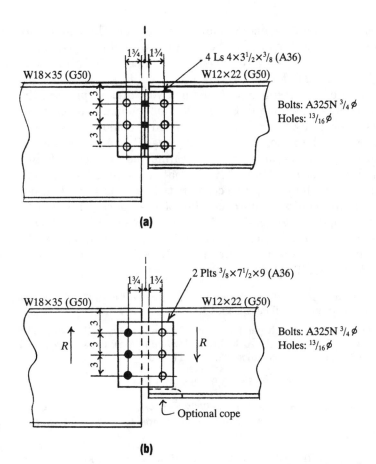

Figure 2.62 Typical shear splices: (a) shear splice with four angles and (b) shear splice with one or two plates.

ter-to-center flange distance for inside and outside plate connections, or the moment divided by the beam depth for outside plate connections. The web connection takes any shear. Two typical shear splices are shown in Fig. 2.62. These are common in cantilever roof construction. Figure 2.62a shows a four-clip angle splice. The angles are shop-bolted (as shown) or shop-welded to the beam webs. The design of this splice is exactly the same as that of a double-angle framing connection. The shear acts at the faying surface of the field connection and each side is designed as a double-angle framing connection. If shop-bolted, all the bolts are in shear only; there is no eccentricity on the shop bolts. If shop-welded, the shop welds see an eccentricity from the location of the shear at the field faying surface to the centroids of the weld groups on each side. This anomaly is historical. The bolted connections derive from riveted connections which were developed before it was considered necessary to satisfy "the niceties of structural mechanics," according to McGuire (1968).

A second type of shear splice uses one or two plates in place of the four angles. This type, shown in Fig. 2.62b, has moment capacity, but has been used for many years with no reported problems. It is generally less expensive than the angle type. Because it has moment capability, eccentricity on the bolts or welds cannot be neglected. It has been shown by Kulak and Green (1990) that if the stiffness on both sides of the splice is the same, the eccentricity is one-half the distance between the group centroids on each side of the splice. This will be the case for a shop-bolted–field-bolted splice as shown in Fig. 2.62b. A good discussion on various shear splice configurations and the resulting eccentricities is given in the AISC LRFD *Manual of Steel Construction,* vol. II, (1994), pp. 9-179 and 9-180.

Example   As an example of the design routine for the Fig. 2.62b splice, its capacity (design strength) will be calculated.

Bolts.   The design strength per bolt in single shear is $\phi r_v = 15.9$ kips. The eccentricity is $e_x = 2.25$. From the AISC LRFD *Manual of Steel Construction,* 2nd edition, Table 8-18, for $e_x = 2.25$ and $n = 3, c = 2.11$, and

$$\phi R_v = 2.11 \times 15.9 \times 2 = 67.1 \text{ kips}$$

Bearing.   Bearing will be critical on the W 12 × 22 web, which has a thickness $t_w = 0.260$ in. Since the edge distances and spacings satisfy the 1½ d and 3d criteria, respectively,

$$\phi R_p = 0.75 \times 2.4 \times 65 \times 0.260 \times 0.75 \times 2.11 = 48.1 \text{ kips}$$

Net shear on splice plates.   This will be more critical than block shear rupture because the edge and end distances are the same and $0.6 \times 58 = 34.8 < 36$.

$$\phi R_{nv} = (9 - 3 \times 0.875)0.375 \times 2 \times 0.75 \times 0.6 \times 58 = 125 \text{ kips}$$

Gross shear on splice plates

$$\phi R_{gv} = 9 \times 0.375 \times 2 \times 0.9 \times 0.6 \times 36 = 131 \text{ kips}$$

Gross shear on beam web (uncoped).   The W 12 × 22 will control. Thus

$$\phi R_{gv} = 0.260 \times 12.31 \times 0.9 \times 0.6 \times 50 = 86.4 \text{ kips}$$

Note that the net shear and block shear are not limit states for the web of either beam unless there is a cope toward which the shear acts. In cantilever construction, the shear, *R*, usually acts as shown in Fig. 2.62b, that is, the W 12 × 22 is pushed down against the bolts so the shear acts away from the cope. Since the shear acts away from the cope, block shear is not a limit state.

The cope will affect the gross shear strength, however. If the cope is 1 in deep

$$\phi R_{gv} = 0.260(12.31 - 1)0.9 \times 0.6 \times 50 = 79.4 \text{ kips}$$

Net section bending on the splice plates. From the AISC LRFD *Manual of Steel Construction,* 2nd edition, Table 12.1, the net section modulus of a splice plate is $S_{net} = 3.75 \text{ in}^2$, $\phi M_n = 0.75 \times 58 \times 3.75 = 163$ kip-in. The design strength for net bending is thus

$$\phi R_{nb} = \frac{163 \times 2}{2.25} = 145 \text{ kips}$$

Gross section bending on the splice plates. For convenience, this is usually considered at the bolt line unless there is a greater moment elsewhere. The section modulus is $Z = 0.25 \times 0.375 \times 9^2 = 7.59$ in, so $\phi M_n = 0.9 \times 36 \times 7.59 = 246$ kip-in, and

$$\phi R_{gb} = \frac{246 \times 2}{2.25} = 219 \text{ kips}$$

Considering all of the limit states, and taking the least capacity, the capacity (design strength) of the splice in Fig. 2.62*b* is 48.1 kips, as determined from bearing on the W 12 × 22 web.

2.5 Miscellaneous Connections

2.5.1 Simple Beam connections under shear and axial load

As its name implies, a simple shear connection is intended to transfer shear load out of a beam while allowing the beam to act as a simply-supported beam. The most common simple shear connection is the double-angle connection with angles shop-bolted or shop-welded to the web of the carried beam and field-bolted to the carrying beam or column. This section, which is from Thornton (1995*a*), will deal with this connection.

Under shear load, the double-angle connection is flexible regarding the simple-beam end rotation, because of the angle leg thickness and the gage of the field bolts in the angle legs. The AISC LRFD *Manual of Steel Construction,* 2d ed., vol. II, p. 9-12, recommends angle thicknesses not exceeding ⅝ in with the usual gages. Angle leg thicknesses of ¼ to ½ in are generally used, with ½-in angles usually being sufficient for the heaviest shear load. When this connection is subjected to axial load in addition to the shear, the important limit states are angle leg bending and prying action. These tend to require that the angle thickness increase or the gage decrease, or both, and these requirements compromise the connection's ability to remain flexible to simple-beam end rotation. This lack of connection flexibility causes a tensile load on the upper field bolts which could lead to bolt fracture and a progressive failure of the connection and the resulting collapse of the beam. It is thought that there has never been a reported failure of this type, but it is perceived to be possible.

Even without the axial load, some shear connections are perceived to have this problem under shear alone. These are the single-plate shear connections (shear tabs) and the tee framing connections. Recent research on the tee framing connections (Thornton, 1996) has led to a formula (AISC *Manual of Steel Construction*, LRFD 2d ed., vol II, p. 9-170) which can be used to assess the resistance to fracture (ductility) of double angle shear connections. The formula is

$$d_{b_{\min}} = 0.163t \ \sqrt{\frac{F_y}{\widetilde{b}}\left(\frac{\widetilde{b}^2}{L^2} + 2\right)}$$

where $d_{b_{\min}}$ = the minimum bolt diameter (A325 bolts) to preclude bolt fracture under a simple-beam end rotation of 0.03 rad

t = the angle leg thickness

$\widetilde{b}$ = the distance from the bolt line to the k distance of the angle (Fig. 2.63)

L = the length of the connection angles

Note that this formula can be used for ASD and LRFD designs in the form given here. It can be used to develop a table (Table 2.1) of angle thicknesses and gages for various bolt diameters which can be used as a guide for the design of double-angle connections subjected to shear and axial tension. Note that Table 2.1 validates AISC's long-standing (AISC, 1970) recommendation (noted previously) of a maximum ⅝-in angle thickness for the "usual" gages. The usual gages would be 4½ to 6½ in. Thus, for a carried beam web thickness of say ½ in, minimum angle gages (GOL) will range from 2 to 3 in. Table 2.1 gives a GOL of 2½ in for ¾-in bolts (the most critical as well as the most common bolt size). Note also that Table 2.1 assumes a significant simple-beam end rotation of 0.03 rad, which is approximately the end rotation that occurs when a plastic hinge forms at the center of the beam. For short beams, beams loaded near their ends, beams with bracing gussets at their end connections, and beams with light shear loads, the beam end rotation will be small and Table 2.1 does not apply.

As an example of a double-clip angle connection, consider the connection of Fig. 2.64. This connection is subjected to a shear load of 33 kips and an axial tensile load of 39 kips.

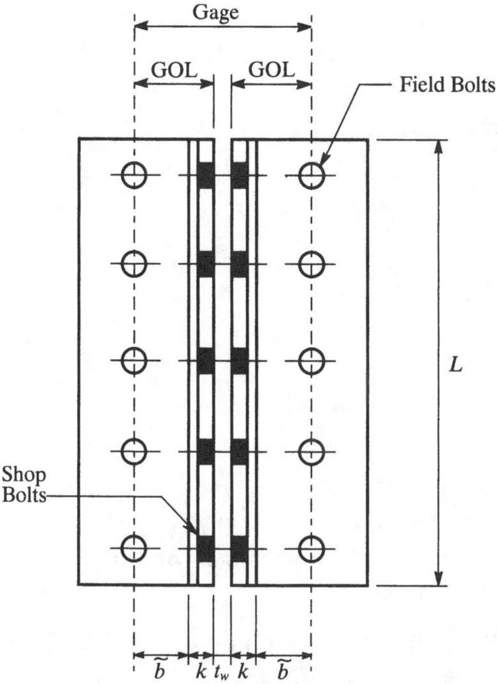

Fgure 2.63 Geometry of double angles (shop-bolted shown).

TABLE 2.1 Estimated Minimum Angle Gages (GOL) for A36 Angles and A325 Bolts for Rotational Flexibility

	Minimum gage of angle (GOL),* in		
Angle thickness, in	¾-in-diam bolt	⅞-in-diam bolt	1-in-diam bolt
⅜	1⅜	1¼	1⅛
½	1⅞	1⅝	1½
⅝	2½	2⅛	1⅞
¾	3¼	2 11/16	2 5/16
1	6	4 5/16	3½

*Driving clearances may control minimum GOL.

Shop bolts. The shop bolts "see" the resultant load, $R = \sqrt{33^2 + 39^2} = 51.1$ kips. The design shear strength of one bolt is $\phi r_v = 15.9$ kips, so these bolts in double shear have a design capacity

$$\phi Rv = 15.9 \times 3 \times 2 = 95.4 \text{ kips} > 51.1 \text{ kips, ok}$$

Beam web. The limit states for the beam web are bearing and block shear rupture (tearout).

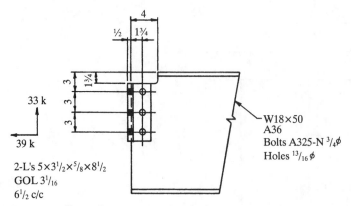

Figure 2.64 Framed connection subjected to axial and shear loads.

Bearing. The edge distances are 1¾ and 1¼ in. The minimum edge distance for the bearing stress to be $2.4F_u$ is $1.5 \times 0.75 = 1.125 < 1.25$, which satisfies the 1½d criterion. The spacing is 3 in and $3 \times 0.75 = 2.25 < 3$, which satisfies the 3d criterion. Thus

$$\phi R_p = 0.75 \times 2.4 \times 58 \times 0.75 \times 0.355 \times 3 = 83.4 \text{ kips} > 51.1 \text{ kips, ok}$$

If the loads of 33 kips shear and 39 kips axial always remain proportional, that is, maintain the bevel of 10⅛ to 12, as shown in Fig. 2.65, the spacing requirement is irrelevant because there is only one bolt in line of force and the true edge distance is 1.94 or 2.29 in rather than 1.25 in as used previously. When there is only one bolt in the line of force

$$R_n = L_e t F_u \le 2.4 \, dt F_u$$

and thus if $L_e \le 2.4d$, the bearing strength will be reduced from the 83.4 kips thus far determined. Since $d = 0.75$, $2.4 \times 0.75 = 1.8$ in,

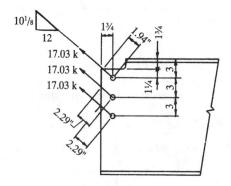

Figure 2.65 Edge distances along the line of action.

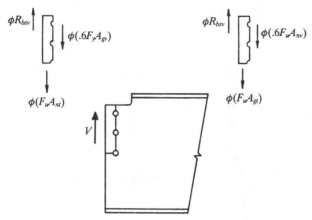

Shear Yield & Tension Fracture   Shear Fracture & Tension Yield

Figure 2.66 Block shear rupture under shear.

which is less than the provided edge distances of 1.94 and 2.29 in, as shown in Fig. 2.65, the bearing strength is not reduced from 83.4 kips, but situations can develop where some reduction occurs. The designer should keep an eye on this.

Block shear rupture (tearout). A simple conservative way to treat block shear when shear and tension are present is to treat the resultant as a shear. Then, from Figs. 2.64 and 2.66,

$$A_{gv} = 7.25 \times 0.355 = 2.57 \text{ in}^2$$
$$A_{nv} = (7.25 - 2.5 \times 0.875)\,0.355 = 1.80 \text{ in}^2$$
$$A_{gt} = 1.75 \times 0.355 = 0.621 \text{ in}^2$$
$$A_{nt} = (1.75 - 0.5 \times 0.875)\,0.355 = 0.466 \text{ in}^2$$
$$0.6F_u A_{nv} = 0.6 \times 58 \times 1.80 = 62.6$$
$$F_u A_{nt} = 58 \times 0.466 = 27.0$$

Since $0.6F_u A_{nv} > F_u A_{nt}$

$$\phi R_{bsv} = 0.75(62.6 + 36 \times 0.621) = 63.7 \text{ kips} > 51.1 \text{ kips, ok}$$

An alternate approach is to calculate a block shear rupture design strength under tensile axial load. From Fig. 2.67:

$$A_{gv} = 0.621 \text{ in}^2$$
$$A_{nv} = 0.466 \text{ in}^2$$
$$A_{gt} = 2.57 \text{ in}^2$$

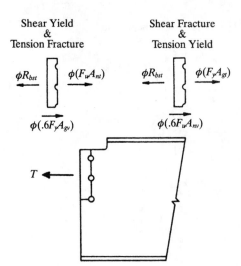

Shear Yield
&
Tension Fracture

ϕR_{bst} $\phi(F_u A_{nt})$

$\phi(.6F_y A_{gv})$

Shear Fracture
&
Tension Yield

ϕR_{bst} $\phi(F_y A_{gt})$

$\phi(.6F_u A_{nv})$

T ←

Figure 2.67 Block shear rupture under tension T.

$$A_{nt} = 1.80 \text{ in}^2$$

$$0.6F_u A_{nv} = 0.6 \times 58 \times 0.466 = 16.2$$

$$F_u A_{nt} = 58 \times 1.80 = 104$$

since $F_u A_{nt} > 0.6F_u A_{nv}$

$$\phi R_{bst} = 0.75(104 + 0.6 \times 36 \times 0.621) = 88.1 \text{ kips}$$

Using an elliptical interaction equation, which is analogous to the von Mises (distortion energy) yield criterion,

$$\left(\frac{V}{\phi R_{bsv}}\right)^2 + \left(\frac{T}{\phi R_{bst}}\right)^2 \leq 1$$

where V is the factored shear and T is the factored tension. Then

$$\left(\frac{33}{63.7}\right)^2 + \left(\frac{39}{88.1}\right)^2 = 0.464 < 1, \text{ ok}$$

This interaction approach is always less conservative than the approach using the resultant $R = \sqrt{V^2 + T^2}$ as a shear because $\phi R_{bst} > \phi R_{bsv}$ for the geometries of the usual bolt positioning in double-angle connections with two or more bolts in a single vertical column. The resultant approach, being much simpler as well as conservative, is the method most commonly used.

Connection angles. Figure 2.64 shows angles 5 × 3½ × ⅝, but assume for the moment that ⅜ angles are to be checked. The shop legs are checked for the limit states of bearing, gross shear and gross tension, and net shear and net tension. Net shear rupture and net tension rupture will control over block shear rupture with the usual connection geometries, that is, 1¼ edge and 1¼ end distances. Since the sum of the clip angle thicknesses = 0.375 + 0.375 = 0.75 > 0.355, the beam web, and not the shop legs, of the clip angles will control.

Connection angles—field legs and field bolts. The field legs of the angles can be checked for gross and net shear using the resultant $R = 51.1$ kips

$$\phi R_{gv} = 0.9 \times 36 \times 8.5 \times 0.375 \times 2 = 207 \text{ kips} > 51.1 \text{ kips, ok}$$

$$\phi R_{nv} = 0.75 \times 58(8.5 - 3 \times 0.875)0.375 \times 2 = 192 \text{ kips} > 51.1 \text{ kips, ok}$$

These limit states seldom control, especially when there is an axial force which, because of the angle leg bending, tends to increase the required angle leg thickness far beyond anything that might be required for shear. Thus, the critical limit state for axially loaded clip angles is leg bending (prying action).

Prying action. The AISC LRFD *Manual of Steel Construction,* 2d ed., has a table to aid in the selection of a clip angle thickness.

The preliminary selection table, Table 11.1, indicates that a ⅝ angle will be necessary. Trying Ls 5 × 3½ × ⅝, and following the procedure of the AISC *Manual of Steel Construction,* 2nd edition,

$$b = \frac{6.5 - 0.355 - 2 \times 0.625}{2} = 2.45$$

$$a = \frac{10.355 - 6.5}{2} = 1.93(<1.25 \times 2.45 = 3.06, \text{ ok})$$

$$b' = 2.45 - \frac{0.75}{2} = 2.08$$

$$a' = 1.93 + \frac{0.75}{2} = 2.31$$

$$\rho = \frac{2.08}{2.31} = 0.90$$

$$p = \frac{8.5}{3} = 2.83$$

$$\delta = 1 - \frac{0.8125}{2.83} = 0.71$$

The shear per bolt $V = 33/6 = 5.5$ kips < 15.9 kips, ok. The tension per bolt $T = 39/6 = 6.5$ kips. Because of interaction,

$$\phi F_t' = 0.75(117 - 1.9f_v) \leq 0.75 \times 90 = 67.5 \text{ ksi}$$

With $f_v = 5.5/0.4418 = 12.5$ ksi

$$\phi F_t' = 0.75(117 - 1.9 \times 12.5) = 70.0 \text{ ksi} > 67.5 \text{ ksi}$$

Use $\phi F_t' = 67.5$ ksi, and $\phi r_t' = 67.5 \times 0.4418 = 29.8$ kips/bolt. Since $T = 6.5$ kips < 29.8 kips, the bolts are satisfactory independent of prying action. Returning to the prying action calculation

$$t_c = \sqrt{\frac{4.44 \times 29.8 \times 2.08}{2.83 \times 36}} = 1.64 \text{ in}$$

$$\alpha' = \frac{1}{0.71 \times 1.90}\left[\left(\frac{1.64}{0.625}\right)^2 - 1\right] = 4.36$$

Since $\alpha' = 4.36$, use $\alpha' = 1$. This means that the strength of the clip angle legs in bending is the controlling limit state. The design strength is

$$T_d = 29.8\left(\frac{0.625}{1.64}\right)^2(1 + 0.71) = 7.4 \text{ kips} > 6.5 \text{ kips, ok}$$

The Ls $5 \times 3\frac{1}{2} \times \frac{5}{8}$ are satisfactory.

Ductility considerations. The $\frac{5}{8}$-in angles are the maximum thickness recommended by AISC for flexible shear connections. Using the formula introduced at the beginning of this section,

$$d_{b_{\min}} = 0.163t\sqrt{\frac{F_y}{\widetilde{b}}\left(\frac{\widetilde{b}^2}{L^2} + 2\right)}$$

with $t = 0.625, F_y = 36, \widetilde{b} = 3.0625 - 1.125 = 1.94, L = 8.5$:

$$d_{b_{\min}} = 0.163 \times 0.625\sqrt{\frac{36}{1.94}\left(\frac{1.94^2}{8.5^2} + 2\right)} = 0.63 \text{ in}$$

Since the actual bolt diameter is 0.75 in, the connection is satisfactory for ductility.

As noted before, it may not be necessary to make this check for ductility. If the beam is short, is loaded near its ends, or for other reasons is not likely to experience very much simple-beam end rotation, this ductility check can be omitted.

This completes the calculations for the design shown in Fig. 2.64.

2.5.2 Reinforcement of axial force connections

It sometimes happens that a simple beam connection, designed for shear only, must after fabrication and erection be strengthened to carry some axial force as well as the shear. In this case, washer plates can sometimes be used to provide a sufficient increase in the axial capacity. Figure 2.68 shows a double-angle connection with washer plates which extend from the toe of the angle to the k distance of the angle. These can be made for each bolt, so only one bolt at a time need be removed, or if the existing load is small, they can be made to encompass two or more bolts on each side of the connection. With the washer plate, the bending strength at the "stem" line, section a-a of Fig. 2.68 is

$$M_n = \tfrac{1}{4}F_y pt^2$$

while that at the bolt line, section b-b, is

$$M_n' = \alpha\delta\tfrac{1}{4}F_y p(t^2 + t_p^2) = \alpha\delta\tfrac{1}{4}F_y pt^2\left(1 + \frac{t_p^2}{t^2}\right) = \alpha\delta\eta M_n$$

where $\eta = 1 + (t_p/t)^2$ and the remaining quantities are in the notation of the AISC LRFD *Manual of Steel Construction* (1994). With the introduction of η, the prying action formulation of the AISC LRFD

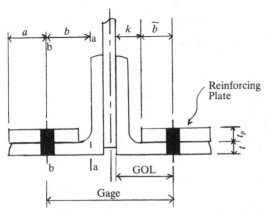

Figure 2.68 Prying action with reinforcing (washer) plate.

Manual of Steel Construction can be generalized for washer plates by replacing δ wherever it appears by the term δη. Thus

$$\alpha' = \frac{1}{\delta\eta(1 + \rho)}\left[\left(\frac{t_c}{t}\right)^2 - 1\right]$$

and

$$T_d = \phi r_t \left(\frac{t}{t_c}\right)^2 (1 + \alpha'\delta\eta)$$

All other equations remain the same.

As an example of the application of this method, consider the connection of Fig. 2.69. This was designed for a shear of 60 kips, but now must carry an axial force of 39 kips when the shear is at 33 kips. Let us check the axial capacity of this connection. The most critical limit state is prying action because of thin angle leg thickness. From Fig. 2.69

$$b = \frac{5.5 - 0.355 - 2 \times 0.375}{2} = 2.20$$

$$a = \frac{8 + 0.355 - 5.5}{2} = 1.43$$

$$1.25 \times 2.20 = 2.25 > 1.43$$

Use $a = 1.43$. Then $b' = 1.82$, $a' = 1.81$, $\rho = 1.01$, $\delta = 0.72$, $V = 33/8 = 4.125$ kips/bolt. The holes are HSSL (horizontal short slots), so $\phi r_v = 8.88$ kips/bolt. Since $4.125 < 8.88$, the bolts are ok for shear (as they

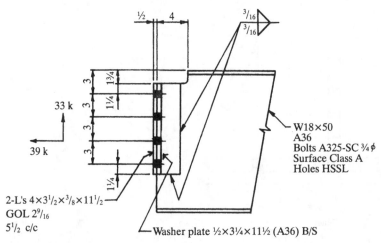

Figure 2.69 A shear connection needing reinforcement to carry axial load of 39 kips.

obviously must be since the connection was originally designed for 60 kips shear). Because this is a shear connection, the shear capacity is reduced by the tension load by the factor $1 - T/(1.13T_b)$, where T is the applied load per bolt and T_b is the specified pretension. Thus, the reduced shear design strength is

$$\phi r_v' = \phi r_v\left(1 - \frac{T}{1.13T_b}\right)$$

This expression can be inverted to a form usable in the prying action equations as

$$\phi r_t' = 1.13T_b\left(1 - \frac{V}{\phi r_v}\right) \le \phi r_t$$

For the present problem

$$\phi r_t' = 1.13 \times 28\left(1 - \frac{4.125}{8.8}\right) = 16.8 \text{ kips} < 29.8 \text{ kips}$$

Use $\phi r_t' = 16.8$ kips. Since $T = 39/8 = 4.875$ kips < 16.8 kips, the bolts are ok for tension/shear interaction exclusive of prying action. Now, checking prying action, which includes the bending of the angle legs,

$$t_C = \sqrt{\frac{4.44 \times 16.8 \times 1.82}{2.875 \times 36}} = 1.15$$

$$\alpha' = \frac{1}{0.72 \times 2.01}\left[\left(\frac{1.15}{0.375}\right)^2 - 1\right] = 5.81$$

Since $\alpha' > 1$, use $\alpha' = 1$, and

$$T_d = 16.8\left(\frac{0.375}{1.15}\right)^2 1.72 = 3.07 \text{ kips} < 4.875, \text{ no good}$$

Thus, the ⅜-in angle legs fail. Try a ½-in washer plate. Then

$$\eta = 1 + \left(\frac{0.5}{0.375}\right)^2 = 2.78$$

$$\alpha' = \frac{1}{0.72 \times 2.78 \times 2.01}\left[\left(\frac{1.15}{0.375}\right)^2 - 1\right] = 2.08$$

Since $\alpha' > 1$, use $\alpha' = 1$

$$T_d = 16.8\left(\frac{0.375}{1.15}\right)^2(1 + 0.72 \times 2.78) = 5.36 \text{ kips} > 4.875 \text{ kips, ok}$$

Therefore, the ½-in washer plates enable the connection to carry $5.36 \times 8 = 42.9$ kips > 39 kips, ok.

If ductility is a consideration, the ductility formula can be generalized to

$$d_{b_{\min}} = 0.163t\sqrt{\eta}\ \sqrt{\frac{F_y}{b}\left(\frac{b^2}{L^2} + 2\right)}$$

With $b = \text{GOL} - k = 2\dfrac{9}{16} - \dfrac{13}{16} = 1.75$

$$d_{b_{\min}} = 0.163 \times 0.375\sqrt{2.78}\ \sqrt{\frac{36}{1.75}\left(\frac{1.75^2}{11.5^2} + 2\right)} = 0.66 \text{ in} < 0.75 \text{ in, ok}$$

2.5.3 Use of ultimate tensile strength to increase axial capacity in the presence of prying action

In the example of the previous section, washer plates were used to increase the axial capacity of the connection. Research (Thornton, 1992, 1997) has shown that the prying action formulation of the AISC *Manual of Steel Construction* is very conservative with respect to strength, and that if F_y, the yield strength, is replaced by F_u, the tensile strength, a much better prediction of strength is achieved. This idea of replacing F_y with F_u in prying calculations was originally suggested by Kato and McGuire (1973), who also showed that it yielded more accurate predictions of strength. The prying action equations are unchanged when F_u replaces F_y, except for

$$t_c = \sqrt{\frac{4.44\phi r_t' b'}{pF_u}}$$

Following the calculations of the preceding section for the example of Fig. 2.69,

$$t_c = \sqrt{\frac{4.44 \times 16.8 \times 1.82}{2.875 \times 58}} = 0.902$$

$$\alpha' = \frac{1}{0.72 \times 2.01}\left[\left(\frac{0.902}{0.375}\right)^2 - 1\right] = 3.31$$

Since $\alpha' > 1$, use $\alpha' = 1$

$$T_d = 16.8\left(\frac{0.375}{0.902}\right)^2(1 + 0.72 \times 1) = 4.99 \text{ kips} > 4.875 \text{ kips, ok}$$

Thus, if the tensile strength approach is used, the $4 \times 3\frac{1}{2} \times \frac{3}{8}$ angles are capable of carrying $4.99 \times 8 = 40.0$ kips > 39 kips and are satisfactory without the reinforcing washer plates.

2.6 References

American Institute of Steel Construction, *Detailing for Steel Construction,* Chicago, IL, 1983.

American Institute of Steel Construction, *Manual of Steel Construction,* 7th ed., Chicago, IL, 1970.

American Institute of Steel Construction, *Manual of Steel Construction,* vol. II—*Connections,* Chicago, IL, 1992.

American Institute of Steel Construction, *Manual of Steel Construction,* 2d ed. (two volumes), LRFD, AISC, Chicago, IL, 1994.

American Institute of Steel Construction, "AISC Initiates Research into K Area Cracking," *Modern Steel Construction,* vol. 37, no. 2, February, 1997a, pp. 23–24.

American Institute of Steel Construction, "Seismic Provisions for Structural Steel Buildings," 1997b.

Anand, S. C., and Bertz, R. F., "Analysis and Design of a Web Connection in Direct Tension," *Engineering Journal,* vol. 18, no. 2, 2nd quarter, 1981, pp. 48–53.

Blodgett, Omer W., *Design of Welded Structures,* The James F. Lincoln Arc Welding Foundation, Cleveland, OH, 1966.

Carter, C. J., Thornton, W. A., and Murray, T. M., Discussion, "The Behavior and Load-carrying Capacity of Unstiffened Seated Beam Connections," *Engineering Journal,* vol. 34, no. 4, 4th quarter, 1997, pp. 151–156.

Fling, R. S., "Design of Steel Bearing Plates," *Engineering Journal,* vol. 7, no. 2, 2d quarter, 1970, pp. 37–40.

Gaylord, E. H., and Gaylord, C. N., *Design of Steel Structures,* McGraw-Hill, New York, 1972, pp. 139–141.

Gross, J. L., "Experimental Study of Gusseted Connections," *Engineering Journal,* vol. 27, no. 3, 3d quarter, 1990, pp. 89–97.

Hardash, S., and Bjorhovde, R., "New Design Criteria for Gusset Plates in Tension," *Engineering Journal,* vol. 23, no. 2, 2d quarter, 1985, pp. 77–94.

Huang, J. S., Chen, W. F., and Beedle, L. S., "Behavior and Design of Steel Beam to Column Moment Connections," *Bulletin 188,* Welding Research Council, New York, October, 1973.

Kato, B. and McGuire, W., "Analysis of T-Stub Flange to Column Connections," *Journal of the Structural Division,* ASCE, vol. 99, no. ST5, May 1973, pp. 865–888.

Kloiber, L., and Thornton, W., "Design Approaches to Shear Connections for Skewed Members in Steel Structures," *Building to Last, Proceedings of Structures Congress XV,* ASCE, Portland, OR, April 13–16, 1997.

Kulak, G. L., Fisher, J. W., and Struik, J. H. A., *Guide to Design Criteria for Bolted and Riveted Joints,* Wiley-Interscience, New York, 1987.

Kulak, G. L., and Green, D. L., "Design of Connections in Wide Flange Beam or Girder Splices," *Engineering Journal,* vol. 27, no. 2, 2d quarter, 1990, pp. 41–48.

Mann, A. P., and Morris, L. J., "Limit Design of Extended End Plate Connection," *Journal of the Structural Division,* ASCE, vol. 105, no. ST3, March 1979, pp. 511–526.

McGuire, W., *Steel Structures,* Prentice Hall, Englewood Cliffs, NJ, 1968, p. 933.

Murray, T., "Design of Lightly Loaded Column Base Plates," *Engineering Journal,* vol. 20, no. 4, 4th quarter, 1983, pp. 143–152.

Richard, R. M., "Analysis of Large Bracing Connection Designs for Heavy Construction," *National Steel Construction Conference Proceedings,* AISC, Chicago, IL, 1986, pp. 31.1–31.24.

Richard, R. M., et al., "Analytical Models for Steel Connections," *Behavior of Metal Structures, Proceedings of the W. H. Munse Symposium,* W. J. Hall and M. P. Gaus, eds., May 17, 1983, ASCE, NY.

Sputo, T., and Ellifrit, D. S. "Proposed Design Criteria for Stiffened Seated Connections to Column Webs," *Proceedings of the AISC National Steel Construction Conference,* Washington, DC, 1991, pp. 8.1–8.26.

Stockwell, F. W., Jr., "Preliminary Base Plate Selection," *Engineering Journal,* vol. 21, no. 3, 3d quarter, 1975, pp. 92–99.

0

0

0

0

0

0

0

0

Struik, J. H. A., and deBack, J., "Tests on T-Stubs With Respect to a Bolted Beam to Column Connections," *Report 6-69-13*, Stevin Laboratory, Delft University of Technology, Delft, The Netherlands, 1969 [as referenced in Kulak, Fisher, and Struik (1987)], pp. 272–282.

Thornton, W. A., "Prying Action—A General Treatment," *Engineering Journal*, vol. 22, no. 2, 2d quarter, 1985, pp. 67–75.

Thornton, W. A., "Design of Small Base Plates for Wide Flange Columns," *Engineering Journal*, vol. 27, no. 3, 3d quarter, 1990a, pp. 108–110.

Thornton, W. A., "Design of Small Base Plates for Wide Flange Columns—A Concatenation of Methods," *Engineering Journal*, vol. 27, no. 4, 4th quarter, 1990b, pp. 173–174.

Thornton, W. A., "On the Analysis and Design of Bracing Connections," *National Steel Construction Conference Proceedings*, AISC, Chicago, IL, 1991, pp. 26.1–26.33.

Thornton, W. A., "Strength and Serviceability of Hanger Connections," *Engineering Journal*, vol. 29, no. 4, 4th quarter, 1992, pp. 145–149.

Thornton, W. A., "Treatment of Simple Shear Connections Subject to Combined Shear and Axial Forces," *Modern Steel Construction*, vol. 35, no. 9, September, 1995a, pp. 9–10.

Thornton, W. A., "Connections—Art, Science and Information in the Quest for Economy and Safety," *Engineering Journal*, vol. 32, no. 4, 4th quarter, 1995b, pp. 132–144.

Thornton, W. A., "Rational Design of Tee Shear Connections," *Engineering Journal*, vol. 33, no. 1, 1st quarter, 1996, pp. 34–37.

Thornton, W. A., "Strength and Ductility Requirements for Simple Shear Connections with Shear and Axial Load," *Proceedings, National Steel Construction Conference*, AISC, Chicago, IL, May 7–9, 1997.

Timoshenko, S. P., and Goodier, J. N., *Theory of Elasticity*, 3d ed., McGraw-Hill, New York, 1970, pp. 57–58.

Vasarhelyi, D. D., "Tests of Gusset Plate Models," *Journal of the Structural Division*, ASCE, vol. 97, no. ST2, February 1971.

Whitmore, R. E., *"Experimental Investigation of Stresses in Gusset Plates,"* University of Tennessee Engineering Experiment Station Bulletin 16, May 1952, University of Tennessee, Knoxville.

Chapter

3

Welded Joint Design and Production

Duane K. Miller
R. Scott Funderburk

The Lincoln Electric Company
Cleveland, OH

(Courtesy of The Steel Institute of New York.)

3.1   Structural Steels for Welded Construction

3.1.1   Introduction

When selecting steels for structural applications, engineers usually select the specific steel based upon the mechanical property of strength, whether yield or tensile. These mechanical properties, along with the modulus of elasticity, in general satisfy the requirements for structural considerations. Additional requirements, such as notch toughness (typically measured by the Charpy V-notch specimen), may be specified to minimize brittle fracture, especially for dynamically loaded structures. With the single exception of weathering steel for uncoated applications exposed to atmospheric conditions, the chemical composition generally is of little concern from a structural point of view.

In terms of weldability, the chemistry of the steel is at least as important as, and arguably more important than, the mechanical properties. During the thermal cutting processes associated with construction, as well as when it is arc-welded, steel is subject to a variety of thermal cycles which can alter its mechanical properties in the area immediately adjacent to the weld metal. This area is known as the *heat-affected zone* and is, by definition, base metal that has been thermally affected by the welding or cutting process. While this region is generally small (typically 2 to 3 mm wide), it may exhibit different strength properties, and the toughness in this region may be dramatically altered. Finally, the base metal composition may have a significant effect on the weld metal composition, particularly for single-pass welds and when welding procedures are used that result in deep penetration. The chemistry of the base material is of the utmost importance in determining the suitability of a steel for welded construction. The mechanical properties of steel cannot be overlooked as they relate to welding, however. The strength of the steel will, in many cases, determine the required strength level of the weld deposit.

Although most steels can be welded, they are not all welded with the same degree of ease. *Weldability* is the term used to describe how readily the steel can be welded. For new construction, it is always advisable to select steels with good weldability since this will inevitably lead to both high quality and economical construction. Steels with reduced weldability may require specialized electrodes, techniques, preheat and postheat treatments, and joint designs. While materials that are difficult to weld are successfully fabricated every day, the use of these materials is generally inappropriate for new construction given the variety of readily available steels with excellent weldability. Occasionally, the engineer is faced with the situation of welding on material with less than optimum weldability, such as when welding on existing structures is necessary. Under these circumstances, it is advisable to proceed with caution, reviewing the principles outlined in this chapter, and, when necessary, to contact a welding engineer with expertise in metallurgy to address the specifics of the situation.

For most applications, however, modern welding codes such as the American Welding Society's *AWS D1.1 Structural Welding Code—Steel* and the American Institute of Steel Construction's *Steel Construction Manual* list weldable steels suitable for construction. These materials have a long history of satisfactory performance, and the codes supply appropriate guidelines as to what precautions or techniques are appropriate for certain materials. For example, AWS D1.1 lists "prequalified steels" that may be used in conjunction with a prequalified welding procedure. The code requirements for the fabrication of these steels are sufficiently justified that the contractor is not required to qualify the welding procedures by test when using this particular material, provided that all the other prequalified requirements were met.

Codes, however, do not necessarily include new developments from the steel producers. An inevitable characteristic of codes is that they will always lag behind industry. Once a particular steel has an acceptable history of performance, it may be incorporated into the applicable specifications. Until that time, the engineer must rely upon research data to determine the suitability of the part for a specific application.

A variety of tests have been devised over the years, each capable of measuring specific aspects of the weldability of the material under different conditions. Some tests measure the heat-affected zone properties, whereas others are more sensitive to weld-metal cracking tendencies. Unproven materials should be carefully reviewed by a competent engineer before used in actual applications, and actual consideration of approximate weldability tests is recommended.

Listed in the following section are typical steels that are used for welded construction today.

3.1.2   Modern base metals for welding

The carbon steels.   Classification of the carbon steels is based principally on carbon content. The groups are: low carbon (to 0.30% carbon), medium carbon (0.30 to 0.45%), and high carbon (more than 0.45%). Mechanical properties of hot finished steels are influenced principally by chemical composition (particularly carbon content), but other factors—finishing temperature, section size, and the presence of residual elements—also affect properties. A ¾-in plate, for example, has higher tensile properties and lower elongation than a 1½-in plate of the same composition, resulting primarily from the higher rate of cooling of the ¾-in plate from the rolling temperature. Medium- and high-carbon steels are not typically used for structural operations and therefore will not be discussed further.

Low-carbon steels.   In general, steels with carbon contents to 0.30% are readily joined by all the common arc-welding processes. These grades account for the greatest tonnage of steels used in welded structures. Typical applications include structural assemblies, as well as many other areas.

Steels with very low carbon content—to 0.13%—are generally good welding steels, but they are not the best for high-speed production welding. The low carbon content and the low manganese content (to 0.30%) tend to produce internal porosity. This condition is usually corrected by modifying the welding procedure slightly—usually by using a slower speed. Steels with very low carbon content are more ductile and easier to form than higher-carbon steels. They are used for applications requiring considerable cold forming, such as stampings or rolled or formed shapes.

Steels with 0.15 to 20% carbon content generally have excellent weldability, and they can be welded with all types of mild-steel electrodes. These steels should be used for maximum production speed on assemblies or structures that require extensive welding.

Steels at the upper end of the low-carbon range—the 0.25 to 030% carbon grades—generally have good weldability, but when one or more of the elements is on the high side of permissible limits, cracking can result, particularly in fillet welds. With slightly reduced speeds and currents, any mild steel type of electrode can be used. In thicknesses up to $\frac{5}{16}$ in, standard procedures apply.

If some of the elements—particularly carbon, silicon, or sulfur—are on the high side of the limits, surface holes may form. Reducing current and speed minimizes this problem.

Although for some welding applications these steels require little or no preheating, heavy sections (2 in or more) and certain joint configurations often require a preheat. In general, steels in the 0.25 to 0.30% carbon range should be welded with low hydrogen processes.

High-strength–low-alloy structural steels. Higher mechanical properties and better corrosion resistance than that of structural carbon steels are characteristics of the high-strength–low-alloy (HSLA) steels. These improved properties are achieved by addition of small amounts of alloying elements. Some of the HSLA types are carbon-manganese steels; others contain different alloy additions, governed by requirements for weldability, formability, toughness, or economy. The strength of these steels is generally between that of structural carbon steels and that of high-strength quenched and tempered steels.

High-strength–low-alloy steels are usually used in the as-rolled condition, although some are available that require heat treatment after fabrication. These steels are produced to specific mechanical property requirements rather than to chemical compositions. Minimum mechanical properties available in the as-rolled condition vary among the grades and, within most grades, with thickness. Ranges of properties available in this group of steels are

1. Minimum yield point from 42,000 to 70,000 psi
2. Minimum tensile strength from 60,000 to 85,000 psi
3. Resistance to corrosion, classified as: equal to that of carbon steels, twice that of carbon steels, or 4 to 6 times that of carbon steels

The high-strength–low-alloy steels should not be confused with the high-strength quenched and tempered alloy steels. Both groups are sold primarily on a trade name basis, and they frequently share the same trade name, with different letters or numbers being used to identify each. The quenched and tempered steels are full-alloy steels that are heat-treated at the mill to develop optimum properties. They are generally martensitic in structure, whereas the HSLA steels are mainly ferritic steels; this is the clue to the metallurgical and fabricating differences between the two types. In the as-rolled condition, ferritic steels are composed of relatively soft, ductile constituents; martensitic steels have hard, brittle constituents that require heat treatment to produce their high-strength properties.

Strength in the HSLA steels is achieved instead by relatively small amounts of alloying elements dissolved in a ferritic structure. Carbon content rarely exceeds 0.28% and is usually between 0.15 and 0.22%. Manganese content ranges from 0.85 to 1.60%, depending on grade, and other alloy additions—chromium, nickel, silicon, phosphorus, copper, vanadium, columbium, and nitrogen—are used in amounts of less than 1%. Welding, forming, and machining characteristics of most grades do not differ markedly from those of the low-carbon steels.

To be weldable, the high-strength steels must have enough ductility to avoid cracking from rapid cooling. Weldable HSLA steels must be sufficiently low in carbon, manganese, and all "deep-hardening" elements to ensure that appreciable amounts of martensite are not formed upon rapid cooling. Superior strength is provided by solution of the alloying elements in the ferrite of the as-rolled steel. Corrosion resistance is also increased in certain of the HSLA steels by the alloying additions.

ASTM specifications. Thirteen ASTM specifications cover the plain-carbon, high-strength–low-alloy, and quenched and tempered structural steels. All of the following steels, except those noted, are prequalified according to D1.1-98:

ASTM A36: Covers carbon steel shapes, plates, and bars of structural quality for use in bolted or welded construction of bridges and buildings and for general structural purposes. Strength requirements are 58- to 80-ksi tensile strength and a minimum of 36-ksi yield strength. In general, A36 is weldable and is available in many shapes and sizes. However, its strength is limited only by a minimum value and could have yield strengths exceeding 50 ksi. Therefore, preheat and low hydrogen electrodes may be required.

ASTM A53: Covers seamless and welded black and hot-dipped galvanized steel pipe. Strength requirements range from 25- to 35-ksi minimum yield and 45- to 60-ksi minimum tensile for three types and two grades. While maximum carbon levels are between 0.25 and 0.30%, pipe is typically manufactured by electric-resistance welding and submerged arc welding. A53 may require low-hydrogen process controls and preheat, especially when the carbon levels are close to the maximum allowed.

ASTM A441: Covers the intermediate-manganese HSLA steels that are readily weldable with proper procedures. The specification calls for additions of vanadium and a lower manganese content (1.25% maximum) than ASTM A440, which is generally not weldable due to its high carbon and manganese leads. Minimum mechanical properties vary from 42- to 50-ksi yield to 63- to 70-ksi tensile. Atmospheric corrosion resistance of this steel is approximately twice that of structural carbon steel, while it has superior toughness at low temperatures. Only shapes, plates, and bars are covered by the specification, but weldable sheets and strip can be supplied by some producers with approximately the same minimum mechanical properties.

ASTM A500: Covers cold-formed welded and seamless carbon steel round, square, rectangular, or special shape structural tubing for welded, riveted, or bolted construction of bridges and buildings and for general structural purposes, and is commonly used in the United States. This tubing is produced in both seamless and welded sizes with a maximum periphery of 64 in and a maximum wall thickness of 0.625 in. Minimum strength properties range from 33-ksi yield and 45-ksi tensile for grade A to 46-ksi yield and 62-ksi tensile for grade C.

ASTM A501: Covers hot-formed welded and seamless carbon steel square, round, rectangular, or special shape structural tubing for welded, riveted, or bolted construction of bridges and buildings and for general structural purposes, and is commonly used in Canada. Square and rectangular tubing may be furnished in sizes from 1 to 10 in across flat sides and wall thicknesses from 0.095 to 1.00 in. Minimum strength requirements are 36-ksi yield and 58-ksi tensile.

ASTM A514: Covers quenched and tempered alloy steel plates of structural quality in thicknesses of 6 in and under intended primarily for use in welded bridges and other structures. Strength requirements range from 100- to 130-ksi tensile and 90- to 100-ksi yield. When welding, the heat input must be controlled and specific minimum and maximum levels of heat input are required. Additionally, a low-hydrogen process is required.

ASTM A516: Covers carbon steel plates intended primarily for service in welded pressure vessels where improved notch toughness is important. These plates are furnished in four grades with strength requirements ranging from 55- to 90-ksi tensile and 30- to 38-ksi yield.

ASTM A572: Includes six grades of high-strength–low-alloy structural steels in shapes, plates, and bars used in buildings and bridges. These steels offer a choice of strength levels ranging from 42- to 65-ksi yields. Proprietary HSLA steels of this type with 70- and 75-ksi yield points are also available. Increasing care is required for welding these steels as the strength level increases.

A572 steels are distinguished from other HSLA steels by their columbium, vanadium, and nitrogen content.

A supplementary requirement is included in the specification that permits designating the specific alloying elements required in the steel. Examples are the type 1 designation, for columbium; type 2, for vanadium; type 3, for columbium and vanadium; and type 4, for vanadium and nitrogen. Specific grade designations must accompany this type of requirement.

ASTM A588: Covers high-strength–low-alloy structural steel shapes, plates, and bars for welded, riveted, or bolted connection. However, it is intended primarily for use in welded bridges and buildings in its unpainted condition, since the atmospheric corrosion resistance in most environments is substantially better than that of carbon steels. When properly exposed to the atmosphere, this steel can be used bare (unpainted) for many applications.

If the steel is to be painted, a low-hydrogen electrode without special corrosion resistance can be used. However, if the steel is to remain bare, then an electrode must be selected that has similar corrosion characteristics.

ASTM A709: Covers carbon and high-strength–low-alloy steel structural shapes, plates, and bars and quenched and tempered alloy steel for structural plates intended for use in bridges. Six grades are available in four yield strength levels of 36, 50, 70, and 100 ksi. Grades 50W, 70W, and 100W have enhanced atmospheric corrosion resistance. From a welding point of view, these grades are essentially the same as A36, A572, A852, and A514, respectively.

ASTM A710: Covers low-carbon age-hardening nickel-copper-chromium-molybdenum-columbium, nickel-copper-columbium, and nickel-copper-manganese-molybdenum-columbium alloy steel plates for general applications. Three different grades and three different conditions provide minimum yield strengths from 50 to 90 ksi. When this steel is to be welded, a welding procedure should be developed for the specific grade of steel and intended service. According to D1.1-98, no preheat is required with SMAW, SAW, GAW, and FCAW electrodes that are capable of depositing weld metal with a maximum diffusible hydrogen content of 8 ml/100 g.

ASTM A852: Covers quenched and tempered high-strength–low-alloy structural steel plates for welded, riveted, or bolted construction. It is intended primarily for use in welded bridges and buildings where savings in weight, added durability, and good notch toughness are important. This steel specification has substantially better atmospheric corrosion resistance than that of carbon structural steels. It has similar chemistry requirements to A588, but has been quenched and tempered to achieve the higher-strength level. Welding technique is important, and a welding procedure suitable for the steel and intended service should be developed. The specification limits the material thickness up to and including 4 in. According to D1.5-96, A852 is an approved bare metal under the A709 specification and D1.1-98 requires welding procedure qualification for this steel.

ASTM 913: Covers high-strength–low-alloy structural steel shapes in grades 60, 65, and 70 produced by the quenching and self-tempering process. The shapes are intended for riveted, bolted, or welded construction of bridges, buildings, and other structures. Although not in D1.5, the maximum yield strengths are 60, 65, and 70 ksi for the respective grades, while the minimum tensile strengths are 75, 80, and 90 ksi. A913 can be welded with a low-hydrogen process, and according to D1.1-98, it must provide a maximum diffusible hydrogen content of 8 ml/100 g. The shapes should not be formed or postweld heat-treated at temperatures exceeding 1100°F (600°C).

3.1.3   Older and miscellaneous base metals

Cast iron.   Cast iron was a popular building material through the late 1800s, and occasionally an engineer is faced with the need to make additions to a cast-iron column, for example. Cast iron may also be encountered in miscellaneous structural applications such as ornate light poles, archways, and other components with decorative functions in addition to accomplishing structural support. Cast iron can be successfully welded but with great difficulty. Unless the welding involves repair of casting defects (voids, slag, or sand pockets), or the reattachment of nonstructural components, it is highly desirable to investigative alternative methods of joining when cast iron is involved. Bolted attachments are generally more easily made than welded ones.

Although there are several types of cast iron, the following are general guidelines to follow when welding cast iron. First, determine what type of cast iron is to be welded (that is, gray, malleable, ductile, or white). If the casting is white iron, which is generally considered unweldable, then an alternative jointing or repairing procedure should be investigated. However, if the casting is gray, malleable, or ductile, an appropriate welding procedure can be developed.

Second, preheating is *almost always* required when welding cast iron. Preheating can be achieved by heating large parts in a furnace or by using a heating torch on smaller parts. In general, the minimum preheat temperature will be around 500°F to sufficiently retard the cooling rate.

Third, any welding processes can be utilized; however, the electrode used should be a cast-iron electrode as specified in AWS A5.15.

Finally, after welding, the casting should be allowed to cool slowly to help reduce the hardness in the heat-affected zone. If the casting is a load-bearing member, caution must be taken to prohibit brittle fracture. The welding procedures should be tested, and a welding expert should be consulted.

Cast steels. Steel castings are often used for miscellaneous structural components such as rockers on expansion joints, miscellaneous brackets, and architectural elements. Welding cast steel is generally similar to welding of rolled steel of similar chemical composition. However, steel castings are often made of significantly different chemical compositions than would be utilized for rolled steel, particularly structural steels expected to be joined by welding. For example, a steel casting made of AISI 4140, a chromium-molybdenum steel with 0.40% carbon, will have reduced weldability because of the alloy content and the high-carbon level. The producers of the steel castings often utilize alloys with reduced weldability simply because it may be easier to obtain the desired mechanical properties with the enriched composition, and welding of these materials is often overlooked.

Steel castings with poor weldability should be joined by bolting when possible, or alternative compositional requirements should be pursued. For example, it may be possible to select material with lower carbon and/or alloy levels for a specific application, increasing the weldability. If the chemistry of the casting is similar to that of a rolled steel, it will have similar weldability characteristics. However, the casting will have associated flow pattern–dependent properties.

Stainless steels. Because of its expense, stainless steel is rarely used for structural applications. The unique characteristics of stainless steel, however, make it ideally suited for applications where the structural material is subjected to corrosive environments, high or low service temperatures, and for applications where the material is to be used in its uncoated state. Certain grades of stainless steel are readily weldable, whereas others are welded with great difficulty.

When stainless steel is used as a structural material, particularly when it is joined to carbon steel elements, it is important to recognize the difference in thermal expansion between the two materials. With stainless expansion rates being 1.5 times that of carbon steel, the differential expansion can cause problems in structures that are subjected to variations in temperature.

The American Welding Society is currently developing a welding code to govern the fabrication of stainless-steel structures. It will be known as AWS D1.6 and, although not complete at the time of the writing of this chapter, it should be available in the near future. This code will provide welding requirements similar to those contained in AWS D1.1, but will deal specifically with stainless steel as the base material.

Metallurgy of stainless steel. Stainless steels are iron-chromium alloys, usually with a low carbon content, containing at least 11.5% chromium, which is the level at which effective resistance to atmospheric corrosion begins.

The American Iron and Steel Institute (AISI) classifies stainless steels by their metallurgical structures. This system is useful because the structure (austenitic, ferritic, or martensitic) indicates the general range of mechanical and physical properties, formability, weldability, and hardenability. The austenitic type generally have good weldability characteristics, high ductility, low yield strength, and high ultimate strength characteristics that make them suitable for forming and deep drawing operations; they also have the highest corrosion resistance of all the stainless steels. Austenitic grades account for the highest tonnage of weldable stainless steels produced.

Ferritic stainless steels are characterized by high levels of chromium and low carbon, plus additions of titanium and columbium. Since little or no austenite is present, these grades do not transform to martensite upon cooling, but remain ferritic throughout their normal operating temperature range. Principal applications of the ferritic types are automotive and appliance trim, chemical processing equipment, and products requiring resistance to corrosion and scaling at elevated temperatures, rather than high strength. Ferritic stainless steels are not easily welded in structural applications. If welding is required, a welding expert should be consulted for structural applications.

Martensitic stainless steels are iron-chromium alloys that can be heat treated to a wide range of hardness and strength levels. Martensitic grades are typically used to resist abrasion. They are not as corrosion resistant as the austenitic and ferritic types. Martensitic stainless steels, like the ferritic, are not easily welded in structural applications, and a welding expert should be consulted when welding.

Aluminum. Aluminum has many characteristics that are highly desirable for engineering applications, including structural, such as high strength-to-weight ratio, and corrosion resistance. Aluminum does not have the high modulus of elasticity associated with steel, but the weight-to-modulus ratio of the two materials is roughly equal. Aluminum is readily welded, but the welded connection rarely duplicates the strength of unwelded base metal. This is because the heat-affected zone (HAZ) in the as-welded state has reduced strength compared to the unaffected base material. This is in stark contrast to the behavior of steel, where the entire welded connection can usually be made as strong as the base material. The degree of strength degradation depends on the particular alloy system used. However, the engineer can conservatively assume that the heat-affected zone will have approximately one-half the strength of the aluminum alloy.

This characteristic is not necessarily a strong impediment to the use of aluminum, however. Creative joint designs and layouts of material can minimize the effect of the reduced strength HAZ. For example, rather than employing a butt joint perpendicular to the primary tensile loading, it may be possible to reorient the joint so that it lies parallel to the stress field, minimizing the magnitude of stress transfer across this interface, and thus reducing the effects of the reduced strength HAZ. Gussets, plates, stiffeners, and increases in thickness of the material at transition points can also be helpful in overcoming this characteristic.

Aluminum cannot be welded to steel or stainless steel by conventional arc-welding processes. It is possible to join aluminum to other materials by alternative welding processes such as explosion bonding. A common approach to welding aluminum to other significantly different materials is to utilize explosion bonding to create a transition member. In the final application, the steel, for example, is welded to the steel portion of the transition member, and the aluminum is welded to the aluminum side. While generally not justified for structural applications, this approach has been used for piping applications, for example. For structural applications, mechanical fasteners are generally employed; however, the galvanic action should be considered.

The requirements for fabrication of aluminum are contained in AWS D1.2 *Structural Welding Code—Aluminum.*

3.2   Weld Cracking/Solutions

Weld cracking is a problem faced occasionally by the fabricator. This section will discuss the various types of cracking and possible solutions for steel alloys.

Several types of discontinuities may occur in welds or heat-affected zones. Welds may contain porosity, slag inclusions, or cracks. Of the three, cracks are by far the most detrimental. Whereas there are acceptable limits for slag inclusions and porosity in welds, cracks are never acceptable. Cracks in, or in the vicinity of, a weld indicate that one or more problems exist that must be addressed. A careful analysis of crack characteristics will make it possible to determine their cause and take appropriate corrective measures.

For the purposes of this section, "cracking" will be distinguished from weld failure. Welds may fail due to overload, underdesign, or fatigue. The cracking discussed here is the result of solidification, cooling, and the stresses that develop due to weld shrinkage. *Weld cracking* occurs close to the time of fabrication. *Hot cracks* are those that occur at elevated temperatures and are usually solidification related. *Cold cracks* are those that occur after the weld metal has cooled to room temperature and may be hydrogen related. Neither is the result of service loads.

Most forms of cracking result from the shrinkage strains that occur as the weld metal cools. If the contraction is restricted, the strains will induce residual stresses that cause cracking. There are two opposing forces: the stresses induced by the shrinkage of the metal and the surrounding rigidity of the base material. The shrinkage stresses increase as the volume of shrinking metal increases. Large weld sizes and deep penetrating welding procedures increase the shrinkage strains. The stresses induced by these strains will increase when higher-strength filler metals and base materials are involved. With a higher yield strength, higher residual stresses will be presented.

Under conditions of high restraint, extra precautions must be utilized to overcome the cracking tendencies which are described in the following sections. It is essential to pay careful attention to welding sequence, preheat and interpass temperatures, postweld heat treatment, joint design, welding procedures, and filler material. The judicious use of peening as an in-process stress relief treatment may be necessary to fabricate highly restrained members.

3.2.1 Centerline cracking

Centerline cracking is characterized as a separation in the center of a given weld bead. If the weld bead happens to be in the center of the joint, as is always the case on a single-pass weld, centerline cracks will be in the center of the joint. In the case of multiple-pass welds, where several beads per layer may be applied, a centerline crack may not be in the geometric center of the joint, although it will always be in the center of the bead (Fig. 3.1).

Centerline cracking is the result of one of the following three phenomena: *segregation-induced cracking, bead shape–induced cracking,* or *surface profile–induced cracking.* Unfortunately, all three phenomena reveal themselves in the same type of crack, and it is often difficult to identify the cause. Moreover, experience has shown that often two or even all three of the phenomena will interact and contribute to the cracking problem. Understanding the fundamental mechanism of each of these types of centerline cracks will help in determining the corrective solutions.

Segregation-induced cracking occurs when low melting-point constituents, such as phosphorus, zinc, copper, and sulfur compounds, in the admixture separate during the weld solidification process. Low melting-point components in the molten metal will be forced to the center of the joint during solidification, since they are the last to solidify and the weld tends to separate as the solidified metal contracts away from the center region containing the low melting-point constituents.

When centerline cracking induced by segregation is experienced, several solutions may be implemented. Since the contaminant usually comes from the base material, the first consideration is to limit the amount of contaminant pickup from the base material. This may be done by limiting the penetration of the welding process. In some cases, a joint redesign may be desirable. The extra penetration afforded by some of the processes is not necessary and can be reduced. This can be accomplished by using lower welding currents.

A buttering layer of weld material (Fig. 3.2) deposited by a low-energy process, such as shielded metal arc welding, may effectively reduce the amount of pickup of contaminant into the weld admixture.

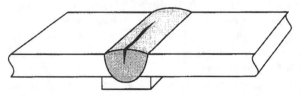

Figure 3.1 Centerline cracking. (*Courtesy of The Lincoln Electric Company.*)

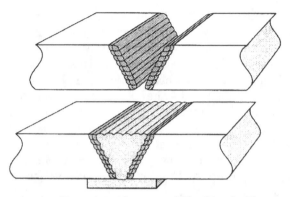

Figure 3.2   Buttering. (*Courtesy of The Lincoln Electric Company.*)

In the case of sulfur, it is possible to overcome the harmful effects of iron sulfides by preferentially forming manganese sulfide. Manganese sulfide (MnS) is created when manganese is present in sufficient quantities to counteract the sulfur. Manganese sulfide has a melting point of 2900°F. In this situation, before the weld metal begins to solidify, manganese sulfides are formed which do not segregate. Steel producers utilize this concept when higher levels of sulfur are encountered in the iron ore. In welding, it is possible to use filler materials with higher levels of manganese to overcome the formation of low melting-point iron sulfide. Unfortunately, this concept cannot be applied to contaminants other than sulfur.

The second type of centerline cracking is known as *bead shape–induced cracking*. This is illustrated in Fig. 3.3 and is associated with deep penetrating processes such as SAW and CO_2-shielded FCAW. When a weld bead is of a shape where there is more depth than width to the weld cross section, the solidifying grains growing perpendicular to the steel surface intersect in the middle, but do not gain fusion across the joint. To correct for this condition, the individual weld beads must have at least as much width as depth. Recom-

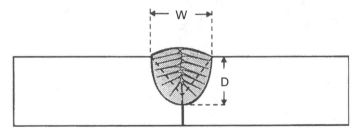

Figure 3.3   Bead shape–induced cracking. (*Courtesy of The Lincoln Electric Company.*)

mendations vary from a 1:1 to a 1.4:1 width-to-depth ratio to remedy this condition. The total weld configuration, which may have many individual weld beads, can have an overall profile that constitutes more depth than width. If multiple passes are used in this situation, and each bead is wider than it is deep, a crack-free weld can be made.

When centerline cracking due to bead shape is experienced, the obvious solution is to change the width-to-depth relationship. This may involve a change in joint design. Since the depth is a function of penetration, it is advisable to reduce the amount of penetration. This can be accomplished by utilizing lower welding amperages and larger-diameter electrodes. All of these approaches will reduce the current density and limit the amount of penetration.

The final mechanism that generates centerline cracks is *surface profile–induced conditions*. When concave weld surfaces are created, internal shrinkage stresses will place the weld metal on the surface into tension. Conversely, when convex weld surfaces are created, the internal shrinkage forces pull the surface into compression. These situations are illustrated in Fig. 3.4. Concave weld surfaces frequently are the result of high arc voltages. A slight decrease in arc voltage will cause the weld bead to return to a slightly convex profile and eliminate the cracking tendency. High travel speeds may also result in this configuration. A reduction in travel speed will increase the amount of fill and return the surface to a convex profile. Vertical-down welding also has a tendency to generate these crack-sensitive concave surfaces. Vertical-up welding can remedy this situation by providing a more convex bead.

3.2.2   Heat-affected zone cracking

Heat-affected zone (HAZ) cracking (Fig. 3.5) is characterized by separation that occurs immediately adjacent to the weld bead. Although it

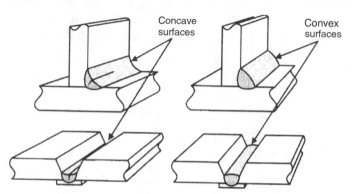

Figure 3.4   Surface profile–induced cracking. (*Courtesy of The Lincoln Electric Company.*)

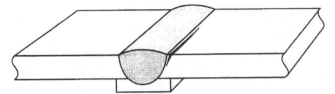

Figure 3.5 Heat-affected zone cracking. (*Courtesy of The Lincoln Electric Company.*)

is related to the welding process, the crack occurs in the base material, not in the weld material. This type of cracking is also known as *underbead cracking, toe cracking,* or *delayed cracking.* Because this cracking occurs after the steel has cooled below approximately 200°C, it can be called *cold cracking,* and because this cracking is associated with hydrogen, it is also called *hydrogen-assisted cracking.*

In order for heat-affected zone cracking to occur, three conditions must be present simultaneously: (1) there must be a sufficient level of hydrogen, (2) there must be a sufficiently sensitive material involved, and (3) there must be a sufficiently high level of residual or applied stress. Sufficient reduction or elimination of one of the three variables will eliminate heat-affected zone cracking. In welding applications, the typical approach is to limit two of the three variables, namely, the level of hydrogen and the sensitivity of the material. Hydrogen can enter into a weld pool through a variety of sources. Moisture and organic compounds are the primary sources of hydrogen. It may be present on the steel, electrode, in the shielding materials, and in atmospheric humidity. Flux ingredients, whether on the outside of electrodes, inside the core of electrodes, or in the form of submerged arc or electroslag fluxes, can adsorb or absorb moisture, depending on storage conditions and handling practices. To limit hydrogen content in deposited welds, welding consumables must be properly maintained, and welding must be performed on surfaces that are clean and dry. The second necessary condition for heat-affected zone cracking is a sensitive microstructure. In the case of heat-affected zone cracking, the area of interest is the heat-affected zone that results from the thermal cycle experienced by the region immediately surrounding the weld nugget. As this area is heated by the welding arc during the creation of the weld pool, it transforms from its room temperature structure of ferrite to the elevated temperature structure of austenite. The subsequent cooling rate will determine the resultant HAZ properties. Conditions that encourage the development of crack-sensitive microstructures include high cooling rates and higher hardenability levels in the steel. High cooling rates are encouraged by lower heat-input welding procedures, greater base material thicknesses, and colder base metal temperatures. Higher hardenability levels result from higher carbon contents and/or alloy levels. For a given steel, the most effective way to reduce the cooling rate is by raising the temperature of the surrounding steel through preheat. This reduces the temperature gradient, slowing cooling rates and limiting the formation of sensitive microstructures. Effective preheat is the primary means by which acceptable heat-affected zone properties are created, although heat input also has a significant effect on cooling rates in this zone.

The residual stresses of welding can be reduced through thermal stress relief, although for most structural applications, this is economically impractical. For complex structural applications, temporary shoring and other conditions must be considered as the steel will have a greatly reduced strength capacity at stress-relieving temperatures. For practical applications, heat-affected zone cracking will be controlled by effective low-hydrogen practice and appropriate preheats.

When sufficient levels of hydrogen, residual stress, and material sensitivity occur, hydrogen cracking will occur in the heat-affected zone. For this cracking to occur, it is necessary for the hydrogen to migrate into the heat-affected zone, an activity that takes time. For this reason, the D1.1 code requires a delay of 48 h for the inspection of welds made on A514 steel, known to be sensitive to hydrogen-assisted heat-affected zone cracking.

With time, hydrogen diffuses from weld deposits. Sufficient diffusion to avoid cracking normally takes place in a few weeks, although it may take many months depending on the specific application. The concentrations of hydrogen near the time of welding are always the greatest, and if hydrogen-induced cracking is to occur, it will generally occur within a few days of fabrication. However, it may take longer for the cracks to grow to a sufficient size to be detected.

Although a function of many variables, general diffusion rates can be approximated. At 450°F, hydrogen diffuses at the rate of approximately 1 in/h. At 220°F, hydrogen diffuses the same 1 in in approximately 48 h. At room temperature, typical diffusible hydrogen rates are 1 in/2 weeks. If there is a question regarding the level of hydrogen in a weldment, it is possible to apply a postweld heat treatment commonly called *postheat*. This generally involves the heating of the weld to a temperature of 400 to 500°F, holding the steel at that temperature for approximately 1 h for each inch of thickness of material involved. At that temperature, the hydrogen is likely to be redistributed through diffusion to preclude further risk of cracking. Some materials, however, will require significantly longer than 1 h/in. This operation is not necessary where hydrogen has been properly controlled, and it is not as powerful as preheat in terms of its ability to prevent underbead cracking. In order for postheat operations to be effective, they must be applied before the weldment is allowed to cool to room temperature. Failure to do so could result in heat-affected zone cracking prior to the application of the postheat treatment.

3.2.3 Transverse cracking

Transverse cracking, also called *cross-cracking,* is characterized as a crack within the weld metal perpendicular to the longitudonal direction (Fig. 3.6). This is the least frequently encountered type of cracking, and is generally associated with weld metal that is higher in strength, significantly overmatching the base material. Transverse cracking is also hydrogen assisted, and like heat-affected zone cracking, is also a factor of excessive hydrogen, residual stresses, and a sensitive microstructure. The primary difference is that transverse cracking occurs in the weld metal as a result of the longitudinal residual stress.

As the weld bead shrinks longitudinally, the surrounding base material resists this force by going into compression. The high strength of the surrounding steel in compression restricts the required shrinkage of the weld material. Due to the restraint of the surrounding base material, the weld metal develops longitudinal stresses which may facilitate cracking in the transverse direction.

When transverse cracking is encountered, a review of the low hydrogen practice is warranted. Electrode storage conditions should be carefully reviewed. If these are proper, a reduction in the strength of the weld metal will usually solve transverse cracking problems. Of course, design requirements must still be met, although most transverse cracking results from weld metal overmatch conditions.

Emphasis is placed upon the weld metal because the filler metal may deposit lower-strength, highly ductile metal under normal conditions. However, with the influence of alloy pickup, it is possible for the weld metal to exhibit extremely high strengths with reduced ductility. Using lower-strength weld metal is an effective solution, but caution should be taken to ensure that the required joint strength is attained.

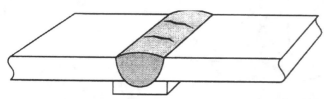

Figure 3.6 Transverse cracking. (*Courtesy of The Lincoln Electric Company.*)

Preheat may have to be applied to alleviate transverse cracking. The preheat will assist in diffusing hydrogen. As preheat is applied, it will additionally expand the length of the weld joint, allowing the weld metal and the joint to contract simultaneously, and reducing the applied stress to the shrinking weld. This is particularly important when making circumferential welds. When the circumference of the materials being welded is expanded, the weld metal is free to contract along with the surrounding base material, reducing the longitudinal shrinkage stress. Finally, postweld hydrogen release treatments that involve holding the steel at 250 to 450°F for extended times will assist in diffusing any residual hydrogen.

3.3 Welding Processes

A variety of welding processes can be used for fabrication in structural applications. However, it is important that all parties involved understand these processes in order to ensure quality and economical fabrication. A brief description of the major processes is provided below.

3.3.1 SMAW

Shielded metal arc welding (SMAW), commonly known as *stick electrode welding* or *manual welding,* is the oldest of the arc-welding processes (Fig. 3.7). It is characterized by versatility, simplicity, and flexibility. The SMAW process is commonly used for tack welding, fabrication of miscellaneous components, and repair welding. There is a practical limit to the amount of current that may be used. The covered electrodes are typically 9 to 18 in long, and if the current is

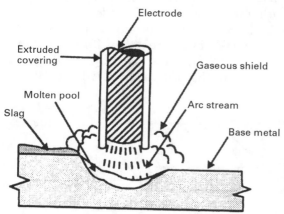

Figure 3.7 SMAW process. (*Courtesy of The Lincoln Electric Company.*)

raised too high, electrical-resistance heating within the unused length of electrode will become so great that the coating ingredients may overheat and "break down," resulting in potential weld quality degradation. SMAW also is used in the field for erection, maintenance, and repairs. SMAW has earned a reputation for depositing high-quality welds dependably. It is, however, slower and more costly than other methods of welding, and is more dependent on operator skill. Consequently, SMAW seldom is used for primary fabrication of structures.

3.3.2 FCAW

Flux-cored arc welding (FCAW) uses an arc between a continuous filler metal electrode and the weld pool. The electrode is always tubular. Inside the metal sheath is a combination of materials that may include metallic powder and flux. FCAW may be applied automatically or semiautomatically.

The flux-cored arc-welding process has become the most popular semiautomatic process for structural steel fabrication and erection. Production welds that are short, change direction, difficult to access, must be done out-of-position (that is, vertical or overhead), or part of a short production run generally will be made with semiautomatic FCAW.

The flux-cored arc-welding process offers two distinct advantages over shielded metal arc welding. First, the electrode is continuous. This eliminates the built-in starts and stops that are inevitable with shielded metal arc welding. Not only does this have an economic advantage because the operating factor is raised, but the number of arc starts and stops, a potential source of weld discontinuities, is reduced.

Another major advantage is that increased amperages can be used with flux-cored arc welding, with a corresponding increase in deposition rate and productivity. With the continuous flux-cored electrodes, the tubular electrode is passed through a contact tip, where electrical energy is transferred to the electrode. The short distance from the contact tip to the end of the electrode, known as *electrode extension* or *electrical stickout*, limits the buildup of heat due to electrical resistance. This electrode extension distance is typically $\frac{3}{4}$ to 1 in for flux-cored electrodes.

Within the category of flux-cored arc welding, there are two specific subsets: self-shielded flux core (FCAW-ss) (Fig. 3.8) and gas-shielded flux core (FCAW-g) (Fig. 3.9). Self-shielded flux-cored electrodes require no external shielding gas. The entire shielding system results from the flux ingredients contained within the core of the tubular electrode. The gas-shielded versions of flux-cored electrodes utilize an externally sup-

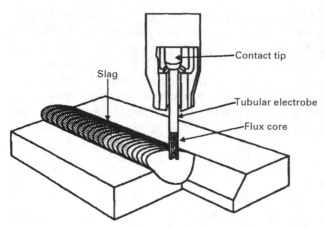

Figure 3.8 Self-shielded FCAW. (*Courtesy of The Lincoln Electric Company.*)

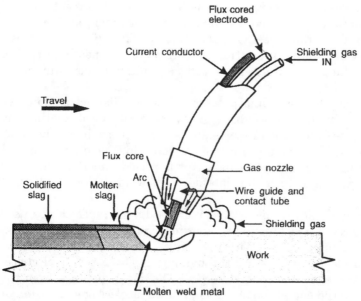

Figure 3.9 Gas-shielded FCAW. (*Courtesy of The Lincoln Electric Company.*)

plied shielding gas. In many cases, CO_2 is used, although other gas mixtures may be used, for example, argon/CO_2 mixtures. Both types of flux-cored arc welding are capable of delivering weld deposits that meet the quality and mechanical property requirements for most structure applications. In general, the fabricator will utilize the process that offers the greatest advantages for the particular environment. Self-shielded flux-cored electrodes are better for field-welding situations. Since no externally supplied shielding gas is required, the process may be used in high winds without adversely affecting the quality of the deposit. With any of the gas-shielded processes, wind shields must be erected to preclude interference with the gas shield in windy weather. Many fabricators have found self-shielded flux core offers advantages for shop welding as well, since it permits the use of better ventilation.

Individual gas-shielded flux-cored electrodes tend to be more versatile than self-shielded flux-cored electrodes, and in general, provide better arc action. Operator appeal is usually higher. While the gas shield must be protected from winds and drafts, this is not particularly difficult in shop fabrication situations. Weld appearance and quality are very good. Higher-strength gas-shielded FCAW electrodes are available, while current technology limits self-shielded FCAW deposits to 90-ksi tensile strength or less.

3.3.3 SAW

Submerged arc welding (SAW) differs from other arc welding processes in that a layer of granular material called *flux* is used for shielding the arc and the molten metal (Fig. 3.10). The arc is struck between

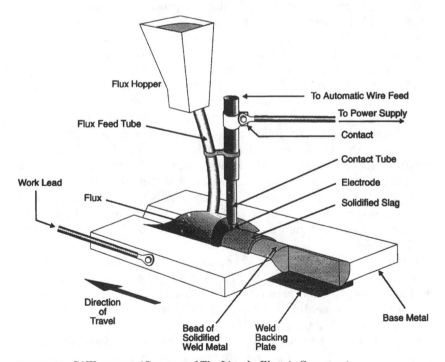

Figure 3.10 SAW process. (*Courtesy of The Lincoln Electric Company.*)

the workpiece and a bare wire electrode, the tip of which is submerged in the flux. Since the arc is completely covered by the flux, it is not visible and the weld is made without the flash, spatter, and sparks that characterize the open-arc processes. The nature of the flux is such that very little smoke or visible fumes are developed.

The process is typically fully mechanized, although semiautomatic operation is often utilized. The electrode is fed mechanically to the welding gun, head, or heads. In semiautomatic welding, the welder moves the gun, usually equipped with a flux-feeding device, along the joint. High currents can be used in submerged arc welding and extremely high-heat input levels can be developed. Because the current is applied to the electrode a short distance above its arc, relatively high amperages can be used on small-diameter electrodes, resulting in extremely high current densities. This allows for high deposition rates and deep penetration.

Welds made under the protective layer of flux (Fig. 3.10) are excellent in appearance and spatter-free. The high quality of submerged arc welds, the high deposition rates, the deep penetration characteristics, and the easy adaptability of the process to full mechanization make it popular for the manufacture of plate girders and fabricated columns.

One of the greatest benefits of the SAW process is freedom from the open arc. This allows multiple arcs to be operated in a tight, confined area without the need for extensive shields to guard the operators from arc flash. Yet this advantage also proves to be one of the chief drawbacks of the process; it does not allow the operator to observe the weld puddle. When SAW is applied semiautomatically, the operator must learn to propel the gun carefully in a fashion to ensure uniform bead contour. The experienced operator relies on the uniform formation of a slag blanket to indicate the nature of the deposit. For single-pass welds, this is mastered fairly readily; however, for multiple-pass welding, the skills required are significant. Therefore, most submerged arc applications are mechanized. The nature of the joint must then lend itself to automation if the process is to prove viable. Long, uninterrupted straight seams are ideal applications for submerged arc. Short, intermittent welds are better made with one of the open-arc processes.

Two electrodes may be fed through a single electrical contact tip, resulting in higher deposition rates. Generally known as *parallel electrode welding,* the equipment is essentially the same as that used for single-electrode welding, and parallel electrode welding procedures may be prequalified under AWS D1.1-98.

Multiple-electrode SAW refers to a variation of submerged arc which utilizes at least two separate power supplies, two separate wire drives, and feeds two electrodes independently. Some applications, such as the manufacture of line pipe, may use up to five independent electrodes in a multiple-electrode configuration. Ac welding is typically used for multielectrode welding. If dc current is used, it is limited usually to the lead electrode to minimize the potentially negative interaction of magnetic fields between the two electrodes.

3.3.4 GMAW

Gas metal arc welding (GMAW) (Fig. 3.11) utilizes equipment similar to that used in flux-cored arc welding. Indeed, the two processes are very similar. The major differences are: gas metal arc uses a solid or metal-cored electrode and leaves no appreciable amount of residual slag.

Gas metal arc has not been a popular method of welding in the typical structural steel fabrication shop because of its sensitivity to mill scale, rust, limited puddle control, and sensitivity to shielding loss. Newer GMAW metal-cored electrodes, however, are beginning to be used in the shop fabrication of structural elements with good success.

A variety of shielding gases or gas mixtures may be used for GMAW. Carbon dioxide (CO_2) is the lowest-cost gas, and while acceptable for welding carbon steel, the gas is not inert but active at elevated temperatures. This has given rise to the term *MAG* (*metal active gas*) for the process when CO_2 is used, and *MIG* (*metal inert gas*) when predominantly argon-based mixtures are used. While shielding

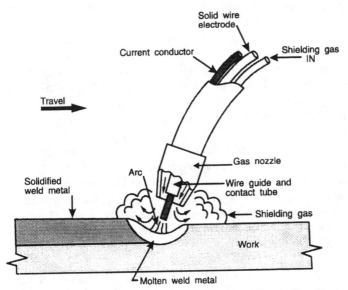

Figure 3.11 GMAW process. (*Courtesy of The Lincoln Electric Company.*)

gas is used to displace atmospheric oxygen, it is possible to add smaller quantities of oxygen into mixtures of argon—generally at levels of 2 to 8%. This helps stabilize the arc and decreases puddle surface tension, resulting in improved wetting. Tri and quad mixes of argon, oxygen, carbon dioxide, and helium are possible, offering advantages that positively affect arc action, deposition appearance, and fume generation rates.

Short arc transfer is ideal for welding on thin-gauge materials. It is generally not suitable for structural steel fabrication purposes. In this mode of transfer, the small-diameter electrode, typically 0.035 or 0.045 in, is fed at a moderate wire feed speed at relatively low voltages. The electrode will touch the workpiece, resulting in a short in the electrical circuit. The arc will actually go out at this point, and very high currents will flow through the electrode, causing it to heat and melt. Just as excessive current flowing through a fuse causes it to blow, so the shorted electrode will separate from the work, initiating a momentary arc. A small amount of metal will be transferred to the work at this time.

The cycle will repeat itself again once the electrode shorts to the work. This occurs somewhere between 60 and 200 times/s, creating a characteristic buzz to the arc. This mode of transfer is ideal for sheet metal, but results in significant fusion problems if applied to heavy materials. A phenomenon known as *cold lap* or *cold casting* may result where the metal does not fuse to the base material. This is unacceptable since the welded connections will have virtually no strength. Great caution must be exercised in the application of the short arc mode to heavy plates. The use of short arc on heavy plates is not totally prohibited, however, since it is the only mode of transfer that can be used out-of-position with gas metal arc welding, unless specialized equipment is used. Weld joint details must be carefully designed when short arc transfer is used. Welders must pass specific qualification tests before using this mode of transfer. Short arc transfer is often abbreviated as GMAW-s, and is not prequalified by the D1.1 code.

Globular transfer is a mode of gas metal arc welding that results when high concentrations of carbon dioxide are used, resulting in an arc that is rough with larger globs of metal ejected from the end of the electrode. This mode of transfer, while resulting in deep penetration, generates relatively high levels of spatter. Weld appearance can be poor and it is restricted to the flat and horizontal position. Globular transfer may be preferred over spray arc transfer because of the low cost of CO_2-shielding gas and the lower level of heat experienced by the operator.

Spray arc transfer is characterized by high wire-feed speeds at relatively high voltages. A fine spray of molten drops, all smaller in diameter than the electrode diameter, is ejected from the electrode toward the work. Unlike short arc transfer, the arc in spray transfer is continuously maintained. High-quality welds with particularly good appearance are the result. The shielding used for spray arc transfer is composed of at least 80% argon, with the balance made up of either carbon dioxide or oxygen. Typical mixtures would include 90-10 argon-CO_2, and 95-5 argon-oxygen. Other proprietary mixtures are available from gas suppliers. Relatively high arc voltages are used with the spray mode of transfer. However, due to the intensity of the arc, spray arc is restricted to applications in the flat and horizontal position, because of the puddle fluidity and lack of a slag to hold the molten metal in place.

Pulsed arc transfer utilizes a background current that is continuously applied to the electrode. A pulsing peak current is optimally applied as a function of the wire-feed speed. With this mode of transfer, the power supply delivers a pulse of current which, ideally, ejects a single droplet of metal from the electrode. The power supply returns to a lower background current which maintains the arc. This occurs between 100 and 400 times/s. One advantage of pulsed arc transfer is that it can be used out-of-position. For flat and horizontal work, it may not be as fast as spray transfer. However, used out-of-position, it is free of the problems associated with the gas metal arc short-circuiting mode. Weld appearance is good and quality can be excellent. The disadvantage of pulsed arc transfer is that the equipment is slightly more complex and more costly. The joints are still required to be relatively clean, and out-of-position welding is still more difficult than with processes that generate a slag that can support the molten puddle.

Metal cored electrodes are a relatively new development in gas metal arc welding. This is similar to flux-cored arc welding in that the electrode is tubular, but the core material does not contain slag-forming ingredients. Rather, a variety of metallic powders is contained in the core. The resulting weld is virtually slag-free, just as with other forms of GMAW. The use of metal-cored electrodes offers many fabrication advantages. They have increased ability to handle mill scale and other surface contaminants.

Finally, metal-cored electrodes permit the use of high amperages that may not be practical with solid electrodes, resulting in potentially higher deposition rates. The properties obtained from metal-cored deposits can be excellent. Appearance is very good. Because of the ability of the filler metal manufacturer to control the composition of the core ingredients, mechanical properties obtained from metal-cored deposits may be more consistent than those obtained with solid electrodes. However, metal-cored electrodes are, in general, more expensive.

3.3.5 ESW/EGW

Electroslag and electrogas welding (ESW/EGW) are closely related processes (Figs. 3.12 and 3.13) that offer high-deposition welding in the vertical plane. Properly applied, these processes offer significant savings over alternative out-of-position methods and, in many cases, savings over flat position welding. Although the two processes have similar applications and mechanical setup, there are fundamental differences in the arc characteristics. Electroslag and electrogas are

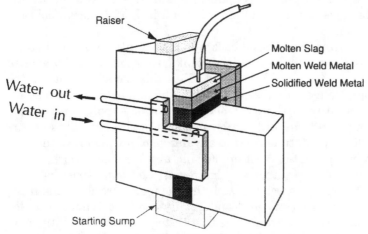

Figure 3.12 ESW process. (*Courtesy of The Lincoln Electric Company.*)

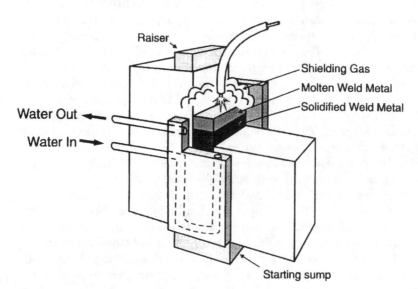

Figure 3.13 EGW process. (*Courtesy of The Lincoln Electric Company.*)

mechanically similar in that both utilize copper dams or shoes that are applied to either side of a square-edged butt joint. An electrode or multiple electrodes are fed into the joint. A starting sump is typically applied for the beginning of the weld. As the electrode is fed into the joint, a puddle is established that progresses vertically. The copper dams, which are commonly water-cooled, chill the weld metal and prevent it from escaping from the joint. The weld is completed in one pass.

These processes may be used for groove welds in butt, corner, and tee joints. Typical applications involve heavier plate, usually 1 in or thicker. Multiple electrodes may be used in a single joint, allowing very heavy plate up to several inches thick to be joined in a single pass. Because of the sensitivity of the process to the number of variables involved, specific operator training is required, and the D1.1-98 code requires welding procedures to be qualified by test.

In building construction, applications for ESW/EGW with traditional connection designs are somewhat limited. However, they can be highly efficient in the manufacture of tree columns. In the shop, the beam flange-to-column welds can be made with the column in the horizontal plane. With the proper equipment and tooling, all four flange welds can be made simultaneously. In addition, continuity plate welds can be made with ESW/EGW. Future connection designs may utilize configurations that are more conducive to these processes.

Another common application is for the welding of continuity plates inside box columns. It is possible to weld three sides of the continuity plate to the interior of the box prior to closing the box with the fourth side. However, once this closure is made, access to the final side of the continuity plate is restricted. This final closure weld can be made by operating through a hole in the outside of the box column. This approach is very popular in the Far East where box columns are widely used.

In electroslag welding, a granular flux is metered into the joint during the welding operation. At the beginning, an arc, similar to that of submerged arc welding, is established between the electrode and the sump.

After the initial flux is melted into a molten slag, the reaction changes. The slag, which is carefully designed to be electrically conductive, will conduct the welding current from the electrode through the slag into the pieces of steel to be joined. As high currents are passed through the slag, it becomes very hot. The electrode is fed through the hot slag and melts. Technically, electroslag welding is not an *arc*-welding process, but a *resistance*-welding process. Once the arc is extinguished and the resistance melting process is stabilized, the weld continues vertically to completion. A small amount of slag is consumed as it chills against the water-cooled copper shoes. In some cases, steel dams instead of copper dams are used to retain the puddle. After completion of the weld, the steel dams stay in place, and become part of the final product. Slag must be replenished, and additional flux is continuously added to compensate for the loss.

One aspect of electroslag welding that must be considered is the very high heat input associated with the process. This causes a large heat-affected zone (HAZ) that may have a lower notch toughness. Electroslag welding is different from electroslag, inasmuch as no flux is used. Electrogas welding is a true arc-welding process and is conceptually more like gas metal arc or flux-cored arc welding. A solid or tubular electrode is fed into the joint, which is flooded with an inert gas shield. The arc progresses vertically while the puddle is retained by the water-cooled dams.

The HAZ performance is dependent not only on the heat input, but also on the nature of the steel. While all processes develop a heat-affected zone, the large size of the electroslag heat-affected zone justifies additional scrutiny. Advances in steel technology have resulted in improved steels, featuring higher cleanliness and toughness, that better retain the HAZ properties in ESW/EGW welds.

3.3.6   GTAW

The *gas-tungsten arc-welding* (GTAW) *process,* colloquially called *TIG welding,* is rarely used in structural applications. However, it may be specified to meet some unique requirements or for a repair welding procedure. GTAW (Fig. 3.14) uses a nonconsumed electrode composed

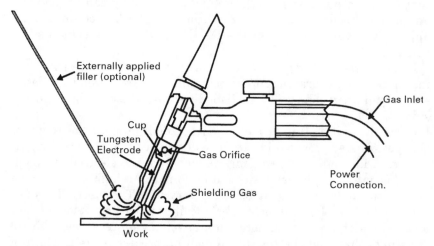

Figure 3.14   Gas-tungsten arc welding. (*Courtesy of The Lincoln Electric Company.*)

of tungsten, a metal with a very high melting point. Between the tungsten and the work, an arc is established that results in heating of the base material. A filler rod may or may not be used. The area is shielded with an inert gas, typically argon, although helium may be used. GTAW is ideally suited for welding on nonferrous materials, such as stainless steel and aluminum. Moreover, it is very effective when joining thin sections.

One area where gas-tungsten arc welding may be used in structural applications is when it is applied for the purpose of "TIG dressing." The TIG dressing technique has been used to extend the fatigue life of fillet welds. With this technique, the gas-tungsten arc process is used to heat and melt the toes of fillet welds, resulting in a new distribution of residual stresses and perhaps improved contour of the toe of the fillet. This has been used to retrofit structures where fatigue cracking is expected. The process is inherently expensive, but may be justified if it extends the life of the structure.

3.4   Welding Process Selection

Any of the common arc-welding processes can be used to achieve the quality required for structural steel applications. While each may have a particular area of strength and/or weakness, the primary consideration as to which process will be used is largely cost-driven. The availability of specialized equipment in one fabrication shop compared to the capabilities of a second shop may dictate significantly different approaches, both of which may prove to be cost-effective. A history of successful usage offers a strong incentive for the fabricator to continue using a given process. The reasons for this go well beyond familiarity and comfort with a specific approach. When welders and procedures are established with a given process, significant costs will be incurred with any change to a new approach.

3.4.1   Joint requirements

Each individual weld-joint configuration and preparation has certain process requirements in order to achieve low-cost welding. Four characteristics must be considered: deposition rate, penetration ability, out-of-position capability, and high travel-speed capacity. Each process exhibits different capabilities in these realms. Once the joint and its associated requirements are analyzed, they should be compared to the various process options and the ability of the process to achieve those requirements. A proper match of weld-joint requirements and process capabilities will lead to dependable and economical fabrication.

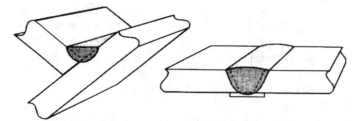

Figure 3.15 Joint requiring substantial fill. (*Courtesy of The Lincoln Electric Company.*)

Some welds, such as large fillet welds and groove welds, require that *high deposition-rate* welding be used (Fig. 3.15) for the most economical fabrication. The cost of making these welds will be determined largely by the deposition rate of the process. The amount of weld material required may be measured in pounds per foot of joint. Once the deposition rate of a process in pounds per hour is known, it is possible to determine the number of feet of weld that can be made in a given hour assuming 100% arc time. This, of course, translates directly to productivity rates.

The second criterion imposed by weld joints is the requirement for *penetration*. Examples are listed under Fig. 3.16 and would include any complete joint-penetration groove weld that has a root face dimension. These joints will be made by welding from one side and back-gouging from the second to ensure complete fusion. With deeper penetration afforded by the welding process, a smaller amount of base metal will have to be removed by back-gouging. Subsequent welding will then be proportionately reduced as well.

While all welding requires fusion, not all joints require deep penetration. For example, simple fillet welds are required by AWS D1.1-98 to have fusion to the root of the joint, but are not required to have penetration beyond the root. This has a practical basis: verification of

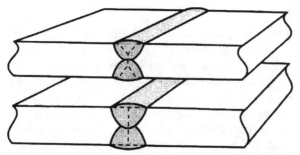

Figure 3.16 Joints requiring substantial penetration. (*Courtesy of The Lincoln Electric Company.*)

penetration beyond the root is impossible with visual inspection. Fusion to the root, and not necessarily beyond, ensures that sufficient strength is generated, provided the weld is properly sized. While penetration can be verified with ultrasonic inspection, fillet welds routinely receive only visual or magnetic particle inspection. Thus, no penetration beyond the root is required, nor is design credit given to deeper penetration in fillet welds if it happens to be present. Figure 3.17 illustrates this requirement.

The *out-of-position* capability of a given welding process refers to the ability to deposit weld metal in the vertical or overhead positions. It is generally more economical to position the work in the flat and horizontal positions. However, this is usually impossible for field erection, and may be impractical under other conditions. The ability to obtain *high travel speeds* is important for small welds. It may not be possible for a high deposition welding process to be used at high travel speeds. The size of the droplet transferred, puddle fluidity, surface tension, and other factors combine to make some processes more capable of high travel speeds than others.

3.4.2 Process capabilities

After the joint is analyzed and specific requirements determined, these are compared to the capabilities of various processes. The process with capabilities most closely matching the requirements typically will be the best and most economical option.

Submerged arc welding and electroslag/electrogas welding have the greatest potential to deliver *high deposition rates*. Multiple-electrode applications of submerged arc extend this capability even further. For joints requiring high deposition rates, submerged arc and electroslag/electrogas welding are ideal processes to contribute to low-cost welding. When the specific conditions are not conducive to SAW but high deposition rates are still required, flux-cored arc welding may be used. The larger-diameter electrodes, which run at higher electrical currents, are preferred.

Deep penetration is offered by the submerged arc-welding process. While electroslag/electrogas also offers deep penetration, the joints on which the electroslag are used typically do not require this capability. Where open-arc processes are preferred, gas-shielded flux-cored arc welding may offer deep penetration.

Out-of-position capability is strongest for the flux-cored and shielded metal arc-welding processes. The slag coatings that are generated by these processes can be instrumental in retaining molten weld metal in the vertical and overhead positions. Submerged arc is not applicable for these joints.

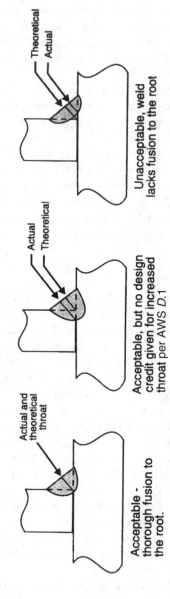

Figure 3.17 Fillet weld requirements. (*Courtesy of The Lincoln Electric Company.*)

The requirement for *high travel speed* capability for welding structural steel members is fairly limited. This typically consists of the travel speed associated with making a ¼-in fillet weld. All of the popular processes, with the exception of electroslag/electrogas, are capable of making ¼-in fillet welds under the proper conditions. Among the variables that need to be considered are electrode size and procedure variables. A common mistake of fabricators is to utilize a process and procedure capable of extremely high deposition rates but limited travel speeds. Oversized welds can result from the inability to achieve high travel speeds. A more economical approach would be to optimize the procedure according to the desired travel speed. This may result in a lower deposition rate but a lower overall cost because overwelding has been eliminated.

3.4.3 Special situations

Self-shielded flux-cored welding is ideal for *outdoor conditions.* Quality deposits may be obtained without the erection of special wind shields and protection from drafts. Shielded metal arc welding is also suitable for these conditions but is considerably slower.

The welding process of choice for field erectors for the last 25 years has been FCAW-ss. It has been the commonly used process for fabrication of steel structures throughout the United States. Its advantages are reviewed in order to provide an understanding of why it has been the preferred process. In addition, its limitations are outlined to highlight areas of potential concern.

The chief advantage of the FCAW-ss process is its ability to deposit quality weld metal under field conditions, which usually involve wind. The code specifically limits wind velocity in the vicinity of a weld to a maximum of 5 mi/h. In order to utilize gas-shielded processes under these conditions, it is necessary to erect windshields to preclude movement of the shielding gas with respect to the molten weld puddle. While tents and other housings can be created to minimize this problem, such activities can be costly and are often a fire hazard. In addition, adequate ventilation must be provided for the welder. The most efficient windshields may preclude adequate ventilation. Under conditions of severe shielding loss, weld porosity will be exhibited. At much lower levels of shielding loss, the mechanical properties (notch toughness and ductility) may be negatively affected, although there will be no obvious evidence that this is taking place.

A variety of other gas-related issues are also eliminated, including ensuring availability of gas, handling of high-pressure cylinders (always a safety concern), theft of cylinders, protection of gas distribution hosing under field conditions, and the cost of shielding gas. Leaks in the delivery system obviously waste shielding gas, but a leak can also allow entry of air into the delivery system. Weld quality can be affected in the same way as shielding loss. Most field erectors have found it advantageous to utilize the self-shielded process and circumvent all such potential problems.

Some projects permit *multiple welding heads* to be simultaneously operated in the same general vicinity. When this is done, submerged arc is an ideal choice. Because of the lack of glare and arc flash, an operator can control multiple arcs that are nearly impossible to control in a situation where the arc intensity from one torch would make it difficult to carefully control another. A typical example would be the use of welding systems that simultaneously make fillet welds on opposing sides of stiffeners.

The easiest way to *control smoke and fumes* in the welding environment is to limit their initial generation. Here, submerged arc is ideal. Smoke exhaust guns are available for the flux-cored arc-welding processes. The most effective process for use with these smoke exhaust guns is FCAW-ss. Because the process is self-shielded, there is no concern about disruption of the gas shielding.

3.5   Welding Procedures

Within the welding industry, the term *welding procedure specification* (or WPS) is used to signify the combination of variables that are to be used to make a certain weld. The terms *welding procedure,* or simply *procedure,* may be used. At a minimum, the WPS consists of the following:

Process (SMAW, FCAW, etc.)

Electrode specification (AWS A5.1, A5.20, etc.)

Electrode classification (E7018, E71T-1, etc.)

Electrode diameter ($\frac{1}{8}$ in, $\frac{5}{32}$ in, etc.)

Electrical characteristics (ac, dc+, dc−)

Base metal specification (A36, A572 GR50, etc.)

Minimum preheat and interpass temperature

Welding current (amperage)/wire-feed speed

Arc voltage

Travel speed

Position of welding

Postweld heat treatment

Shielding gas type and flowrate

Joint design details

The welding procedure is somewhat analogous to a cook's recipe. It outlines the steps required to make a quality weld under specific conditions.

3.5.1 Effects of welding variables

The effects of the variables are somewhat dependent on the welding process being employed, but general trends apply to all the processes. It is important to distinguish the difference between constant current (CC) and constant voltage (CV) electrical welding systems. Shielded metal arc welding is always done with a CC system. Flux-cored welding and gas metal arc welding generally are performed with CV systems. Submerged arc may utilize either.

Amperage is a measure of the amount of current flowing through the electrode and the work. It is a primary variable in determining heat input. Generally, an increase in amperage means higher deposition rates, deeper penetration, and more admixture. The amperage flowing through an electric circuit is the same, regardless of where it is measured. It may be measured with a tong meter or with the use of an electric shunt. The role of amperage is best understood in the context of heat input and current density considerations. For CV welding, an increase in wire-feed speed will directly increase amperage. For SMAW on CC systems, the machine setting determines the basic amperage, although changes in the arc length (controlled by the welder) will further change amperage. Longer arc lengths reduce amperage.

Arc voltage is directly related to arc length. As the voltage increases, the arc length increases, as does the demand for arc shielding. For CV welding, the voltage is determined primarily by the machine setting, so the arc length is relatively fixed in CV welding. For SMAW on CC systems, the arc voltage is determined by the arc length, which is manipulated by the welder. As arc lengths are increased with SMAW, the arc voltage will increase and the amperage will decrease. Arc voltage also controls the width of the weld bead, with higher voltages generating wider beads. Arc voltage has a direct effect on the heat-input computation.

The voltage in a welding circuit is not constant, but is composed of a series of voltage drops. Consider the following example: Assume the power source delivers a total system voltage of 40 V. Between the power source and the welding head or gun, there is a voltage drop of perhaps 3 V associated with the input-cable resistance. From the point of attachment of the work lead to the power source work terminal, there is an additional voltage drop of, say, 7 V. Subtracting the 3 V and the 7 V from the original 40 V, this leaves 30 V for the arc. This example illustrates how important it is to ensure that the voltages used for monitoring welding procedures properly recognize any losses in the welding circuit. The most accurate way to determine arc voltage is to measure the voltage drop between the contact tip and the workpiece. This may not be practical for semiautomatic welding, so voltage is typically read from a point on the wire feeder (where the gun and cable connection is made) to the workpiece. For SMAW, welding voltage is not usually monitored, since it is constantly changing and cannot be controlled except by the welder. Skilled workers hold short arc lengths to deliver the best weld quality.

Travel speed, measured in inches per minute, is the rate at which the electrode is moved relative to the joint. All other variables being equal, travel speed has an inverse effect on the size of the weld beads. As the travel speed increases, the weld size will decrease. Extremely low travel speeds may result in reduced penetration, as the arc impinges on a thick layer of molten metal and the weld puddle rolls ahead of the arc. Travel speed is a key variable used in computing heat input; reducing travel speed increases heat input.

Wire-feed speed is a measure of the rate at which the electrode is passed through the welding gun and delivered to the arc. Typically measured in inches per minute (in/min), the wire-feed speed is directly proportional to deposition rate and directly related to amperage. When all other welding conditions are maintained constant (for example, the same electrode type, diameter, electrode extension, arc voltage, and electrode extension), an increase in wire-feed speed will directly lead to an increase in amperage. For slower wire-feed speeds, the ratio of wire-feed speed to amperage is relatively constant and linear.

For higher levels of wire-feed speed, it is possible to increase the wire-feed speed at a disproportionately high rate compared to the increase in amperage. When these conditions exist, the deposition rate per amp increases but at the expense of penetration.

Wire-feed speed is the preferred method of maintaining welding procedures for constant-voltage wire-feed processes. The wire-feed speed can be independently adjusted, and measured directly, regardless of the other welding conditions. It is possible to utilize amperage as an alternative to wire-feed speed although the resultant amperage for a given wire-feed speed may vary, depending on the polarity, electrode diameter, electrode type, and electrode extension. Although equipment has been available for 20 y that monitors wire-feed speed, many codes such as AWS D1.1 continue to acknowledge amperage as the primary method for procedure documentation. D1.1 does permit the use of wire-feed speed control instead of amperage, providing a wire-feed speed–amperage relationship chart is available for comparison.

Electrode extension, also known as *stickout,* is the distance from the contact tip to the end of the electrode. It applies only to the wire-feed processes. As the electrode extension is increased in a constant-voltage system, the electrical resistance of the electrode increases, causing the electrode to be heated. This is known as *resistance heating* or I^2R heating. As the amount of heating increases, the arc energy required to melt the electrode decreases. Longer electrode extensions may be employed to gain higher deposition rates at a given amperage. When the electrode extension is increased without any change in wire-feed speed, the amperage will decrease. This results in less penetration and less admixture. With the increase in electric stickout, it is common to increase the machine voltage setting to compensate for the greater voltage drop across the electrode.

In constant-voltage systems, it is possible to simultaneously increase the electric stickout and wire-feed speed in a balanced manner so that the current remains constant. When this is done, higher deposition rates are attained. Other welding variables such as voltage and travel speed must be adjusted to maintain a stable arc and to ensure quality welding. The ESO variable should always be within the range recommended by the manufacturer.

Electrode diameter means larger electrodes can carry higher welding currents. For a fixed amperage, however, smaller electrodes result in higher deposition rates. This is because of the effect on current density discussed in the following.

Polarity is a definition of the direction of current flow. Positive polarity (reverse) is achieved when the electrode lead is connected to the positive terminal of the direct-current (dc) power supply. The work lead is connected to the negative terminal. Negative polarity (straight) occurs when the electrode is connected to the negative terminal and the work lead to the positive terminal. Alternating current (ac) is not a polarity, but a current type. With ac, the electrode is alternately positive and negative. Submerged arc is the only process that commonly uses either electrode positive or electrode negative polarity for the same type of electrode. Ac may also be used. For a fixed wire-feed speed, a submerged arc electrode will require more amperage on positive polarity than on negative. For a fixed amperage, it is possible to utilize higher wire-feed speeds and deposition rates with negative polarity than with positive. Ac exhibits a mix of both positive and negative polarity characteristics.

The magnetic field that surrounds any dc conductor can cause a phenomenon known as *arc blow,* where the arc is physically deflected by the field. The strength of the magnetic field is proportional to the square of the current value, so this is a more significant potential problem with higher currents. Ac is less prone to art blow, and can sometimes be used to overcome this phenomenon.

Heat input is proportional to the welding amperage, times the arc voltage, divided by the travel speed. Higher heat inputs relate to larger weld cross-sectional areas and larger heat-affected zones, which may negatively affect mechanical properties in that region. Higher heat input generally results in slightly decreased yield and tensile strength in the weld metal, and generally lowers notch toughness because of the interaction of bead size and heat input.

Current density is determine by dividing the welding amperage by the cross-sectional area of the electrode. For solid electrodes, the current density is therefore proportional to I/d^2. For tubular electrodes where current is conducted by the sheath, the current density is related to the area of the metallic cross section. As the current density increases, there will be an increase in deposition rates, as well as penetration. The latter will increase the amount of admixture for a given joint. Notice that this may be accomplished by either the amperage or decreasing the electrode size. Because the electrode diameter is a squared function, a small decrease in diameter may have a significant effect on deposition rates and plate penetration.

Preheat and interpass temperature are used to control cracking tendencies, typically in the base materials. Regarding weld metal properties, for most carbon-manganese-silicon systems, a moderate interpass temperature promotes good notch toughness. Preheat and interpass temperatures greater than 550°F may negatively affect notch toughness. Therefore, careful control of preheat and interpass temperatures is critical.

3.5.2 Purpose of welding procedure specifications (WPSs)

The particular values for the variables discussed previously have a significant effect on weld soundness, mechanical properties, and productivity. It is therefore critical that those procedural values used in the actual fabrication and erection be appropriate for the specific requirements of the applicable code and job specifications. Welds that will be architecturally exposed, for example, should be made with procedures that minimize spatter, encourage exceptional surface finish, and have limited or no undercut. Welds that will be covered with fireproofing, in contrast, would naturally have less restrictive cosmetic requirements.

Many issues must be considered when selecting welding procedure values. While all welds must have fusion to ensure their strength, the required level of penetration is a function of the joint design in the weld type. All welds are required to deliver a certain yield and/or tensile strength, although the exact level required is a function of the connection design. Not all welds are required to deliver minimum specified levels of notch toughness. Acceptable levels of undercut and porosity are a function of the type of loading applied to the weld. Determination of the most efficient means by which these conditions can be met cannot be left to the welders, but should be determined by knowledgeable welding technicians and engineers who create written welding procedure specifications and communicate those requirements to welders by the means of these documents. The WPS is the primary tool that is used to communicate to the welder, supervisor, and inspector how a specific weld is to be made. The suitability of a weld made by a skilled welder in conformance with the requirements of a WPS can only be as good as the WPS itself. The proper selection of procedure variable values must be achieved in order to have a WPS appropriate for the application. This is the job of the welding expert who generates or writes the WPS. The welder is generally expected to be able to follow the WPS, although the welder may not know how or why each particular variable was selected. Welders are expected to ensure welding is performed in accordance with the WPS. Inspectors do not develop WPSs, but should ensure that they are available and are followed.

The *D1.1-98 Structural Welding Code—Steel* requires written welding procedures for all fabrication performed. The inspector is obligated to review the WPSs and to make certain that production welding parameters conform to the requirements of the code. These WPSs are required to be written, regardless of whether they are prequalified or qualified by test. Each fabricator or erector is responsible for the development of WPSs. Confusion on this issue apparently still exists since there continue to be reports of fabrication being performed in the absence of written welding procedure specifications. One prevalent misconception is that if the actual parameters under which welding will be performed meet all the conditions for "prequalfied" status, written WPSs are not required. This is not true; according to the code, the requirement is clear.

The WPS is a communication tool, and it is the primary means of communication to all the parties involved regarding how the welding is to be performed. It must therefore be readily available to foremen, inspectors, and the welders.

The code is not prescriptive in its requirements regarding availability and distribution of WPSs. Some shop fabricators have issued each welder employed in their organization with a set of welding procedures that are typically retained in the welder's locker or tool box. Others have listed WPS parameters on shop drawings. Some company bulletin boards have listings of typical WPSs used in the organization. Some suggest that WPSs should be posted near the point where welding is being performed. Regardless of the method used, WPSs must be available to those authorized to use them.

It is in the contractor's best interest to ensure that efficient communication is maintained with all parties involved. Not only can quality be compromised when WPSs are not available, but productivity can suffer as well. Regarding quality, the limits of suitable operation of the particular welding process and electrode for the steel, joint design, and position of welding must be understood. It is obvious that the particular electrode employed must be operated on the proper polarity, proper shielding gases must be used, and amperage levels must be appropriate for the diameter of electrode and for the thickness of material on which welding is performed. Other issues are not necessarily so obviously apparent. The required preheat for a particular application is a function of the grade(s) of steel involved, the thickness(es) of material, and the type of electrode employed (whether low hydrogen or nonlow hydrogen). The required preheat level can be communicated by means of the written WPS.

Lack of conformance with the parameters outlined in the WPS may result in the deposition of a weld that does not meet the quality requirements imposed by the code or the job specifications. When an unacceptable weld is made, the corrective measures to be taken may necessitate weld removal and replacement, an activity that routinely increases the cost of that particular weld tenfold. Avoiding these types of unnecessary activities by clear communication has obvious ramifications in terms of quality and economics.

There are other economic issues to be considered as well. In a most general way, the cost of welding is inversely proportional to the deposition rate. The deposition rate, in turn, is directly tied to the wire-feed speed of the semiautomatic welding processes. If it is acceptable, for example, to make a given weld with a wire-feed speed of 200 in/min, then a weld made at 160 in/min (which may meet all the quality requirements) would cost approximately 25% more than the weld made with the optimum procedure. Conformance with WPS values can help ensure that construction is performed at rates that are conducive to the required weld quality and are economical as well. Some wire feeders have the ability to preset welding parameters, coupled with the digital LED display or analog meters that indicate operational parameters, which can assist in maintaining and monitoring WPS parameters.

The code imposes minimum requirements for a given project. Additional requirements may be imposed by contract specifications. The same would hold true regarding WPS values. Compliance with the minimum requirements of the code may not be adequate under all circumstances. Additional requirements can be communicated through the WPS, such as recommendations imposed by the steel producer, electrode manufacturer, or others can and should be documented in the WPS.

3.5.3 Prequalified welding procedure specifications

The AWS D1.1 code provides for the use of prequalified WPSs. Prequalified WPSs are those that the AWS D1 Committee has determined to have a history of acceptable performance, and so does not subject them to the qualification testing imposed on all other welding procedures. The use of prequalified WPSs does not preclude their need to be in a written format. The use of prequalified WPSs still requires that the welders be appropriately qualified. All the workmanship provisions imposed in the fabrication section of the code apply to prequalified WPSs. The only code requirement exempted by prequalification is the nondestructive testing and mechanical testing required for qualification of welding procedures.

A host of restrictions and limitations imposed on prequalified welding procedures do not apply to welding procedures that are qualified by test. Prequalfied welding procedures must conform with all the prequalified requirements in the code. Failure to comply with a single prequalified condition eliminates the opportunity for the welding procedure to be prequalified. The use of a prequalified welding procedure does not exempt the engineer from exercising engineering judgment to determine the suitability of the particular procedure for the specific application.

In order for a WPS to be prequalified, the following conditions must be met:

- The welding process must be prequalified. Only SMAW, SAW, GMAW (except GMAW-s), and FCAW may be prequalified.
- The base metal/filler metal combination must be prequalified.
- The minimum preheat and interpass temperatures prescribed in D1.1-98 must be employed.
- Specific requirements for the various weld types must be maintained. Fillet welds must be in accordance with D1.1-98, Sec. 3.9, plug and slot welds in accordance with D1.1-98, and groove welds in accordance with D1.1-98, Secs. 3.11 and 3.12, as applicable. For the groove welds, whether partial-joint penetration or complete-joint penetration, the required groove preparation dimensions are shown in the code.

Even if prequalified joint details are employed, the welding procedure must be qualified by test if other prequalified conditions are not
met. For example, if a prequalified detail is used on an unlisted steel,
the welding procedures must be qualified by test.

Prequalified status requires conformance to a variety of procedural
parameters. These include maximum electrode diameters, maximum
welding current, maximum root-pass thickness, maximum fill-pass
thickness, maximum single-pass fillet weld sizes, and maximum single-pass weld layers.

In addition to all the preceding requirements, welding performed
with a prequalified WPS must be in conformance with the other code
provisions contained in the fabrication section of AWS *D1.1-98
Structural Welding Code.*

The code does not imply that a WPS that is prequalified will automatically achieve the quality conditions required by the code. It is the
contractor's responsibility to ensure that the particular parameters
selected within the requirements of the prequalified WPS are suitable
for the specific application. An extreme example will serve as an illustration. Consider the following example of a hypothetical proposed
WPS for making a $\frac{1}{4}$-in fillet weld on $\frac{3}{8}$-in A36 steel in the flat position.
The weld type and steel are prequalified. SAW, a prequalified process,
is selected. The filler metal selected is F7A2-EM12K, meeting the
requirements of D1.1-98. No preheat is specified since it would not be
required. The electrode diameter selected is $\frac{3}{32}$ in, less than the $\frac{1}{4}$-in
maximum specified. The maximum single-pass fillet weld size in the
flat position, according to D1.1-98 is unlimited, so the $\frac{1}{4}$-in fillet size
can be prequalified. The current level selected for making this particular fillet weld is 800 A, less than the 1000-A maximum specified.

However, the amperage level imposed on the electrode diameter for
the thickness of steel on which the weld is being made is inappropriate.
It would not meet the requirements of the fabrication chapters which
require that the size of electrode and amperage be suitable for the
thickness of material being welded. This illustration demonstrates the
fact that compliance with all prequalified conditions does not guaran-
tee that the combination of selected variables will always generate an
acceptable weld.

Most contractors will determine preliminary values for a prequalified
WPS based upon their experience, recommendations from publications
such as the AWS *Welding Handbooks,* from AWS *Welding Procedures
Specifications* (AWS B2.1), or other sources. It is the responsibility of
the contractor to verify the suitability of the suggested parameters
prior to the application of the actual procedure on a project, although
the verification test need not be subject to the full range of procedure
qualification tests imposed by the code. Typical tests will be made to

determine soundness of the weld deposit (for example, fusion, tie-in of weld beads, freedom from slag inclusions). The plate could be nondestructively tested or, as is more commonly done, cut, polished, and etched. The latter operations allow for examination of penetration patterns, bead shapes, and tie-in. Welds that are made with prequalified WPSs that meet the physical dimensional requirements (fillet weld size, maximum reinforcement levels, and surface profile requirements) and are sound (that is, adequate fusion, tie-in, and freedom from excessive slag inclusions and porosity) should meet the strength and ductility requirements imposed by the code for welding procedures qualified by test. Weld soundness, however, cannot be automatically assumed just because the WPS is prequalified.

3.5.4 Guidelines for preparing prequalified WPSs

When developing prequalified WPSs, the starting point is a set of welding parameters appropriate for the general application being considered. Parameters for overhead welding will naturally vary from those required for down-hand welding. The thickness of material involved will dictate electrode sizes and corresponding current levels. The specific filler metals selected will reflect the strength requirements of the connection. Many other issues must be considered. Depending on the level of familiarity and comfort the contractor has with the particular values selected, welding a mock-up may be appropriate. Once the parameters that are desired for use in production are established, it is essential to check each of the applicable parameters for compliance with the D1.1-98 code.

To assist in this effort, Annex H has been provided in the D1.1-98 code. This contains a checklist that identifies prequalified requirements. If any single parameter deviates from these requirements, the contractor is left with two options: (1) the preliminary procedure can be adjusted to conform with the prequalified constraints or (2) the WPS can be qualified by test. If the preliminary procedure is adjusted, it may be appropriate to reexamine its viability by another mock-up.

The next step is to document, in writing, the prequalified WPS values. A sample form is included in Annex E of the code. The fabricator may utilize any convenient format. Also contained in Annex E are a series of examples of completed WPSs that may be used as a pattern.

3.5.5 Qualifying welding procedures by test

Conducting qualification tests. There are two primary reasons why welding procedures may be qualified by test. First, it may be a contractual requirement. Secondly, one or more of the specific conditions to be used in the production may deviate from the prequalified requirements. In either case, a test weld must be made prior to the establishment of the final WPS. The first step in qualifying a welding procedure by test is to establish the procedure that is desired to be qualified. The same sources cited for the prequalified WPS starting points could be used for WPSs qualified by test. These will typically be the parameters used for fabrication of the test plate, although this is not always the case, as will be discussed later. In the simplest case, the exact conditions that will be encountered in production will be replicated in the procedure qualification test. This would include the welding process, filler metal, grade of steel, joint details, thickness of material, preheat values, minimum interpass temperature level, and the various welding parameters of amperage, voltage, and travel speed. The initial parameters used to make the procedure qualification test plate beg for a name to define them, although there is no standard industry term. It has been suggested that "TWPS" be used where the "T" could alternatively be used for temporary, test, or trial. In any case, it would define the parameters to be used for making the test plate since the validity of the particular parameters cannot be verified until successfully passing the required test. The parameters for the test weld are recorded on a procedure qualification record (PQR). The actual values used should be recorded on this document. The target voltage, for example, may be 30 V but, in actual fact, only 29 V were used for making the test plate. The 29 V would be recorded.

After the test plate has been welded, it is allowed to cool and the plate is subjected to the visual and nondestructive testing as prescribed by the code. The specific tests required are a function of the type of weld being made and the particular welding consumables. The types of qualification tests are described in D1.1-98, paragraph 4.4.

In order to be acceptable, the test plates must first pass visual inspection followed by nondestructive testing (NDT). At the contractor's option, either RT or UT can be used for NDT. The mechanical tests required involve bend tests (for soundness) macroetch tests (for soundness), and reduced section tensile tests (for strength). For qualification of procedures on steels with significantly different mechanical properties, a longitudinal bend specimen is possible. All weld metal tensile tests are required for unlisted filler metals. The nature of the bend specimens, whether side, face, or root, is a function of the thickness of the steel involved. The number and type of tests required are defined in D1.1-98, Table 4.2, for complete joint penetration groove welds; D1.1-98, Table 4.3, for partial joint penetration groove welds; and D1.1-98, Table 4.4, for fillet welds.

Once the number of tests has been determined, the test plate is sectioned and the specimens machined for testing. The results of the tests are recorded on the PQR. According to D1.1-98, if the test results meet all the prescribed requirements, the testing is successful and welding procedures can be established based upon the successful PQR. If the test results are unsuccessful, the PQR cannot be used to establish the WPS. If any one specimen of those tested fails to meet the test requirements, two retests of that particular type of test may be performed with specimens extracted from the same test plate. If both of the supplemental specimens meet the requirements, the D1.1-98 allows the tests to be deemed successful. If the test plate is over 1½-in thick, failure of a specimen necessitates retesting of all the specimens at the same time from two additional locations in the test material.

It is wise to retain the PQRs from unsuccessful tests as they may be valuable in the future when another similar welding procedure is contemplated for testing.

The acceptance criteria for the various tests are prescribed in the code. The reduced section tensile tests are required to exceed the minimum specified tensile strength of the steel being joined. Specific limits on the size, location, distribution, and type of indication on bend specimens are prescribed in D1.1-98, paragraph 4.8.3.3.

Writing WPSs from successful PQRs.   When a PQR records the successful completion of the required tests, welding procedures may be written from that PQR. At a minimum, the values used for the test weld will constitute a valid WPS. The values recorded on the PQR are simply transcribed to a separate form, now known as a WPS rather than a PQR.

It is possible to write more than one WPS from a successful PQR. Welding procedures that are sufficiently similar to those tested can be supported by the same PQR. Significant deviations from those conditions, however, necessitate additional qualification testing. Changes that are considered significant enough to warrant additional testing are considered essential variables, and these are listed in D1.1-98, Tables 4.5, 4.6, and 4.7. For example, consider an SMAW welding procedure that is qualified by test using an E8018-C3 electrode. From that test, it would be possible to write a WPS that utilizes E7018 (since this is a decrease in electrode strength) but it would not be permissible to write a WPS that utilizes E9018-G electrode (because Table 4.5 lists an increase in filler metal classification strength as an essential variable). It is important to carefully review the essential variables in order to determine whether a previously conducted test may be used to substantiate the new procedure being contemplated.

D1.1-98, Table 4.1, defines the range of weld types and positions qualified by various tests. This table is best used, not as an after-the-fact evaluation of the extent of applicability of the test already conducted, but rather for planning qualification tests. For example, a test plate conducted in the 2G position qualifies the WPS for use in either the 1G or 2G position. Even though the first anticipated use of the WPS may be for the 1G position, it may be advisable to qualify in the 2G position so that additional usage can be obtained from this test plate.

In a similar way, D1.1-98, Table 4.7, defines what changes can be made in the base metals used in production versus qualification testing. An alternative steel may be selected for the qualification testing simply because it affords additional flexibility for future applications.

If WPS qualification is performed on a nonprequalified joint geometry, and acceptable test results are obtained, WPSs may be written from that PQR utilizing any of the prequalified joint geometries (D1.1-98, Table 4.5, item 32).

3.5.6 Approval of WPSs

After a WPS is developed by the fabricator or erector, it is required to be reviewed in accordance to D1.1 requirements. For prequalified WPSs, the inspector is required to review the WPSs to ensure that they meet all the prequalified requirements. For WPSs that are qualified by test, the AWS D1.1-98 code requires these to be submitted to the engineer for review.

The apparent logic behind the differences in approval procedures is that while prequalified WPSs are based upon well-established time-proven, and documented welding practices, WPSs that have been qualified by test are not automatically subject to such restrictions. Even though the required qualification tests have demonstrated the adequacy of the particular procedure under test conditions, further scrutiny by the engineer is justified to ensure that it is applicable for the particular situation that will be encountered in production.

In practice, it is common for the engineer to delegate the approval activity of all WPSs to the inspector. There is a practical justification for such activity: the engineer may have a more limited understanding of welding engineering, and the inspector may be more qualified for this function. While this practice may be acceptable for typical projects that utilize common materials, more scrutiny is justified for unusual applications that utilize materials in ways that deviate significantly from normal practice. In such situations, it is advisable for the engineer to retain the services of a welding expert to evaluate the suitability of the WPSs for the specific application.

3.6 Weld Size Determination

3.6.1 Strength of welded connections

A welded connection can be designed and fabricated to have a strength that matches or exceeds that of the steel it joins. This is known as a *full-strength connection* and can be considered 100% efficient. Welded connections can be designed so that if loaded to destruction, failure would occur in the base material. Poor weld quality, however, may adversely affect weld strength. Porosity, slag inclusions, and lack of fusion may decrease the capacity of a complete joint penetration (CJP) groove weld.

A connection that duplicates the base metal capacity is not always necessary and when unwarranted, its specification unnecessarily increases fabrication costs. In the absence of design information, it is possible to design welds that have strengths equivalent to the base material capacity. Assuming the base plate has been properly selected, a weld sized around the base plate will be adequate as well. This, however, is a very costly approach. Economical welded structures cannot be designed on this basis. Unfortunately, the overuse of the CJP detail and the requirement of "matching filler metal" serves as evidence that this is often the case.

3.6.2 Variables affecting welded connection strength

The strength of a welded connection is dependent on the weld metal strength and the area of weld that resists the load. Weld metal strength is a measure of the capacity of the deposited weld metal itself, measured in units such as ksi (kips per square inch). The connection strength reflects the combination of weld metal strength and cross-sectional area, and would be expressed as a unit of force, such as kips. If the product of area times the weld metal strength exceeds the load applied, the weld should not fail in static service.

The area of weld metal that resists fracture is the product of the theoretical throat multiplied by the length. The *theoretical weld throat* is defined as the minimum distance from the root of the weld to its theoretical face. For a CJP groove weld, the theoretical throat is assumed to be equal to the thickness of the plate it joins. Theoretical throat dimensions of several types of welds are shown in Fig. 3.18.

For fillet welds or partial-joint penetration groove welds, using filler metal with strength levels equal to or less than the base metal, the theoretical failure plane is through the weld throat. When the same weld is made using filler metal with a strength level greater than that of the base metal, the failure plane may shift into the fusion boundary or heat-affected zone. From a design perspective, this is an undesirable condition and may lead to performance problems.

Complete-joint penetration groove welds that utilize weld metal with strength levels exactly equal to the base metal will theoretically fail in either the weld or the base metal. Since the weld metal is generally slightly higher in strength than the base metal, the theoretical failure plane for transversely loaded connections is assumed to be in the base metal.

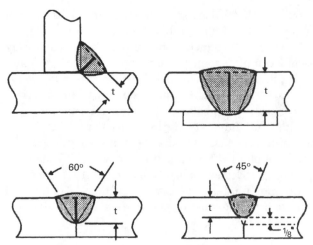

Figure 3.18 Theoretical throats. (*Courtesy of The Lincoln Electric Company.*)

In review, connection strength is governed by three variables: weld metal strength, weld length, and weld throat. The highest value of weld metal strength to be used for these calculations is a value comparable to the base metal. The weld length is often fixed, due to the geometry of the parts being joined, leaving one variable to be determined, namely, the throat dimension.

3.6.3 Determining throat size for tension or shear loads

For tension or shear loads, the required capacity the weld must deliver is simply the force divided by the length of the weld. The result, in units of force per length (such as kips per inch) can be divided by the weld metal capacity, in units of force per area (such as kips per square inch). The final result would be the required throat, in inches. Weld metal allowables which incorporate factors of safety can be used instead of the actual weld metal capacity. This directly generates the required throat size.

To determine the weld size, it is necessary to consider what type of weld is to be used. Assume the preceding calculation determined the need for a 1-in throat size. If a single fillet weld is to be used, a throat of 1.0 in would necessitate a leg size of 1.4 in, shown in Fig. 3.19. For double-sided fillets, two 0.7-in leg size fillets could be used. If a single PJP groove weld is used, the effective throat would have to be 1.0 in. The actual depth of preparation of the production joint would be 1.0 in or greater, depending on the welding procedure and included angle used. A double PJP groove weld would require two effective throats of

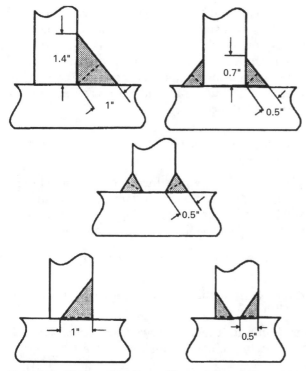

Figure 3.19 Weld combinations with equal throat dimensions.
(*Courtesy of The Lincoln Electric Company.*)

0.5 in each. A final option would be a combination of partial-joint penetration groove welds and external fillet welds. As shown in Fig. 3.11,
a 60° included angle was utilized for the PJP groove weld and an
unequal leg fillet weld applied externally. This acts to shift the effective throat from the normal 45° angle location to a 30° throat.

If the plates being joined are 1.0 in thick, a CJP groove weld is the
only type of groove weld that will effectively transfer the stress, since
the throat on a CJP weld is equal to the plate thickness. PJP groove
welds would be incapable of developing adequate throat dimensions
for this application, although the use of a combination PJP fillet weld
would be a possibility.

3.6.4 Determining throat size for compressive loads

When joints are only subject to compression, the unwelded portion of
the joint may be milled-to-bear, reducing the required weld throat.
Typical of these types of connections are column splices where partial-

joint penetration (PJP) groove welds frequently are used for static structures. In dynamic structures subject to many compression cycles, CJP groove welds often are required for fatigue reasons.

In theory, compression joints require no welds, providing the base metals will bear on another bearing surface. Some horizontal shearing forces may be present and the use of a weld with a throat equal to 50% of the base metal thickness is common.

3.6.5   Practical approach to determine weld size for bending or torsional loads

The following is a simple method to determine the correct amount of welding required for adequate strength for a bending or torsional load. This is a method in which the weld is treated as a line, having no area, but a definite length and outline. This method has the following advantages:

1. It is not necessary to consider throat areas because only a line is considered.

2. Properties of weld are easily found from a table without knowing weld leg size.

3. Forces are considered on a unit length of weld instead of stresses, thus eliminating the knotty problem of combining stresses.

4. Actual test values of welds are given as force per unit length of weld instead of unit stress on throat of weld.

3.6.6   Treat weld as a line

Visualize the welded connection as a single line, having the same outline as the connection, but no cross-sectional area. Notice (Fig. 3.20)

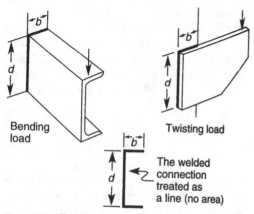

Bending load

Twisting load

The welded connection treated as a line (no area)

Figure 3.20   Treating weld as line.

that the area of the welded connection now becomes just the length of the weld.

Instead of trying to determine the stress on the weld (this cannot be done unless the weld size is known), the problem becomes a much simpler one of determining the force on the weld.

3.6.7   Use standard formulas to find force on weld

When the weld is treated as a line by inserting the property of the welded connection into the standard design formula used for that particular type of load (see Table 3.1), the force on the weld may be found in terms of pounds per lineal inch of weld.

For example, for bending:

Standard design formula   Same formula used for weld
(bending stress)   (treating weld as a line)

$$\sigma = \frac{M}{S} \qquad\qquad f = \frac{M}{S_w}$$

Normally the use of these standard design formulas results in a *unit stress,* in pounds per square inch; however, when the weld is treated as a line, these formulas result in a *force* on the weld, in pounds per lineal inch.

For secondary welds, the weld is not treated as a line, but standard design formulas are used to find the force on the weld, in pounds per lineal inch.

In problems involving bending or twisting loads, Table 3.2 is used. It contains the section modulus, S_w, and polar moment of inertia, J_w, of some 13 typical welded connections with the weld treated as a line.

For any given connection, two dimensions are needed, width b and depth d.

Section modulus S_w is used for welds subjected to bending loads, and polar moment of inertia J_w for twisting loads.

Section moduli S_w from these formulas are for maximum force at the top as well as the bottom portions of the welded connections. For the unsymmetrical connections shown in the table, maximum bending force is at the bottom.

If there is more than one force applied to the weld, these are found and combined together. All forces which are combined (vectorially added) must occur at the same position in the welded joint.

Calculating weld size for longitudinal welds.   Longitudinal welds constitute the majority of the welding performed in many types of construction and hence justify special emphasis. These include the web-to-

TABLE 3.1   Treating a Weld as a Line

Type of loading	Standard design formula	Treating the weld as a line
	Stress lb/in²	Force, lb/in
Primary welds transmit entire load at this point		
Tension or compression	$\sigma = \dfrac{P}{A}$	$f = \dfrac{P}{A_w}$
Vertical shear	$\sigma = \dfrac{V}{A}$	$f = \dfrac{V}{A_w}$
Bending	$\sigma = \dfrac{M}{S}$	$f = \dfrac{M}{S_w}$
Twisting	$\sigma = \dfrac{TC}{J}$	$f = \dfrac{TC}{J_w}$
Secondary welds hold section together—low stress		
Horizontal shear	$\tau = \dfrac{VA_y}{It}$	$f = \dfrac{VA_y}{In}$
Torsion horizontal shear*	$\tau = \dfrac{T}{2At}$	$f = \dfrac{T}{2A}$

A = area contained within median line.
* Applies to closed tubular section only.

b = width of connection, in
d = depth of connection, in
A = area of flange material held by welds in horizontal shear, in²
y = distance between center of gravity of flange material and N.A. of whole section, in
I = moment of inertia of whole section, in.⁴
C = distance of outer fiber, in
t = thickness of plate, in
J = polar moment of inertia of section, in.⁴
P = tensile or compressive load, lb
V = vertical shear load, lb

M = bending moment, in·lb
T = twisting moment, in·lb.
A_w = length of weld, in
S_w = section modulus of weld, in²
J_w = polar moment of inertia of weld, in³
N_x = distance from x axis to face
N_y = distance from y axis to face
S = stress in standard design formula, lb/in²
f = force in standard design formula when weld is treated as a line, lb/in
n = number of welds

TABLE 3.2 Treating a Weld as a Line

Outline of welded joint b = width, d = depth	Bending (about horizontal axis $x - x$)	Twisting
	$S_w = \dfrac{d^2}{6}$ in²	$J_w = \dfrac{d^3}{12}$ in³
	$S_w = \dfrac{d^2}{3}$	$J_w = \dfrac{d(3b^2 + d^2)}{6}$
	$S_w = bd$	$J_w = \dfrac{b^3 + 3bd^2}{6}$
$Ny = \dfrac{b^2}{2(b+d)}$ $Nx = \dfrac{d^2}{2(b+d)}$	$S_w = \dfrac{4bd + d^2}{6} = \dfrac{d^2(4b + d)}{6(2b + d)}$ top bottom	$J_w = \dfrac{(b + d)^4 - 6b^2d^2}{12(b + d)}$
$Ny = \dfrac{b^2}{2b+d}$	$S_w = bd + \dfrac{d^2}{6}$	$J_w = \dfrac{(2b + d)^3}{12} - \dfrac{b^2(b + d)^2}{2b + d}$
$Nx = \dfrac{d^2}{b+2d}$	$S_w = \dfrac{2bd + d^2}{3} = \dfrac{d^2(2b + d)}{3(b + d)}$ top bottom	$J_w = \dfrac{(b + 2d)^3}{12} - \dfrac{d^2(b + d)^2}{b + 2d}$
	$S_w = bd + \dfrac{d^2}{3}$	$J_w = \dfrac{(b + d)^3}{6}$
$Ny = \dfrac{d^2}{b+2d}$	$S_w = \dfrac{2bd + d^2}{3} = \dfrac{d^2(2b + d)}{3(b + d)}$ top bottom	$J_w = \dfrac{(b + 2d)^3}{12} - \dfrac{d^2(b + d)^2}{b + 2d}$
$Ny = \dfrac{d^2}{2(b+d)}$	$S_w = \dfrac{4bd + d^2}{3} = \dfrac{4bd^2 + d^3}{6b + 3d}$ top bottom	$J_w = \dfrac{d^3(4b + d)}{6(b + d)} + \dfrac{b^3}{6}$
	$S_w = bd + \dfrac{d^2}{3}$	$J_w = \dfrac{b^3 + 3bd^2 + d^3}{6}$
	$S_w = 2bd + \dfrac{d^2}{3}$	$J_w = \dfrac{2b^3 + 6bd^2 + d^3}{6}$

228   Chapter Three

TABLE 3.2   Treating a Weld as a Line (*Continued*)

	$S_w = \dfrac{\pi d^2}{4}$	$J_w = \dfrac{\pi d^3}{4}$
	$I_w = \dfrac{\pi_d}{2}\left(D^2 + \dfrac{d^2}{2}\right)$ $S_w = \dfrac{I_w}{c}$ where $c = \dfrac{\sqrt{D^2 + d^2}}{2}$	

flange welds on I-shaped girders, and the welds on the comers of the
box girders. These welds primarily transmit horizontal shear forces
resulting from the change in moment along the member. To determine
the force between the members being joined, the following equation
may be used:

$$f = \frac{Vay}{In}$$

where f = force on weld
  V = total shear on section at a given position along the beam
  a = area of flange connected by the weld
  y = distance from the neutral axis of the whole section to the
    center of gravity of the flange
  I = moment of inertia of the whole section
  n = number of welds joining the flange to webs per joint

The resulting force is then divided by the allowable stress in the
weld metal and the weld throat is attained. This particular procedure
is emphasized because the resultant value for the weld throat is nearly always less than the minimum allowable weld size. The minimum
size then becomes the controlling factor.

3.6.8   Filler metal strength requirements

Filler metal strength may be classified as "matching," "undermatching," or "overmatching." *Matching filler metal* has the same or slightly higher minimum specified yield and tensile strength compared to
the minimum specified properties of the base material. Emphasis is
placed on minimum specified properties because actual properties are
routinely higher. Matching filler metal for A572 GR50 would be E70

material, where the minimum specified filler metal/base metal properties for yield are 60/50 ksi and for tensile are 70/65 ksi. Even though the filler metal has slightly higher properties than the base metal, this is considered to be a matching combination.

Many see the filler metal recommendations provided in codes that reference "matching" combinations for various grades of steel and assume that is the only option available. While this will never generate a nonconservative answer, it may eliminate better options. Matching filler metal tables are designed to give recommendations for the one unique situation where matching is required (for example, CJPs in tension). Other alternatives should be considered, particularly when the residual stresses on the welded connection can be reduced in crack-sensitive or distortion-prone configurations.

Matching filler metal is required for CJP groove welds loaded in tension. In order to achieve a full-strength welded connection, the filler metal must have a strength that is at least equal to that of the material it joins.

Undermatching weld metal may be used for all weld types and loading types except one: complete-joint penetration groove welds loaded in tension. For all other joints and other loading types, some degree of undermatching is permitted. For example, CJPs in compression may be made with weld metal that has a strength of up to 10 ksi less than matching. CJPs in shear or loading parallel to the longitudinal axis may be made with undermatching filler material. All PJPs, fillet welds, and plug or slot welds may be made with undermatching weld metal. Design of the weld sizes, however, must incorporate the lower filler metal strength in order to ensure the welded connection has the proper capacity.

Undermatching may be used to reduce the concentration of stresses in the base material. Lower-strength filler material generally will be more ductile than higher strength weld metal. In Fig. 3.21, the first weld was made with matching filler material. The second design utilizes undermatching weld metal. To obtain the same capacity in the

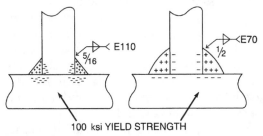

100 ksi YIELD STRENGTH

Figure 3.21 Matching and undermatching filler metal. (*Courtesy of The Lincoln Electric Company.*)

second joint, a larger fillet weld has been specified. Since the residual stresses are assumed to be of the order of the yield point of the weaker material in the joint, the first example would have residual stresses in the weld metal and the base metal of approximately 100 ksi. In the second example, the residual stresses in the base material would be approximately 60 ksi, since the filler metal has the lower yield point. These lower residual stresses will reduce cracking tendencies.

In situations where the weld size is controlled by the minimum permitted size, undermatching is a particularly desirable option. If a $\frac{1}{4}$-in fillet weld is required because of the minimum fillet weld size, it may be made of undermatching weld material without increasing the weld size due to the undermatching requirement.

Overmatching weld metal should be discouraged. It offers no advantages, and will increase residual stresses and distortion. Higher yield-strength weld metal generally is less ductile and more crack sensitive. Exceptions to this guideline are the filler materials used to join A588 weathering steel. In the process of adding alloys for atmospheric corrosion resistance, most filler metals for weathering steel will deposit 80-ksi tensile strength weld metal. Compared to the 70-ksi tensile strength weathering steel, this is an overmatch. The combination, however, performs well and because of the limited alternatives, this slight overmatch is permitted.

Caution must be exercised when overmatching filler metal is deliberately used. The strength of fillet and PJP groove welds is controlled by the throat dimension, weld length, and capacity of the weld metal. In theory, overmatching filler metal would enable smaller weld sizes to be employed and yet create a weld of equal strength. However, the strength of a connection is dependent not only on the weld strength but also on the strength of the fusion zone. As weld sizes are reduced, the fusion zone is similarly decreased in size. The capacity of the base metal is not affected by the selection of filler metal, so it remains unchanged. The reduction in weld size may result in an overstressing of the base metal.

Consider three tee joints containing PJP groove welds and illustrated in Fig. 3.22. A load is applied parallel to the weld, that is, the weld is subject to shear. The allowable stress on the groove weld is 30% of the nominal strength of the weld metal, that is, the "E" number (for example, E60, E70, etc.). Allowable stress on the base metal shall not exceed 40% of the yield strength of the base metal. The first combination employs a very close match of weld metal to base metal, namely, A572 GR50 welded with E70 electrode. The second example examines the same steel welded with undermatching E60 electrode, and the final illustration shows an example of overmatching with E80 electrode.

The weld capacity, in kips per inch, has been determined by multiplying the weld throat by the allowable stress. In the undermatching

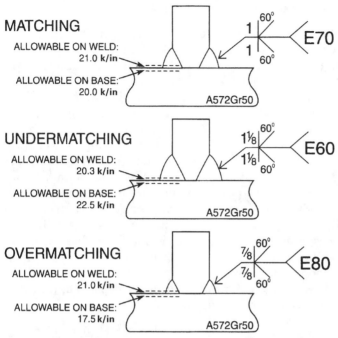

Figure 3.22 Effect of filler metal strength level. (*Courtesy of The Lincoln Electric Company.*)

case, notice that the weld controls. If the weld is properly designed, the base metal will not be overstressed. With matching weld metal, the loading on both the weld and the base metal is essentially the same. But, in the case of the overmatching combination, the weld has 20% more capacity than the base metal. If a designer overlooked loading on the base metal, the connection could easily be overlooked.

It should be noted, however, that all filler metal combinations will overmatch A36. Particular caution should be taken when sizing PJP groove welds on this steel to ensure that the base metal allowables are not exceeded.

Another area of potential problem is the slightly overmatched alloy filler metals used on A588. For PJP groove welds subject to shear loading, weld sizes that are determined based upon E80 filler metal will result in an overstressing of the base metal. However, the problem is eliminated when design calculations are made based on E70 filler metal. This is the recommended approach since some acceptable filler metals for weathering applications are classified E70. The base metal will not be overstressed, and the fabricator will have the flexibility of employing either E70 or E80 filler metal.

3.7   Welding Cost Analysis

Welding is a labor-intensive technology. Electricity, equipment depreciation, electrodes, gases, and fluxes constitute a very small portion of the total welding cost. Therefore, the prime focus of cost control will be on reducing the amount of time required to make a weld.

The following example is given to illustrate the relative costs of material and labor, as well as to assess the effects of proper process selection. The example to be considered is the groove weld of beam flange-to-column connections. Since this is a multiple-pass weld, the most appropriate analysis method is to consider the welding cost per weight of weld metal deposited, such as dollars per pound. Other analysis methods include cost per piece, ideal for manufacturers associated with the production of identical parts on a repetitive basis, and cost per length, appropriate for single-pass welds of substantial length. The two welding processes to be considered are shielded metal arc welding and flux-cored arc welding. Either would generate high-quality welds when properly used.

To calculate the cost per weight of weld metal deposited, an equation taking the following format is used:

$$\text{Cost per weight} = \frac{\text{Electrode cost}}{\text{Efficiency}} + \frac{\text{Labor} + \text{Overhead rate}}{(\text{Deposition rate})(\text{Operating factor})}$$

The cost of the electrode is simply the purchase cost of the welding consumable used. Not all of this filler metal is converted directly to deposited weld metal. There are losses associated with slag, spatter, and in the case of SMAW, the stub loss (the end portion of the electrode that is discarded). To account for these differences, an efficiency factor is applied. The following efficiency factors are typically used for the various welding processes:

Process	Efficiency, %
SMAW	60
FCAW	80
GMAW	90 (CO_2 shielding)
	98 (mixed gas)
SAW	100 (flux not included)

The cost to deposit the weld metal is determined by dividing the applicable labor and overhead rate by the deposition rate, that is, the amount of weld metal deposited in a theoretical, continuous 1 h of production. This cannot be maintained under actual conditions since welding will be interrupted by many factors, including slag removal, replacement of electrode, repositioning of the work or the welder with respect to the work, etc. To account for this time, an "operating factor" is used which is defined as the *arc-on* time divided by the total time associated with welding activities. For SMAW, replacement of electrodes takes place approximately every minute because of the finite length of the electrodes used. The following operating factors are typically used for the various processes and method of application:

Method	Operating factor, %
Manual SMAW	30
Semiautomatic	40
Mechanized	50

Operating factors for any given process can vary widely, depending on what a welder is required to do. In shop situations, a welder may receive tacked assemblies and be required only to weld and clean them. For field erection, the welder may "hang iron," fit, tack, bolt, clean the joint, reposition scaffolding, and perform other activities in addition to welding. Obviously, operating factors will be significantly reduced under these conditions.

The following examples are the actual procedures used by a field erector. The labor and overhead costs do not necessarily represent actual practice. The operating factors are unrealistically high for a field erection site, but have been used to enable comparison of the relative cost of filler metals versus the labor required to deposit the weld metal, as well as the difference in cost for different processes. Once the cost per deposited pound is known, it is relatively simple to determine the quantity of weld metal required for a given project, and multiply it by the cost per weight to determine the cost of welding on the project.

Process	SMAW	FCAW
Electrode classification	E7018	E70TG-K2
Electrode diameter, in	$\frac{3}{16}$	$\frac{7}{64}$
Amperage	225	430
Voltage	NA	27
Electrode efficiency, %	60	80
Electrode cost, $/lb	1.23	2.27
Operating factor, %	30	40
Deposition rate, lb/h	5.5	14.5
Labor and overhead rate, $/h	50	50

For SMAW:

$$\text{Cost per weight} = \frac{\$1.23}{60\%} + \frac{\$50.00}{(5.5)(30\%)} = \$2.05 + \$30.30 = \$32.35/\text{lb}$$

For FCAW:

$$\text{Cost per weight} = \frac{\$2.27}{80\%} + \frac{\$50.00}{(14.5)(40\%)} = \$11.46 + \$8.62 = \$11.46/\text{lb}$$

In the SMAW example, the electrode cost is approximately 6% of the total cost. For the FCAW example, primarily due to a decrease in the labor content, the electrode cost is 25% of the total. By using FCAW, the total cost of welding was decreased approximately 65%. While the FCAW electrode costs 85% more than the SMAW electrode, the higher electrode efficiency reduces the increase in electrode cost to only 39%. The first priority that must be maintained when selecting welding processes and procedures is the achievement of the required weld quality. For different welding methods which deliver the required quality, it is generally advantageous to utilize the method that results in higher deposition rates and higher operating factors. This will result in reduced welding time with a corresponding decrease in the total building erection cycle, which will generally translate to a direct savings for the final owner, not only lowering the cost of direct labor, but also reducing construction loan costs.

3.8 Techniques to Limit Distortion

3.8.1 Why distortion occurs

Distortion occurs due to the nonuniform expansion and contraction of weld metal and adjacent base material during the heating and cooling cycles of the welding process. At elevated temperatures, hot, expanded weld and base metal occupies more physical space than it will at room temperatures. As the metal contracts, it induces strains that result in stresses being applied to the surrounding base materials. When the surrounding materials are free to move, distortion results. If they are not free to move, as in the case of heavily restrained materials, these strains can induce cracking stresses. In many ways, distortion and cracking are related. It should be emphasized that not only the weld metal, but also the surrounding base material, is involved in this contraction process. For this reason, welding processes and procedures that introduce high amounts of energy into the surrounding base material will cause more distortion. Stresses resulting from material shrinkage are inevitable in welding. Distortion, however, can be minimized, compensated for, and predicted. Through efficient planning, design, and fabrication practices, distortion-related problems can be effectively minimized.

3.8.2 Control of distortion

Design concepts to minimize distortion. The engineer who is aware of the effects of distortion can design measures into the welded assemblies that will minimize the amount of distortion. These concepts include the following:

Minimize the amount of weld metal: Any reduction in the amount of weld metal will result in a decrease in the amount of distortion:

- Use the smallest acceptable weld size
- Use intermittent welds where acceptable
- Utilize double-sided joints versus single-sided joints where applicable
- Use groove details that require the minimum volume of weld per metal per length

Fabrication practices that minimize distortion.   Fabricators can use techniques that will minimize distortion. These include the following:

Use as few weld passes as possible: Fewer passes are desirable inasmuch as they limit the number of heating and cooling cycles to which the joint will be subjected. The shrinkage stresses of each pass tend to accumulate, increasing the amount of distortion when many passes are used. Note that this is in direct contrast with the criterion of maximizing notch toughness.

Avoid overwelding: Overwelding results in more distortion than is necessary. Holding weld sizes to print requirements will help avoid unnecessary distortion.

Obtain good fit-up: Poor fit-up, resulting in gaps and larger included angles for bevel preparations, means more weld metal is placed in the joint than is required, contributing to excessive distortion.

Use high-productivity, low-heat input welding procedures: Generally speaking, high-productivity welding procedures (those using high amperages and high travel speeds) result in a lower net heat input than low-productivity procedures. At first, high-amperage procedures may seem to be high-heat input procedures. However, for a given weld size, the high-amperage procedures are high travel-speed procedures. This will result in a decreased amount of heat-affected zone and reduced distortion.

Use clamps, strongbacks, or fixtures to restrict the amount of distortion: Any tooling or restraints that avoid rotation of the part will reduce the amount of distortion experienced. In addition, fixturing may be used to draw heat away, particularly if copper chill bars and clamps are used in the vicinity of the joint. The arc should never impinge on copper as this could cause cracking.

Use a well-planned welding sequence: A well-planned welding sequence is often helpful in balancing the shrinkage forces against each other.

Preset or precamber parts before welding: Parts to be joined may be preset or precambered before welding. When weld shrinkage causes distortion, the parts will be drawn back into proper alignment.

3.9 References

1. American Institute of Steel Construction, *Manual of Steel Construction,* 2d ed., vol. I, LRFD, Chicago, IL, 1994.
2. American Welding Society, *Structural Welding Code—Steel,* D1.1, Miami, FL, 1998.

Partially Restrained Connections

Roberto T. Leon, P.E.
Georgia Institute of Technology
Atlanta, GA

4.1 Introduction

The American Institute of Steel Construction (AISC) specification has recognized semirigid (type 3) or partially restrained (PR) construction since the 1940s (AISC, 1947). Because the design of PR connections is predicated on a set of forces obtained from a careful advanced structural analysis of the proposed frame and because few, if any, design texts address this issue, this chapter will begin with a thorough discussion of PR connection and frame behavior. Once these issues are

(*Courtesy of The Steel Institute of New York.*)

understood, the connection design can proceed as for any other steel connection. Design examples for several types of PR connections, including welded and bolted plates, T stubs, and end plates will then be discussed. It should be noted that PR connections may benefit significantly from using reinforcing bars in the floor slabs to carry negative moments over the supports. The design of this type of composite PR connection has been covered in detail in several recent publications (Leon 1996, 1997), and will not be repeated here.

4.2 Connection Classification

From the 5th to the 8th edition of the allowable stress specification (AISC, 1947, 1978), PR connections were categorized as Type 3 construction. Type 3 design was predicated on the assumption that "connections of beams and girders possess a dependable and known moment capacity intermediate in degree between the rigidity of Type 1 (rigid) and the flexibility of Type 2 (simple)." This definition is confusing since it mixes strength and stiffness concepts, and was generally interpreted as referring to the initial stiffness, K_i, of the connection as characterized by the slope of its moment-rotation curve.

Moment-rotation (M-θ) curves are generally assumed to be the best characterization of connection behavior. These M-θ curves are generally derived from experiments on cantilever-type specimens (Fig. 4.1a).

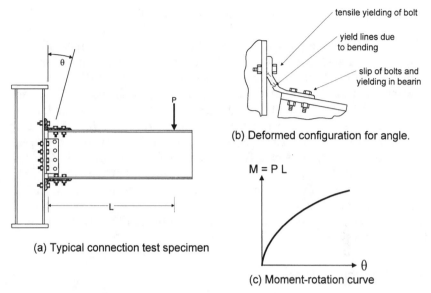

(b) Deformed configuration for angle.

(a) Typical connection test specimen

(c) Moment-rotation curve

Figure 4.1   Derivation of M-θ curves from experiments.

The moments are calculated directly from the statics of the specimen, while the rotations are measured over a distance typically equal to the beam depth. The rotation reported thus includes most deformation components occurring in the joint region. For the case of a top-and-seat angle shown in Fig. 4.1, these components include, among others, the elastic deformations due to the pullout of the angle, the rotation due to yield-line formation in the leg bolted to the column due to bending, yielding of the angle leg attached to the beam in tension, slip of the bolts, and hole elongation due to bearing (Fig. 4.1b).

Because little data on M-θ curves has been available in the published literature until recently and because the specification provided no guidelines on how to implement this concept in practice, Type 3 construction has seldom been explicitly used. However, extensive use of PR connections was made through the artifice of Type 2 "wind" construction, where the connections were assumed to be simple for gravity loads and rigid for wind loads. While extensive research (Ackroyd and Gerstle, 1982) has shown this procedure to be generally safe, the final forces and deformations computed from a simplified analysis can be different from those using an advanced analysis program that incorporates the entire nonlinear M-θ relationship shown in Fig. 4.1c.

The description of Type 3 construction used in previous versions of the steel specification cannot properly account for the effect of connection flexibility at the serviceability, ultimate strength, or stability limit states. The first LRFD specification (AISC, 1986) recognized these limitations and changed the types of construction to fully restrained (FR) and partially restrained (PR) to more realistically recognize the effects of the connection flexibility on frame performance. The definition of PR connections in the first two LRFD versions of the specification (AISC, 1986, 1993), however, conformed to that used in previous allowable stress design (ASD) versions. Research in this area has led to more comprehensive proposals for connection classification (Gerstle, 1985; Nethercot, 1985; Bjorhovde et al., 1992; Eurocode 3, 1992) that clarify the combined importance of stiffness, strength, and ductility in connection design. The next version of the LRFD specification will probably carry a much more detailed connection classification scheme in the commentary. The discussion here is in substantial agreement with the main concepts that will appear in that commentary.

4.2.1   Connection stiffness

As noted earlier, the *connection stiffness* can be taken as the slope of the M-θ curve. Since the curves are nonlinear from the start, it is possible to define this stiffness based on the tangent approach (such as for K_i in Fig. 4.2) or on a secant approach (such as K_{conn} in Fig. 4.2). A tangent approach is viable only if the analysis programs available can handle a continuous, nonlinear rotational spring. Even in this case,

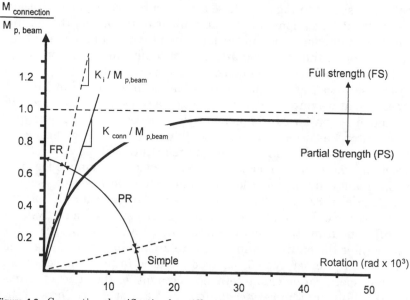

Figure 4.2 Connection classification by stiffness, strength, and ductility.

however, the computational overhead can be large and this option is recommended only for verification of the seismic performance of irregular structures. In most designs for regular frames, a secant approach will probably yield a reasonable solution at a fraction of the calculation effort required by the tangent approach. In this case, the analysis can be carried out in two steps by using linear springs. For service loads, a K_{serv} can be used for deflections and drift checks. The service secant stiffness can be taken at 0.0025 rad. A K_{ult}, based on a secant stiffness to a rotation of 0.02 rad, can be used for checks related to ultimate strength. Clearly, the deformations computed for the service load level will be fairly accurate, since the deviation of K_{serv} from the true curve is typically small. On the other hand, the deformations computed for the ultimate strength case will probably not be very accurate, since there can be very large deviations and the linear spring K_{ult} can only be interpreted as an average. However, this approximation is probably sufficient for design purposes. Designers should be conscious that no theoretical proof exists that a secant stiffness such as K_{ult} will provide a conservative result.

The stiffness of the connection is meaningful only when compared to the stiffness of the connected members. For example, a connection can be classified as rigid (type FR) if the ratio, α, of the connection secant stiffness at service level loads K_{serv} to the beam stiffness, EI/L, is greater than about 18 for unbraced frames. Generally connections with $\alpha \leq 2$ are regarded as pinned connections. Limits on the ranges of α cannot be established uniquely because they will vary depending on the limit state used to derive them. For regular frames, for example, one commonly used criterion to establish an upper limit is that the reduction in elastic buckling capacity due to the flexibility of the connections should not exceed 5% from that given by an analysis assuming rigid connections (Eurocode 3, 1992). Because this reduction in buckling capacity is tied to whether the frame is braced or unbraced, the value of 20 is suggested for unbraced frames, while a value of 8 is sufficient for braced frames. For continuous beams in braced frames, on the other hand, limits based on achieving a certain percentage of the fixed-end moment or reaching a deflection limit seem more reasonable (Leon, 1994).

4.2.2 Connection strength

A connection can also be classified in terms of strength as either a full-strength (FS) connection or a partial-strength (PS) connection. An FS connection develops the full plastic moment capacity, M_p, of the beam framing into it, while a PS connection can only develop a portion of it. For classifying connections according to strength, it is common to nondimensionalize the vertical axis of the M-θ curve by the beam plastic moment capacity, $M_{p,\text{beam}}$, as is shown in Fig. 4.2. Connections not capable of transmitting at least $0.2M_p$ at a rotation of 0.02 rad are considered to have no flexural strength. Because many PR connections do not exhibit a plateau in their strength even at large rotations, an arbitrary rotation value must be established to compare connection strength $M_{p,\text{conn}}$ to the capacity of the beam. For this purpose a rotation of 0.02 rad is recommended.

4.2.3 Connection ductility

Connection ductility is a key parameter either when the deformations are concentrated in the connection elements, as is the typical case in PR connections, or when large rotations are expected in the areas adjacent to the connections, as in the case of ductile moment frames with welded connections. The ductility required will depend on the flexibility of the connections and the particular application (for example, a braced frame in a nonseismic area versus an unbraced frame in a high seismic area). A connection can be classified as ductile based on both its absolute and its relative rotation capacity. Figure 4.3 shows the proposed absolute rotation criteria for special moment frames (SMF, 0.03 rad) and intermediate moment frames (IMF, 0.02 rad) in seismic areas (AISC, 1997).

The figure also shows a relative ductility index ($\mu = \theta_u/\theta_y$) that can be used for comparing the rotation capacity of connections with similar moment-rotation characteristics. In order to compute a relative

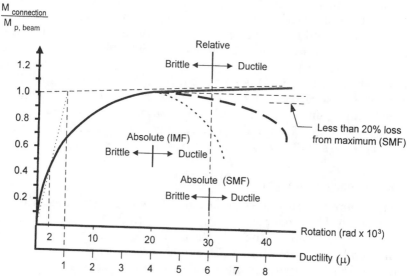

Figure 4.3 Possible connection ductility classification.

ductility, μ, a yield rotation, θ_y, must be defined. For PR connections, such as the one shown in Figure 4.3, this definition is troublesome since a yield moment is difficult to determine. In this case, for the hardening connection (solid line) and for illustrative purposes only, the yield rotation is defined as the rotation at the intersection of the service and hardening stiffnesses of the connection. In general, relative ductilities of 6 or more have been associated with ductile connections.

Both the absolute and relative rotation capacities, however, need to take into account any strength degradation that may occur as a result of local buckling or slip, particularly under cyclic loads. The behavior of the connections shown by the solid (hardening), dashed (gradually softening), and dotted (degrading) lines in Figure 4.3 can lead to significant differences in frame behavior, especially with respect to strength and stability. For classification purposes, a simple rule requires that the strength degradation in ductile connections subjected to cyclic loads be limited to 20% of the maximum capacity when the relative or absolute rotation limits are reached (AISC, 1997).

Limits for ductility criteria, such as those described previously, are only now being developed, but this issue is highlighted here to remind designers that analysis assumptions (unlimited rotational ductility, in general) must be consistent with the detailing of the connection. This is true for both PR and FR frames.

4.2.4 Derivation of M-θ curves

As noted earlier, M-θ curves have typically been derived from experiments. Many of these tests have been collected into databases (Goverdhan, 1984; Ang and Morris, 1984; Nethercot, 1985; Kishi and Chen, 1986, for example). Using these databases, equations for the complete M-θ curves for different types of connections have been proposed. However, numerous important variables, such as the actual yield strength of the materials and the torque in the bolts, are generally poorly documented or missing for many of these tests. Thus, many of the M-θ curves and equations available from these databases cannot be considered as reliable. In addition, care should be exercised when utilizing tabulated moment-rotation curves not to extrapolate to sizes or conditions beyond those used to develop the database since other failure modes may control (ASCE, 1997).

Two approaches have recently become practical alternatives and/or complements to experimental testing in developing M-θ curves. The first alternative is a detailed, nonlinear finite element analysis of the connection. While time-consuming because of the extensive parametric studies required to derive reliable M-θ curves, this approach has gone from a pure research tool to an advanced design office tool in just a few years, thanks to the tremendous gains in computational power available in new desktop workstations.

The other approach is the one proposed by the Eurocodes and commonly labeled the *component approach*. In this case each deformation mechanism in a joint is identified and individually quantified through a series of small component tests. These tests are carefully designed to measure one deformation component at the time. Each of these components is then represented by a spring with either linear or nonlinear characteristics. These springs are arranged in series or in parallel and the overall M-θ curve is derived with the aid of simple computer programs that conduct the analysis of the spring system. Figure 4.4 shows a typical model for a T-stub connection. Here the $K1$ and $K2$ springs model the panel zone deformation due to shear, while springs $K3$ and $K4$ model the bending deformations of the T stubs. Spring $K3$ and $K4$ are made up of the contributions of several other springs that model different deformation components (Fig. 4.4b).

4.2.5 Analysis

For many types of connections, the stiffness at the service load level falls somewhere in between the fully restrained and simple limits, and thus designers need to account for the PR behavior. The M-θ characteristic can be obtained from experiments or models as described in the previous section. The effect of PR connections on both force distribution and deformations in simple systems will be illustrated with two short examples.

Figure 4.5 shows the moments and deflections in a beam subjected to a uniformly distributed load. The horizontal axis is logarithmic and shows the ratio of the connection to beam stiffness ($\alpha = K_{\text{serv}} L/EI$).

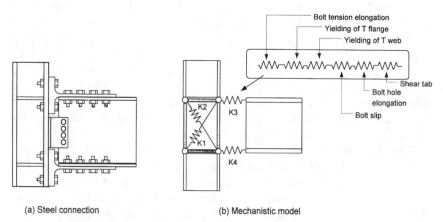

(a) Steel connection   (b) Mechanistic model

Figure 4.4 Component model for a T-stub connection.

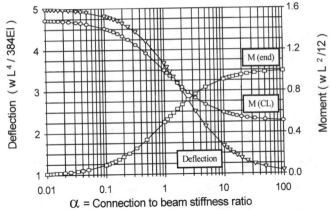

Figure 4.5 Moments and deflections for a beam under uniformly distributed load with PR connections at its ends.

The deformations range from that of a simply supported beam ($\Delta = 5wL^4/384EI$) for a very flexible connection ($\alpha \to 0$) to that of a fixed beam ($\Delta = wL^4/384EI$) for a very stiff connection ($\alpha \to \infty$). From both the deflection and force distribution standpoints, for a range of $15 < \alpha < \infty$ the behavior of the connection is essentially that of a fixed beam. Similarly, for a range of $0 < \alpha < 0.3$, the beam is essentially simply supported. Note that the ranges given here were selected arbitrarily, and that they will vary somewhat with the loading condition. This is why, as was noted earlier in the discussion of connection stiffness, the selection of limits for α to separate FR, PR, and simple behavior are not straightforward. It is important to note, however, that the horizontal axis of Fig. 4.5 is logarithmic. This means that apparently large changes in connection stiffness actually result in much smaller changes in forces or deformations. This lack of sensitivity is actually what allows us to design PR connections by simplified methods, since it means that the connection stiffness does not need to be known with great precision.

Figure 4.6 shows the results of an analysis for the general case of a one-story, one-bay frame with springs both at the connections to the beam, K_{conn}, and at the base of the structure, K_{base}. A simple formula for the drift cannot be written for this general case. The figure shows the drifts for five levels of base fixity ($\alpha_{base} = K_{base} H_e/EI_{col} = 0, 1, 2, 5,$ 5, 10, and ∞) versus a varying $\alpha_{beam} = (K_{conn} L/EI)$. The calculations are for a frame with an $I_{beam} = 2000$ in^4, $L = 288$ in, $I_{col} = 500$ in^4, $H = 144$ in, a concentrated horizontal load at the top of $P = 2.4$ kips, and a distributed load on the beam of $w = 0.08333$ kip/in. The vertical axis gives the deflection as a multiplier, τ, of the fully rigid case, where $K_{conn} = K_{base} = \infty$. The drift value for the latter is 0.025 in. For the case

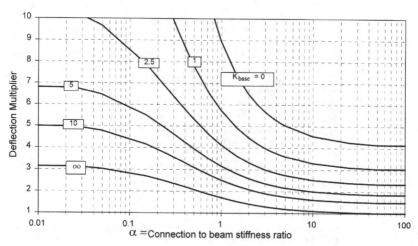

Figure 4.6 Drifts of a simple frame with various degrees of base fixity and connection stiffness.

of $K_{base} = \infty$, as the connection stiffness decreases, the deflection reduces to that of a cantilever subjected to $P/2$ ($\tau = 3.25$). For the other extreme ($K_{base} = 0$), the deflections increase rapidly from $\tau = 4.06$ as the stiffness of the connection is decreased since we are approaching the unstable case of a frame with pins at all connections as $\alpha \to 0$.

Another very important lesson to be drawn from Fig. 4.6 is the large effect of the base fixity on frame drift. While it is common to assume in the analysis that the column bases are fixed, such degree of fixity is difficult to achieve in practice even if the column is embedded into a large concrete footing. Most footings are not perfectly rigid or pinned, with the practical range probably being $1 < K_{base} < 10$. As can be seen from Fig. 4.6, the difference in drift between the assumption of $K_{base} = \infty$ (perfect base fixity) and a realistic assumption ($K_{base} = 10$) ranges from about 50% when K_{conn} is ∞ to about 300% when α is 0.

Figure 4.6 indicates that there are infinite combinations of K_{base} and K_{conn} for a given deflection multiplier. Consider the case of a one-story, one-bay frame with the properties given for Fig. 4.6. Thus, for a target deflection multiplier of, say, 3, one can design the frame with a pinned base and a K_{conn} approaching infinity ($\alpha = 0$), or one can design a rigid footing with a connection having an $\alpha = 2$. This flexibility in design is what makes PR connection design both attractive and somewhat disconcerting. It is attractive because it provides the designer with a wide spectrum of possibilities in selecting the structural members and their connections. It is disconcerting because most designers do not have extensive experience with PR analysis and PR frame behavior.

There are currently numerous good texts that address the analysis and design of PR frames (Bjorhovde et al., 1988a, 1988b, 1992, 1996; Chen and Lui, 1991; CTBUH, 1993; Chen and Toma, 1994; Chen et al., 1995; Leon et al., 1996). There is a considerable range in the complexity of the analysis approaches proposed in the literature. Clearly, the appropriate degree of sophistication of the analysis depends on the problem at hand. The designer should, when incorporating connection restraint into the design, take into account the effect of reduced connection stiffness on the stability of the structure and the effect of connection deformations on the magnitude of second-order effects (ASCE, 1997). Usually design for PR construction requires separate analysis to determine the serviceability limit state and the ultimate limit state because of the nonlinear nature of the M-θ curves.

4.3   Design of Bolted PR Connections

The design of a connection must start from a careful assessment of its intended performance. This requires the designer to determine the performance criteria with respect to stiffness (FR, PR, or simple), the strength (FS or PS), and the ductility. The stiffness is critical with respect to serviceability, while strength and ductility are critical with respect to life safety issues. These criteria must be consistent with the model assumed for analysis. From Fig. 4.2, if an assumption of a rigid connection (FR) was made in the analysis, the resulting connection will typically be fully welded, welded-bolted, or a thick end-plate type. Similarly, if the connection was assumed as simple, then a shear plate welded to the column and bolted to the beam or angles bolted to both column and beam are appropriate.

If explicit use of PR behavior was made in the analysis, in the form of a rotational spring with a given K_{serv}, then a wide variety of connections can be chosen, ranging from an end plate (close to FR/FS performance) to top-and-seat angles (close to simple performance). The key here is to match the K_{serv} of the connection as designed to that assumed in the analysis. The matching should be done at the service level because drift and deflection criteria will probably govern the design in modern steel frames. The stiffness of the connection should be checked with at least the component model approach (Fig. 4.4). Since the stiffness of the connection will be dependent on the actual configuration of the connecting elements and the size of the framing members, it is possible to adjust the stiffness to match that assumed in design.

The ultimate strength and ductility of the connection as designed must also be compatible with that assumed in design. In this case it is imperative to identify all possible failure modes for the connection

TABLE 4.1   Failure Modes for Bolted Connections*

Connection type	CW-FB	CB-FB	EP	TS
Strength (FS or PS)	FS	FS	FS/PS	PS
Stiffness (FR or PR)	FR	FR/PR	PR	PR
Ductile:				
Slippage of slip-critical (friction) bolts	1	1		
Flexural beam yielding adjacent to nodal zone	2	2	1	
Yielding of connecting elements in tension	3	3		2
Formation of yield lines in connecting elements		4	2	1
Yielding of slab reinforcement in tension				
Panel zone yielding	4	5	3	3
Limited local buckling	5	6	4	4
Semiductile:				
Elongation of bolt holes due to bearing	6	7		5
Yielding of bolts to column flange in tension	8	9	5	6
Shear yielding of bolts to beam flange	7	8		7
Severe local buckling of beam flange	9	10	6	8
Brittle:				
Fracture of welds between column and plate	A			
Fracture/failure of shear connection to web	A	A	A	A
Bearing/crushing failure of concrete				
Fracture of shear studs and rebar				
Fracture of beam flange due to local buckling	A	A	A	
Shear failure of bolts	A	A	A	A
Tensile failure of bolts (including prying action)		A	A	A
Fracture of beam through net section		A	A	A
Fracture of connecting element through net section		A	A	A
Column web failure (yielding, crippling, buckling)	A	A	A	A
Edge distance or spacing failure of bolts	A	A	A	A
Block shear	A	A	A	A

NOTE: A indicates a brittle failure mode that should be carefully checked in design; CW-FB = column-welded–flange-bolted connections; CB-FB = column bolted–flange bolted; EP = end plate; TS = top-and-seat angles with double web angles; and PR-CC = partially restrained composite connection.
*From FEMA (1997a).

as designed. Moreover, it is necessary to understand the hierarchy of failure modes so that modes are excluded. Table 4.1 (FEMA, 1997a) shows a proposed hierarchy for seismic design of a variety of connections: column welded-beam bolted (CW-BB), column- and beam-bolted or T stub (CB-BB), end plates (EP), top-and seat connections (TS), and partially restrained composite connections (PR-CC). The table indicates the type of failure associated with each mechanism (ductile, semiductile, or brittle), and lists the ductile and semiductile mechanisms in descending order of desirability. This table is arbitrary and reflects the biases of the author. Designers are encouraged to develop their own lists and rankings based on their experience and regional preferences of fabricators and erectors.

Special note should be made of the fact that the material properties play an important role in connection performance. As our understanding of the failures in steel frames during the 1994 Northridge earth-quake improves, it is clear that material performance played a key role in some of the failures encountered. Issues related to the ductility and toughness of the base materials for both welds and bolts, installation procedures, QA/QC in the field, and need for new, tighter material specifications are receiving considerable attention (FEMA, 1997). Designers should strive to obtain the latest information in this area so that future failures are avoided.

The design process outlined previously is probably markedly different from that adopted for most routine design today. It places a heavy additional burden on designers both in terms of professional responsibility and continuing education, not to mention substantial additional design time. Two important points need to be made with respect to these issues. First, as our designs become more optimal with respect to both strength and stiffness, many of the traditional assumptions made in design need to be carefully reexamined. These include, for example, serviceability criteria based on substantially different partition and cladding systems than those used today. Second, these optimized systems are far more sensitive to the assumptions about connection behavior since typically far fewer moment-resisting connections are used in steel frames today than 20 y ago.

In this section the fundamentals of design for full-strength, fully restrained (FS/FR) bolted connections will be discussed first, followed by that for partial-strength, partially restrained (PS/PR) ones. The design for both seismic and nonseismic cases will be discussed. In the case of seismic design the proposed provisions for AISC (1998) and UBC (2000) will be employed since they are far superior to the current codes (AISC, 1992). The emphasis will be on understanding the basic steps in connection design and developing an understanding of the crucial mechanisms governing their behavior.

4.3.1 Column-welded–beam-bolted connections

The design of column-welded–beam-bolted (CW-BB) connections (Fig. 4.7) has been discussed extensively by Astaneh-Asl (Astaneh-Asl, 1995). The mechanistic model for this type of connection, labeled column-welded–beam-welded (CW-BB), is essentially the same as that shown in Fig. 4.4 for a T-stub connection. The main differences are that the springs representing the tension elongation of the bolts and the yielding in the flange have to be replaced by a spring that represents the behavior of the weld between the column flange and the beam flange.

Table 4.1 lists the main failure modes for this type of connection. In general the desired failure mechanisms will be slip of the bolts fol-

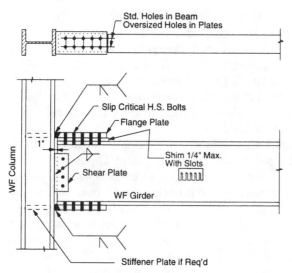

Figure 4.7 Typical CW-BB connection (Astaneh-Asl, 1995).

lowed by yielding of the beam and the connection plate. The main failure modes to avoid are brittle failure of the welds, shear failure of the bolts, and a net section failure in the connecting plate or beam. With this hierarchy established, it is possible to develop a design strategy, as outlined in the steps shown in the following paragraphs, for the design of these connections under monotonic loads.

The design of any connection subjected to seismic loads is similar in principle to the static design, except that a capacity design approach must be followed. In this context, capacity design implies that the connection must be designed to behave in a ductile manner under the maximum expected forces that can be introduced by the framing members. Thus, for WC-BB connections, the welds need to be strong and tough enough such that the weld strength does not control and fracture problems related to the welding procedures and materials are eliminated. For WC-BB connections, yielding should be limited to the connection plate or the beam flange. This requires a careful assessment of the minimum and maximum capacities associated with each of the springs in Fig. 4.4, since the forces are inertial rather than gravity type. In the new version of the seismic specification for building (ASCE, 1997b), there is an Ω_0 factor that intends to address the overstrength from the system standpoint. This is not the same as the local overstrengths at an individual connection that are being discussed here. This design approach is different from the static (that is, nonseismic) case where the connection can be designed for forces derived from the structural analysis, and without regard to what the actual ultimate capacity and failure mode of each of the connection components is.

Before looking at examples of CW-BB connections for both static and seismic loading cases, a number of important design issues, as identified by Astaneh-Asl (1995) need to be understood. These issues, discussed in detail in the following list, are of particular significance for CW-BB connections, but the principles involved are applicable to most strong PR connections. The reader should be aware that extensive changes are in the offing for the upcoming AISC seismic provisions (AISC, 1997). At the time this chapter was written, the final draft of those provisions was not available. Therefore, there may be inconsistencies between what is written here and the new AISC seismic specification:

1. *Proportioning of flange connection:* Whenever possible, the yield strength of the connection elements (top and bottom plates) should be matched to that of the beam flange. This will ensure that distributed yielding takes place and that severe local buckling will not ensue. Severe local buckling can result in an early fracture of the beam flanges if cyclic loads are present. Astaneh-Asl (1995) recommends that for yielding on the gross section:

$$b_p\, t_p\, F_{yp} \cong b_f\, t_f\, F_{yf} \tag{4.1}$$

where b and t = the width and thickness, F_y = the expected yield strength, and the subscripts p and f refer to the plate and beam flange, respectively. Usually, the expected yield strength of the materials is not known when the design is done. For designs not involving seismic forces, the average material properties, as opposed to the nominal ones, can be used throughout. For the case of seismic forces the same assumptions can be made with regard to sizing the plate, but an overstrength factor must be applied to undesirable modes of failure. This overstrength factor accounts both from the difference between nominal and actual yield strengths and the effects of strain hardening that are likely to occur as the rotations exceed 0.02 rad. For the design of the bolt groups and for checking failures of the net sections, a yield overstrength factor, R_y, of 1.5 has been suggested in the new AISC seismic specification for A36 steel and A572 grade 42 (AISC, 1997). For all other materials, a yield overstrength factor of 1.1 is suggested. Note that R_y is similar, but of larger magnitude, than the ß factor used in the interim FEMA guidelines (FEMA 1997a). It is expected that ongoing statistical studies will lead to a definitive value for these factors. In order to ensure a ductile failure, the ratio of the effective area, A_e, to the gross area, A_g, of the plate should be

$$\frac{A_e}{A_g} \geq \frac{R_y F_y}{F_u} \tag{4.2}$$

2. *Design of welds:* For the case of seismic loads another key issue is the design of the welds to the column flange. In this area there are recent, detailed guidelines proposed by SAC (FEMA 1995, 1997a) and AISC (1997). The AISC provisions require that a Welding Procedure Specification (WPS) be prepared as required by AWS D1.1 (AWS, 1997). AWS D1.1 provides detailed procedures for welding (see Chap. 3) and this standard should become familiar to all structural engineers. In addition, a minimum Charpy N-notch test (CVN) toughness of 20 ft-lb at $-20°F$ is required of all filler metal by the seismic AISC specification.

3. *Local buckling criteria:* The current limits suggested by AISC (1997) of $65/\sqrt{F_y}$ for b/t in beam flanges in compression and $970/\sqrt{F_y}$ for webs in flexural compression seem to provide a reasonable limit to ensure that the nominal plastic moment capacity of the section is reached. For seismic applications, these limits have been tightened somewhat, to $52/\sqrt{F_y}$ for b/t in beam flanges and $520/\sqrt{F_y}$ for webs in flexural compression to ensure not only that the capacity can be reached, but also that sufficient rotational ductility is available. The typical buckle that forms when these criteria are met is a smooth, small local buckle. This precludes the development of a sharp buckle that may lead to fracture under reversed inelastic loading. The current limits on web slenderness also seem to provide reasonable limits although the actual performance will be tied to the detailing of the web connections and whether composite action is expected. The slenderness of the connection plates, measured between the weld to the column flange and the centerline of the first row of bolts, should also be kept as low as practicable to prevent the formation of a local or global buckle in this area. Current criteria for unsupported compression elements are applicable in this case.

4. *Bolts:* The bolt group should be designed not only to prevent a shear failure of the connectors but also to provide adequate performance during the slipping phase of the moment-rotation behavior. Since slip provides a good energy dissipation mechanism, it is prudent to design the connection such that the slip occurs well above the service load but also below the ultimate strength of the connection. To meet this criteria, Astaneh-Asl (1995) recommends that the nominal slip resistance, $F_{slippage}$, be such that:

$$1.25F_{service} \leq F_{slippage} \leq 0.80F_{ultimate}$$

where $F_{service}$ corresponds to the nominal slip strength of the bolt group and $F_{ultimate}$ corresponds to the nominal shear strength of the bolts.

5. *Web connection design:* The design of the web connection is usual-

ly made without much regard to the contribution of this part of the connection to the flexural strength of the joint unless the flange connections carry less than 70% of the total moment (AISC, 1992). It is clear from the performance of MRFs during the Northridge earthquake that careful attention should be paid to ensure that the web connection is detailed to provide rotational ductility and strength that are compatible with the action of the flanges. Astaneh-Asl (1995) suggests that the shear plates be designed to develop the plastic moment strength of the web:

$$h_p t_p (0.6 F_{yp}) \geq h_{gw} t_{gw} (0.6 F_{ygw})$$

$$h_p^2 t_p (F_{yp}) \geq h_{gw}^2 t_{gw} (F_{ygw})$$

where h and t are the depth and thickness, F_y is the yield strength, and the subscripts p and gw refer to the shear plate and the beam web, respectively. Here again, allowances should be made for the steel overstrength (say $R_y = 1.1$ to 1.5). Failure modes to be avoided include bolt shear, block shear, net area fractures, and weld fractures.

Design Example 4.1. Design a full-strength connection between a W 21 × 62 girder and a W 14 × 120 column. Both sections are A572, GR 50. Design for static loads assuming the analysis shows a maximum moment, M_u, of 425 ft-kip (5100 kip-in) and a maximum shear, Vu, of 75 kips. The service moment, M_{serv}, is 180 kip-ft (2160 kip-in). Assume A325X ⅞-in-diameter bolts.

1. Check beam moment capacity:

$$\phi M_n = \phi Z_x F_y = (0.9)(144\ \text{in}^3)(50\ \text{ksi}) = 6480\ \text{kip-in, ok}$$

Check local buckling:

Flange:

$$\frac{65}{\sqrt{F_y}} = 9.2 > 6.1,\ \text{ok}$$

Web:

$$\frac{970}{\sqrt{F_y}} = 137 > 46.9,\ \text{ok}$$

2. Check net area fracture versus gross section yielding of the girder flange:

Gross section:

$$\phi A_{fg} F_y = (0.9)(8.24 \times 0.615)(50) = 228.0\ \text{kips}$$

Net section:

$$\phi A_{fn} F_u = (0.75)[(8.24 - 2) \times 0.615)](65) = 187.1\ \text{kips, governs}$$

Since the net section governs, the effective flange area, A_{fe}, is

$$A_{fe} = \frac{5}{6} \frac{F_u}{F_y} A_{fn} = \frac{5}{6} \frac{65}{50} \, 3.838 = 4.15 \text{ in}^2$$

This is an 18% decrease from the gross flange area, so that the effective plastic section modulus, Z_e, is

$$Z_e = Z_x - 2\left(0.18 A_{fg} \frac{d}{2}\right) = 144 - 19.2 = 124.7 \text{ in}^3$$

Thus the moment capacity is

$$\phi M_n = \phi Z_e F_y = (0.9)(124.7 \text{ in}^3)(50 \text{ ksi}) = 5611 \text{ kip-in, ok}$$

The force nominal force in the flanges and plates to carry this moment is

$$F_{\text{flange}} = F_{\text{plate}} = \frac{5611 \text{ kip-in}}{20.99 \text{ in}} = 267.3 \text{ kips}$$

3. Determine the size of the flange plate, assuming that the plate thickness, t_p, will be ⅝ in and that the plate is 50 ksi. Balancing the plastic capacity of the plate against that of the beam and leaving out the ϕ factor, gives a plate width, b_p, of

$$b_p = \frac{(124.7 \text{ in}^3)(50)}{21 \times 0.625 \times 50} = 9.50 \text{ in}$$

4. Check gross (A_{pg}) and net (A_{pn}) areas for the plate:

Gross section:

$$\phi A_{pg} F_y = (0.9)(9.50 \times 0.625)(50) = 267.2 \text{ kips}$$

Net section:

$$\phi A_{pn} F_u = (0.75)[(9.50 - 2) \times 0.625](65) = 228.5 \text{ kips} < F_{\text{plate}}, \text{ no good}$$

Increase plate thickness to 0.75 in, ok (net section = 274.2 kips > F_{plate}).

5. Determine number of A325X⅞-in bolts required for shear in the flanges:

$$N_{\text{bolts}} = \frac{267.3}{27.1} = 9.8 \rightarrow 10 \text{ bolts}$$

Assuming a gage of 5.0 in, this means the edge distance for a 9.5-in-wide plate about are equal to the minimum required. Assuming a bolt spacing of $3d$ and a distance between the last bolt and the weld equal to 4 in, the length of the plate is

$$L_{\text{plate}} = \text{end distance} + 4 \text{ spacings} + 4 \text{ in} = 16 \text{ in}$$

6. Check bearing on the beam flange:

$$F_{bearing} = N_{bolts} (\phi R_n) = (10)(0.75 \times 102) = 767 \text{ kips, ok}$$

7. Check service load slip capacity:

$$M_{slip} = (10 \text{ bolts})(10.2 \text{ kips/bolt})(20.99 \text{ in}) = 2141 \text{ kip-in}$$
$$\cong 2160 \text{ kip-in, say ok}$$

8. Check block shear. Assume shear failure along the bolts and tensile failure across bolt gage:

$$A_{gv} = 2\left(16 - 4 - \frac{1}{2}\right)(0.75) = 17.25 \text{ in}^2$$

$$A_{gt} = (5)(0.75) = 3.75 \text{ in}^2$$

$$A_{nv} = 17.34 - 2(3.5 \times 1)(0.75) = 12.00 \text{ in}^2$$

$$A_{nt} = 3.75 - 2(0.5 \times 0.75) = 3.00 \text{ in}^2$$

$$\phi R_n = \phi [0.6 F_u A_{nv} + F_y A_{gt}] = 491.6 \text{ kips, ok}$$

9. Determine weld size: Assume a full-penetration weld. The weld thickness, based on a 70-ksi electrode, is

$$t_{weld} = \frac{F_{plate}}{(0.6 F_{EXX} \times b_p)}$$

$$= \frac{267.3 \text{ kips}}{0.6 \times 70 \text{ ksi} \times 9.5 \text{ in}} = 0.67 \text{ in say } \tfrac{3}{4} \text{ in}$$

10. Detail the shear connection to the web. The design of the shear connection for this case will not be carried out in detail here (see Chap. 2 for design of shear connections). From the *AISC Manual*, Part II, a $\tfrac{5}{16}$-in pair of angles with $4\tfrac{7}{8}$-in A325X bolts provide 137 kips of shear resistance. This is larger than the 75 kips required for design.

11. Check connection stiffness. For a quick check assume the full force on the plate and a linear distribution of the force along the plate. Compute the elongation of the plate and add $\tfrac{1}{16}$ in for the slip of the bolts:

$$\Delta_{conn} = \frac{F_{plate} (L_{plate}/2)}{EA_{plate}} + \frac{1}{16} = 0.0129 + 0.063 = 0.0754 \text{ in}$$

Assume that the connection rotates about the center of the beam. The connection rotation is

$$\theta_{conn} = \frac{\Delta_{conn}}{(d/2)} = \frac{0.0754}{(20.99/2)} = 0.0072 \text{ rad}$$

The connection stiffness is

$$k_{conn} = \frac{M}{\theta_{conn}} = \frac{267.3 \times 20.99}{0.0072} = 780{,}570 \text{ kip-in/rad}$$

The relative stiffness, assuming the beam is 24 ft long is

$$\alpha = \frac{k_{conn} L_{beam}}{EI_{beam}} = \frac{780{,}570 \times (24 \times 12)}{(29{,}000 \times 1330)} = 5.82$$

Note that this will put us into the weak PR range. This calculation is misleading since much of the Δ_{conn} comes from the slip. At the service load range, if we assume no slip, the actual α is close to 34.0. Thus the moment-rotation curve for this connection indicates rigid behavior at the service level, and PR behavior between the service and ultimate load level.

12. Other checks related to Chap. N of the Specification are shown in Examples 4.3 and 4.4.

The final configuration for the connection is shown in Fig. 4.8. Figures 4.7, and 4.9 through 4.10 show some typical details and variations proposed by Astaneh-Asl for this type of connections. Figure 4.9 shows a variation where the bottom flange is welded rather than bolted, while Fig. 4.10 shows a connection to the weak axis of the column.

It is important to note that in the previous example it was assumed that the loads were well known. In Example 4.2, which follows, it will be shown that while a connection can be designed to connect similar size members in a frame located in a high seismic zone, the requirements can be very different. A complete design example for this type of connection, similar to Example 4.2, is also included as App. B in Astaneh-Asl (1995) and the reader is referred to that publication for complete details.

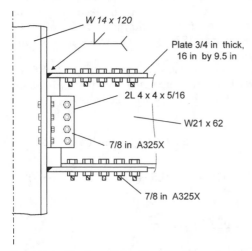

Figure 4.8 Final configuration for Example 4.1 connection.

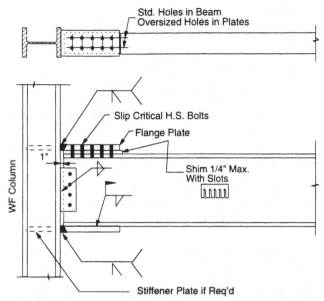

Figure 4.9 Typical CW-BB at the top and CW-BW at bottom connection (Astaneh-Asl, 1995).

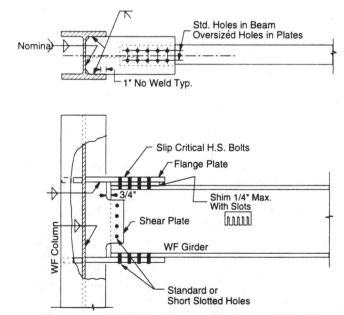

Figure 4.10 Typical CW-BB connection to weak axis of the column (Astaneh-Asl, 1995).

Design Example 4.2.   Design a full-strength connection between a W 21 $\times$ 62 girder and a W 14 $\times$ 120 column. Both sections are A572, Gr 50. Design for seismic loads in UBC Zone 3.

1. Determine maximum beam moment capacity:

$$M_n = Z_x R_y F_y = 144 \text{ in}^3 (1.1 \times 50 \text{ ksi}) = 7920 \text{ kip-in}$$

If we assume that all bending forces are transmitted through the beam flange, the force in the beam flange consistent with this moment would be:

$$F_{\text{flange}} = \frac{7920 \text{ kip-in}}{20.99 \text{ in}} = 377.3 \text{ kips}$$

This force is well beyond what the flange can carry. This difference arises because the flanges of the W 21 $\times$ 62 have a plastic capacity of about 103.2 in³ compared to the overall section plastic section of 144 in³. Thus the flange can carry only:

$$F_{\text{flange}} = \frac{377.3 \text{ kips} \times 103.3 \text{ in}^3}{144 \text{ in}^3} = 270.5 \text{ kips}$$

The connection must be designed for this force, which is 22% larger than in the previous example. The local buckling criteria for seismic design are somewhat stricter than those in the regular specification (AISC, 1997):

Flange:

$$\frac{52}{\sqrt{F_y}} = 7.3 > 6.1, \text{ ok}$$

Web:

$$\frac{520}{\sqrt{F_y}} = 73.5 > 46.9, \text{ ok}$$

The maximum unbraced length, L_b, for seismic design is

$$L_b < 2500 \, r_y F_y = \frac{2500(1.87)}{50} = 93.5 \text{ in} = 7.8 \text{ ft, say } 7.5 \text{ ft for design}$$

2. Check net area fracture versus gross section yielding of the girder flange. At this stage it is necessary to carefully check the minimum and maximum forces that can be introduced by the beam. To ensure ductile behavior, the maximum expected force required to yield the beam flange through its gross area has to be smaller than the minimum expected force required to fracture the beam through its net area. The maximum force that can be transmitted to the joint though the gross area of the beam is the same as for the previous example, but with the ϕ factor removed and the yield strength increased by R_y. Thus the maximum load through the gross area is 1.22 (228 kips) = 278.7 kips. This is somewhat larger but consistent with the flange capacity computed in part 1. On the other hand, the ultimate strength through the net area could be as small as that computed in the previous example. For this connection that is 205.8 kips, which

includes the ϕ factor but also an R_u factor to account for the increase in ultimate strength. This R_u factor is taken as 1.1 to be consistent with R_y since little or no data is available to determine the increase in ultimate strength. The maximum force that can be transmitted through the net section is irrelevant because the design is governed by the smallest force that can be expected to cause brittle fracture.

The problem with this design approach is that for most beams with narrow flanges, the difference between the net and gross areas is so large that it is almost impossible to ensure that the minimum capacity related to fracture is larger than the maximum capacity related to yielding. In general, it will be necessary to weld a small plate to the underside of the beam flanges in the connection area to ensure that yielding governs over fracture. A plate 0.25 in thick will increase the net section capacity to 308.0 kips, which will fulfill the required strength ratio.

3. Determine the size of the flange plate, assuming that the plate thickness, t_p, will be 1 in and that the plate is 50 ksi. Balancing the plastic capacity of the plate against that of the beam and leaving out the ϕ factor, gives a plate width, b_p, of

$$b_p = \frac{278.7 \text{ kips}}{0.75 \times 50} = 7.43 \text{ in, say } 7.50 \text{ in}$$

4. Check neat and gross areas for the plate:

Gross section:

$$A_{pg} F_y R_y = (7.50 \times 0.75)(50 \times 1.1) = 309.4 \text{ kips}$$

Increase plate width to 9.0 in and thickness to ⅞ in:

Net section:

$$\phi A_{pn} F_u R_u = (0.75)[(9.0 - 2) \times 0.875](65 \times 1.1)$$
$$= 328.4 \text{ kips} > F_{\text{plate}} = 309.4 \text{ kips, ok}$$

5. Determine number of A325X ⅞-in bolts required for shear:

$$N_{\text{bolts}} = \frac{309.4 \text{ kips}}{27.1 \text{ kips/bolt}} = 11.4, \text{ say } 12 \text{ bolts}$$

4.3.2 Column-bolted–beam-bolted connections (T stubs)

Bolted T-stub connections were a popular connection in moment-resisting frames before field-welded connections became economical, and still represent the most efficient kind of column-bolted–beam-bolted (CB-BB) connection. The mechanistic model for this type of connection is shown in Fig. 4.4. The important conceptual difference between a CW-BB and a CB-BB is that for T stubs the springs that represent the connection to the column flange have lower strength and stiffness. This is because they represent the flexural deformations that can take place in the flanges of the tee as well as any axial deformation of the bolts to the column flange. Both of these are flexible when compared to the axial stiffness of a weld, which can be considered to be an almost rigid element. In addition for the CB-BB connections, the spring representing the bolts needs to include the prying action, which can significantly increase the force in the bolts at ultimate. Figure 4.11 shows prying action in a very flexible T stub. In this case the flexibility of the flange of the stub results in an additional prying force, Q, at the tip of the stub flange. This force increases the nominal force in the bolts above its nominal pretension value, T.

For the case of the T stub, the springs shown in Fig. 4.4 can have a wide range of strength and stiffnesses, depending primarily on the thickness of the flanges and the location and size of the bolts to the column. The big advantage of this type of connection over a CW-BB one is that these springs can provide a much larger deformation capacity than a weld would. A T-stub connection can thus provide a good balance between strength, stiffness, and ductility.

The design of a T-stub connection essentially follows the same steps as for the CW-BB connections described previously, with important additional design provisions for prying action, bolt tensile elongation capacity, local effects on the column flange, and bolt shear strength. The strength of the connection to the column, taking into account prying action, is limited by

- *The bending strength of the flanges of the T:* This depends primarily on the thickness of the flanges and the exact location of the bolt holes.

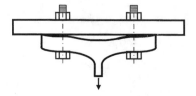

(a) Flexible T-stub deformed due to a force on stem of the T.

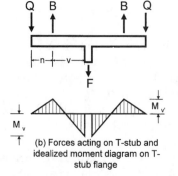

(b) Forces acting on T-stub and idealized moment diagram on T-stub flange

Figure 4.11 Prying action in T stub, showing the case of a flexible flange.

- *The ultimate tensile strength of the stem of the T:* The net area generally governs over the gross area criteria because the width of the stem at the critical section for net area is not too different from that of the critical section for gross area.

- *The tensile strength of the bolts:* This is influenced primarily by the prying action.

- *The shear strength of the bolts:* It is difficult to fit more than 8 to 10 bolts in the stem of a conventional T (cut from a W shape) and thus large bolts may be needed.

Each of these failure modes must be checked individually and the lowest strength taken as the controlling value. Guidelines for these calculations are given in the AISC manual (AISC, 1993), textbooks [see, for example, pp. 848–856 of Salmon and Johnson (1996)], and in the standard references (Kulak, et al., 1987). An excellent review of the design, including some of the numerical problems that can be encountered, is given by Thorton (1985).

The effect of reversed cyclic loading on these connections is to progressively decrease the tension in the bolts to the column flange. Because of prying action, the stress range in these bolts is probably significantly larger than that calculated based on the simplified models used for design. This can result in either low-cycle fatigue failures or in fracture of the bolt due to excessive elongation.

Design Example 4.3. A rigid connection is to be designed to transfer a factored moment of 260.6 kip-ft and a factored shear of 112 kip from a W 21 × 57 beam to the flange of a W 14 × 82 column. The connection consists of tee sections for moment transfer and web angles for shear transfer. All materials are A36 steel. Bolts are to be 1-in A325N bolts.

1. If all bending moment is carried by the tees, the force of the internal couple is

$$F \cong \frac{M_u}{d_b} = \frac{260.6 \times 12}{21.06} = 148.4 \text{ kips}$$

2. Determine the minimum number of bolts, N, required to carry the tensile force to the column flange. Ignore the prying forces for now and check later.

$$\phi R_n = \phi F_t A_b = (0.75)(90)(0.785) = 53.0 \text{ kips/bolt}$$

$$N = \frac{148.4}{53} = 2.8, \text{ say 4 bolts}$$

The bolt capacity is consistent with that given in the AISC *Manual of Steel Construction*, LRFD, vol. II, Table 8-15, p. 8-27. Note that because prying forces can be large in this type of connection, it is best to have a very conservative number of bolts to the column flange. This check is used here mostly to ensure that a reasonable number of bolts are needed (that is, 4 to 8 bolts rather than more which would be hard to accommodate).

3. Determine the number of bolts, M, required to transmit the forces from the tee to the beam flanges through shear (bolts are in single shear):

$$\phi R_n = 0.75\, F_v A_b = 0.75(48)(0.7856) = 28.3 \text{ kips/bolt}$$

This capacity is consistent with that given in the AISC *Manual of Steel Construction,* LRFD, vol. II, Table 8-11, p. 8-24.

Check bearing strength:

$$\phi R_u = \phi\,(2.4)dtF_u = 0.75(2.4)(1)(t)(58) \geq 104.4t$$

Shear will govern if the thickness of the plate is greater than:

$$t = \frac{28.3}{104.4} = 0.27 \text{ in}$$

This is small so bearing probably will not govern; this will be checked later. Thus

$$M \geq \frac{148.4 \text{ kips}}{28.3 \text{ kips}} \geq 5.2 \qquad \text{Use 6 bolts}$$

4. Determine thickness t_w required to transmit tension on the stem of the tee (AISC *Manual of Steel Construction,* LRFD, vol. I, D1):

   a. Assume that the plate width at the critical section is about 9 in (total column flange width is 10.13 in). Then the required tee stem thickness for capacity controlled by gross area yielding is

$$A_g \geq \frac{T_u}{0.90F_y} \geq \frac{148.4}{0.9 \times 36} \geq 4.58 \text{ in}^2 \qquad \text{so } t_w = \frac{4.58}{9.00} = 0.51 \text{ in}$$

   b. Assume that the net area is given by the total width (9.00 in) minus two bolt holes ($2 \times 1.125 = 2.25$) or 6.75 in. Then the required tee stem thickness for capacity governed by net area fracture is

$$A_n = \frac{T_u}{0.75F_u} \geq \frac{148.4}{0.75 \times 58} \geq 3.41 \text{ in}^2 \qquad \text{so } t_w = \frac{3.41}{6.75} = 0.51 \text{ in}$$

Both of them are close enough to say that t_w should be just over 0.5 in.

5. Determine the flange thickness, t_f, for the tee section. This needs to take prying action into account. A simplified mechanism for computing the additional forces due to prying action is shown in Fig. 4.12. The prying forces, Q, arise from the additional forces developed at the end of the T flanges as the T stub is pulled. Assuming that each side of the flange can be modeled as a two-span beam with one end fixed (at the web) and one end free to rotate (edge of T stub), the maximum forces can be calculated based on the formation of plastic hinges at both the web and the edge of the bolt. For details see Salmon and Johnson (1996, pp. 905–909).

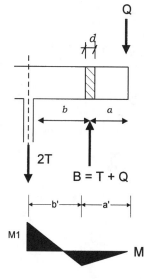

Figure 4.12 Prying action mechanism (Salmon and Johnson, 1995).

From this type of model, equations for the required plate thickness can be derived. One such equation is that proposed by Thornton (1992):

$$t_t \geq \sqrt{\frac{4Tb'}{\phi_b\, wF_y\, (1 + \alpha\delta)}} \qquad \phi_b \geq 0.9$$

where T = force in the bolt, kips
  b' = distance from the web centerline to the inside edge of the bolt, in
  a' = distance from inside edge of the bolt to edge of T stub, in
  w = tributary width of the flange, in
  F_y = yield strength of the T stub, ksi
  α, δ, β = constants as defined in the following

Salmon and Johnson, following Thornton, recommend to compute ß as a function of α and δ, where

$$\beta = \left(\frac{B}{T} - 1\right)\frac{a'}{b'}$$

where $\alpha = M_1/\delta M_2$
  δ = ratio of net area at bolt line to gross area at M_1
  B = maximum bolt resistance, kips

For our example, assuming the gage, g, is about 4 in:

$$b = \frac{g}{2} - \frac{t_w}{2} = 1.75 \text{ in}$$

$$b' = b - \frac{d}{2} = 1.25 \text{ in}$$

$$\beta = \left(\frac{53.0}{37.1} - 1\right)(1.25) = 0.54 \qquad \left(\text{assume } \frac{a'}{b'} = 1.25\right)$$

For purposes of this calculation the force, B, can be taken, as an upper bound, as the bolt capacity in tension (53.0 kips). The force, T, is the part of the total tension force going to each bolt (148.3/4 bolts = 37.1 kips). The value of a' is a guess since we have not chosen a tee yet. Use α as follows:

$$\text{if } \beta \geq 1, \text{ use } \alpha = 1 \rightarrow \text{large prying force}$$

$$\text{if } \beta < 1, \text{ use } \alpha = \text{lesser of } \frac{1}{\delta}\left(\frac{\beta}{1-\beta}\right) \text{ and } 1.0$$

estimate:

$$\delta = \frac{\text{net width at bolt line}}{\text{gross width at critical section near webface}}$$

$$\delta = \frac{4.5 - [1 + (1/16)]}{4.5} = 0.763$$

$$\frac{1}{\delta}\left(\frac{\beta}{1-\beta}\right) = \frac{1}{0.763}\left(\frac{0.54}{1-0.54}\right) = 1.53 \rightarrow x = 1$$

$$t_f \geq \sqrt{\frac{4(148.4/2)1.25}{0.9 \times 9 \times (36) \times (1 + 1.0 \times 0.76)}} = 0.72 \text{ in}$$

Try a WT 12 × 47, $t_f = 0.875$ in, $t_w = 0.515$ in.

6. Check the prying force using the formula proposed by Salmon and Johnson:

$$Q \geq T\left(\frac{\alpha\delta}{1 + \alpha\delta}\right)\left(\frac{b'}{a'}\right)$$

$$a' = a + \frac{d}{2} = \frac{b_f - g}{2} + \frac{d}{2} = \frac{9.065 - 4}{2} + 0.5 = 3.03 \text{ in}$$

$$b' = b - \frac{d}{2} = \frac{g}{2} - \frac{t_w}{2} - \frac{d}{2} = \frac{4}{2} - \frac{0.51}{2} - 0.5 = 1.25 \text{ in}$$

Taking the width of the tee section as 9 in with two holes deducted;

$$\delta \geq \frac{9 - 2(1 + 1/16)}{9} \geq 0.76$$

$$\alpha \geq 1 \rightarrow \alpha\,\delta = 0.763$$

$$Q = T\left(\frac{0.763}{1 + 0.763}\right)\left(\frac{1.25}{2}\right) = 0.27T$$

$$T + Q = 1.27T$$

$$1.27T_u = 1.27 \times 37.1 = 47.1 \leq \phi R_n$$

7. Recheck the thickness, t_f, required

$$t_f \geq \sqrt{\frac{4.T\,b'}{\phi\,WF_{y(1-8)}}} = 0.72 \text{ in}$$

Design is satisfactory; use WT 12 × 47 to carry tensile and compression forces.

8. Design the angles for shear transfer:

$$P_u = 112 \text{ kips}$$

The angles are in single shear so:

$$\phi\,R_n \geq 0.75\,(48)\,1\,(0.7854) \geq 28.3 \text{ kips/bolt}$$

There are six bolts so the capacity is 6 × 28.3 = 169 kips > P_u, so shear does not govern. Bearing will control in the angles and not in the column flange which is much thicker. The minimum angle thickness is therefore:

$$\phi\,R_n \text{ (bearing)} \geq \phi\,(2.4F_u)\,dt \geq 28.3 \text{ kips}$$

$$t_{f,\,\min} \geq \frac{28.3}{(0.75)(2.4)(58)(1)} \geq 0.27 \text{ in} \rightarrow \text{use } 2L\ 4 \times 4 \times \tfrac{5}{16} \times 1 \text{ ft } 0 \text{ in}$$

Number of required bolts:

$$N = \frac{112}{28.3/\text{bolts}} = 3.95 \text{ bolts} \Rightarrow \text{use 4 bolts}$$

Check shear in net section:

$$A_{ns} = t[12 \text{ in} - 5(1 + \tfrac{1}{16})] = 6.6875t$$

$$\text{required } t \geq \frac{112/2}{0.75 \times 0.6 \times 58 \times 6.6575} \geq 0.32 \text{ in} \Rightarrow \text{use } t = \tfrac{3}{8} \text{ in}$$

$$\rightarrow \text{use } 2L\ 4 \times 4 \times \tfrac{3}{8} \times 1 \text{ ft } 0 \text{ in}$$

9. Check if stiffeners in column are required. To avoid stiffeners, the column web must be checked for:
   a. Compression zone:
   (1) K.1.3—local web yielding:

$$P_{bf} = \phi(5k + t_{fb})F_{yc}\,t_{wc} \geq (1.0)\,(5 \times 1.625 + 0.65)\,(36)\,(0.51) = 161.1 \text{ kips}$$

   (2) K1.4—web crippling:

$$P_{bf} = \phi135t_{wc}^2\left[1 + 3\left(\frac{t_{fb}}{d}\right)\left(\frac{t_{wc}}{t_{fc}}\right)^{1.5}\right]\frac{F_{yc}t_{yc}}{t_{wc}}$$

$$= 0.75 \times 135\,(0.51)^2\left[1 + 3\left(\frac{0.65}{14.31}\right)\left(\frac{0.51}{0.855}\right)^{1.5}\right]\frac{36 \cdot 0.55}{0.51}$$

1086 kips > 174.4 kips o.k.

(3) K1.5—compression buckling of the web:

$$R_n = \frac{4100tw^3 \sqrt{F_{yc}}}{dc} = \frac{4100 \,(0.51)^3 \sqrt{36}}{11} = 267 \text{ kips}$$

Thus, none of the capacities are exceeded and no stiffeners are required!

b. Tension zone:

(1) K1.2—local flange bending:

$$P_{bf} = \phi_b \, 6.25 \, t_{fc}^2 \, F_{yc} = 0.9 \times 6.25 \,(0.855)^2 \times 36 = 148.0 \text{ kips}$$

Thus, P_{bf} equals the required strength and no tension stiffener are required!

Design Example 4.4. Design an interior connection between W 24 × 94 A36 beams and a W 14 × 283 A572/50 column. Assume all bolts are 1-in A490X or 1⅛-in A325. The connection consists of T stubs for moment transfer and web angles for shear transfer, and is to be designed for seismic forces.

1. Check beam compactness criteria:

   Flange:

   $$\frac{52}{\sqrt{F_y}} = 8.5 > 7.8, \text{ ok}$$

   Web:

   $$\frac{253}{\sqrt{F_y}} = 42.1 > 41.9, \text{ ok}$$

2. If all bending moment is carried by the tees and assuming that the beam can develop its plastic moment capacity:

   $$F \cong \frac{M_u}{d_b} = \frac{(1.5)(254 \text{ in}^3)(36 \text{ ksi})}{24.31 - 0.875} = 585.3 \text{ kips}$$

   Note that an R_y of 1.5 was assumed for this calculation since the beam is A36 steel. The actual flange force, as before, is dependent on the effective plastic capacity of the beam. In this case the ratio of the effective to total plastic capacity is 0.74 so that the actual design force for the beam flange is:

   $$F_{\text{flange}} = 585.3 \times 0.74 = 433.1 \text{ kips}$$

3. Determine the minimum number of bolts, N, required to carry the tensile force to the column flange. Assume that prying forces will be large and will use about 20% of the tensile capacity. Assuming 1-in A490X bolts:

   $$N = \frac{433.1 \,(1.2)}{66.6} = 7.8, \text{ say 8 bolts}$$

Because prying forces can be large in this type of connection, it is best to have a conservative number of bolts to the column flange. It will be difficult, however, to place more than the eight bolts required by this calculation. This is because the bolts to the column should be in multiples of four, to fit two in each side of the column web. Most tees will be able to accommodate only one row of bolts, and if this calculation indicates that more than eight bolts are needed, it will be necessary to carefully assess the practicality of the connection type. In addition, the design assumed A490 bolts which are generally assumed to be less ductile than A325. If possible, therefore 1⅛-in A325 bolts, with a capacity of 67.1 kips/bolt, should be used.

4. Determine the number of bolts, M, required to transmit the forces from the tee to the beam flanges through shear (bolts are in single shear). The shear capacity of the 1⅛-in A325X bolts is 44.7 kips. First, check the bearing strength assuming the T stub has a stem thickness equal to t:

$$\phi R_u = \phi(2.4)dtf_u R_u = 0.75(2.4)(1)(t)(58 \times 1.1) \geq 114.8t$$

Shear will govern if the thickness of the T stub is greater than:

$$t_w \text{ (stem of T stub)} = \frac{44.7 \text{ kips}}{114.8} = 0.39 \text{ in}$$

This is small so bearing probably will not govern; this will be checked later. Thus

$$M \geq \frac{433.1 \text{ kips}}{44.7 \text{ kips}} \geq 9.7 \qquad \text{use 10 bolts}$$

5. Check beam flange capacities:

For gross area:

$$R_y A_g F_y = 1.5 \, (9.065 \times 0.875) \, (36) = 428.3 \text{ kips}$$

For effective area:

$$\phi \, A_e F_u R_u = 0.75 \, [9.065 - 2(1.25)] \, 0.875 \, (58 \times 1.1) = 274.3 \text{ kips}$$

The value of R_u, which is the equivalent of R_y but reflects the increase at ultimate, was taken as 1.1 since it is not expected that the increase in ultimate strength for A36 steel will be as large as that for the yield strength. The beam-flange thickness will need to be augmented with a plate. The plates should be ⅜ in thick by 4.5 in wide on each side of the web to bring the net area resistance to 469.3 kips.

6. Determine thickness t_w required to transmit tension on the stem of the tee:
   a. The connection is to a heavy W 14, so assume that the T stub will be 16 in at the column face and taper to 9 in at its end. The width of the beam flange is just over 9 in so this configuration is reasonable. With this taper, and using five rows of 1⅛-in bolts, the minimum length of the stem is

$l_s = 4$ (minimum bolt spacing) + 2 (minimum edge distance)
+ clearance + flange thickness

$l_s = 4\,(3 \times 1.125) + 2\,(1.5) + 0.5 + 1.5 = 18.5$ in will be needed

The required tee stem thickness for capacity controlled by gross area yielding, assuming the thickness at the critical section is 12.9 in, is

$$A_g \geq \frac{T_u}{R_y F_y} \geq \frac{428.5}{1.5 \times 36} \geq 7.93 \text{ in}^2 \qquad \text{so } t_w = \frac{7.93}{12.9} = 0.62 \text{ in}$$

b. Assume that the net area is given by the total width at the location of the first line of bolts (12.5 in) minus two bolt holes ($2 \times 1.25 = 2.5$) or 10 in. Then the required tee stem thickness for capacity governed by net area fracture is

$$A_n = \frac{T_u}{R_u F_u} \geq \frac{469.3}{1.1 \times 58} \geq 7.33 \text{ in}^2 \qquad \text{so } t_w = \frac{7.33}{10} = 0.73 \text{ in}$$

These checks are intended to ensure that yielding on the gross area will occur before fracture in the net area. The preceding calculations indicate that it is possible to obtain a good balance between yielding and fracture with a tee stem thickness of around ¾ in.

7. Determine the flange thickness, t_f, for the tee section following the same procedure as in Example 4.3. The minimum gage, g, that can be used is about 4 in:

$$b = \frac{g}{2} - \frac{t_w}{2} \approx 1.75 \text{ in}$$

$$b' = b - \frac{d}{2} = 1.25 \text{ in}$$

Then

$$\text{ß} = \left(\frac{B}{T} - 1\right)\frac{a'}{b'} = \left(\frac{53.0}{43.0} - 1\right)(1.25) = 0.29$$

Estimate:

$$\delta = \frac{\text{net width at bolt line}}{\text{gross width at critical section near web face}}$$

$$\delta = 16 - \frac{(4 \times 1.125)}{16} = 0.719$$

$$\frac{1}{\delta}\left(\frac{\text{ß}}{1 - \text{ß}}\right) = \frac{1}{0.719}\left(\frac{0.29}{1 - 0.29}\right) = 0.57 \rightarrow \alpha = 0.57$$

$$t_f \geq \sqrt{\frac{4(433.1/2)1.75}{16\,(1.5 \times 36)\,(1 + 0.57 \times 0.719)}} = 1.43 \text{ in}$$

A quick check of this value can be made from Table 11-1 in the AISC *Manual of Steel Construction,* LRFD, vol. II. For a thickness of 1.25 in and a *b* of 2.25 in, the table gives 22.5 kips/lin in. In this case we have about 16 in which gives a total force of 360 kips. This is very close to the 433.1 kips if we multiply it by the ϕ factor of 0.9 and the R_y of 1.5, so the design appears reasonable.

Try a WT 20 × 1056:

$$t_w = 0.75 \text{ in}$$

which is slightly greater than 0.73 in required for net area failure in tension

$$t_f = 1.41 \text{ in}$$

which is slightly less than the 1.43 in required

The flanges of the tee must be cut to meet the required distances. For this size WT the minimum gage will be 4.25 in, so the assumed values of *a* and *b* can be achieved. The total width of the flange should be about $[(2 \times 2.25 \times 1.75) + t_w]$ or 8.6 in. Thus we need to trim about 3.5 on each flange. Note that several design trials assuming that the flanges were not trimmed did not produce satisfactory results. This is due to the large increase in prying forces and required flange thickness if the values of *a* and *b* are not kept as low as possible. Note that this reduction in flange width is also required so that an angle can be fit for carrying the shear forces in the beam web.

8. Check the prying force using the formula proposed by Salmon and Johnson:

$$Q \geq T \left(\frac{\alpha \, \delta}{1 + \alpha \, \delta} \right) \left(\frac{b'}{a'} \right)$$

$$Q = T \left[\frac{0.57 \times 0.719}{1 + (0.57 \times 0.719)} \right] \frac{1.40}{2.54} = 0.16T$$

$$T + Q = 1.16 \, T$$

$$1.16 \, T_u = 1.16 \times 43.0 = 49.9 \text{ kips} \leq \phi R_n$$

The entire design must now be rechecked (not shown).

9. Design the angles for shear transfer for either the maximum shear from lateral loads (assume beam length = 28 ft):

$$V_u = 1.25 \, \frac{2M_n}{L} = 78.4 \text{ kips}$$

or the maximum reaction (from the load tables, p. 4-44 of the AISC *Manual of Steel Construction,* LRFD, vol. II):

$$V_u = 256 \text{ kips (controls)}$$

Note that this latter number is probably too conservative since it is unlikely that the gravity loads will govern the shear design for this

beam. However, since this is the worst case it will be used in these calculations. From clearing considerations, the available length for the
angle is about 16.4 in (24.31 + 0.71 + −8.6). Assuming 3-in gages, this
implies that only five rows of bolts can be used. From p. 9-58 of the
LRFD manual, five rows of ⅞-in-diameter A325N bolts with a ½ angle
will give a capacity of 248 kips (close enough). The beam web capacity
is 457 × 0.515 = 235 kips which is also close enough.

→ Use 2L 4 × 4 × ½ × 1 ft 4 in with 5⅞-in A325X bolts

10. Check for moment capacity of the web angles (must transfer at least
    30% of M_p). Provide a fillet weld around the angles to transfer this
    additional moment. Alternatively, and probably better, increase the
    angle thickness to ½ in and number of the bolts to 7 rows of A490X, 1 in
    diameter.
11. Check service level forces. Design resistance at service = 16.9 kips/bolt
    (AISC *Manual of Steel Construction,* LRFD, vol. II, p. 8-29)

$$F_{service} = 169.0 \text{ kips}$$

Check that this is greater than 1.25 $M_{service}$ (not specified for this
problem) and less *b* than 0.8 M_p (= 220 kips, ok)
12. Check panel zone shear:

$$V_u \geq \frac{0.9\Sigma\phi M_p}{d_c} = 884.9 \text{ kips}$$

The ultimate shear capacity is

$$V_u = (0.75)(0.6)(36)(16.74)(1.29)(1.24) = 435 \text{ kips}$$

Doubler plates are needed. Use a ⅝-in plate on each side of the web.
13. Continuity plates:

$$P_{bf} = \phi_b\, 6.25\, t_{fc}^2\, F_{yc} = 0.9 \times 6.25\, (2.07)^2\, 36 = 867.7 \text{ kips}$$

Check versus Sec. 8.5 of AISC (1992):

$$P_{required} = 1.8\, F_{yb} b_f t_{bf} = 1.8 \times 36 \times 16.11 \times 0.71 = 741 \text{ kips}$$

No continuity plates are required.
14. Check if stiffeners in column are required for gravity loads. To avoid
    stiffeners, the column web must be checked for:
    a. Compression zone:
       (1) K.1.3—local web yielding:

$$P_{bf} = \phi(5k + t_{fb})F_{yc} t_{wc} \geq (1.0) (5 \times 2.75 + 0.875)(36)(1.29)$$
$$= 679.2 \text{ kips}$$

(2) K1.4—web crippling:

$$P_{bf} = \phi 135 t_{wc}^2 \left[1 + 3\left(\frac{t_{fb}}{d}\right)\left(\frac{t_{wc}}{t_{fc}}\right)^{1.5} \right] \sqrt{\frac{f_{yc} t_{yc}}{t_{wc}}}$$

$$= 0.75 \times 135 \,(1.29)^2 \left[1 + 3\left(\frac{0.875}{16.74}\right)\left(\frac{1.29}{2.07}\right)^{1.5} \right] \sqrt{\frac{36 \bullet 0.875}{1.29}}$$

$$= 897 \text{ kips}$$

(3) K1.5—compression buckling of the web:

$$R_n = \frac{4100 \, t_w^3 \, \sqrt{F_{yc}}}{d_c} = \frac{4100 \,(1.29)^3 \, \sqrt{36}}{16.74 - (2 \times 1.25)} = 3708 \text{ kips}$$

Thus none of the capacities are exceeded and no stiffeners are required!
b. Tension zone:
(1) K1.2—local flange bending:

$$P_{bf} = \phi_b \, 6.25 \, t_{fc}^2 \, F_{yc} = 0.9 \times 6.25 \,(2.07)^2 \, 36 = 867.7 \text{ kips}$$

Thus, P_{bf} equals the required strength and no tension stiffeners are required! The final design for Example 4.4 is shown in Fig. 4.13.

4.3.3 End-plate connections

End-plate connections are common in some areas of the country and very popular in prefabricated metal buildings. The mechanistic behavior of an end-plate connection is very similar to that of a T stub, with the difference being that the size of the plate is longer than that of the flange of a T stub. If the plate is thin or of moderate thickness compared to the column flange, yield lines will form between the holes in the plate resulting in a plastic mechanism. Because the pattern of yield lines can be complex, the computation of the strength of the plate is not as simple as for a T stub. In the latter case, only two yield lines occur on each half of the stub, one at the bolts and one at the intersection of the flange and web (Fig. 4.12). Two typical yield-line patterns for some common end-plate configurations are shown in Fig. 4.14 (Murray and Meng, 1996; Murray and Watson, 1996). The group patterns can be very complex and not easy to determine for cases with multiple bolts in one row. Yield lines around each individual bolt, in addition to the group patterns shown in Fig. 4.12, are also possible. If the end plate is thick, the behavior will shift to that of a thick T stub. In this case the failure will be either by tension in the bolts to the column or bolt shear in the connection to the beam. In all cases, care should be exercised in not overstressing the column flanges. The

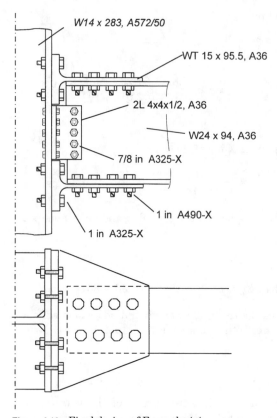

Figure 4.13 Final design of Example 4.4.

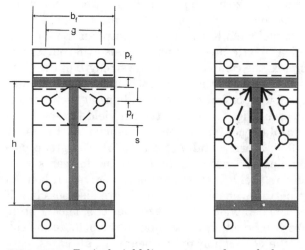

Figure 4.14 Typical yield-line patterns for end plates (Murray and Meng, 1996 and Murray and Watson, 1996).

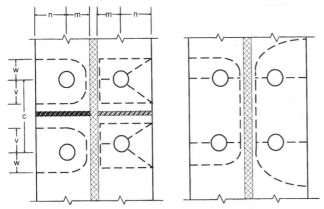

Figure 4.15 Typical yield-line patterns for column flanges with and without stiffeners (Nader and Astaneh-Asl, 1992).

strength of the column flange can be checked by a yield-line approach (Nader and Astaneh, 1992), just as for the plate itself (Fig. 4.15).

An excellent review of the development of end-plate connections is given by Griffiths (1984), and detailed design guidelines and design aids for their design under monotonic loading are available (Murray, 1990). Recently Murray and Meng (1996) have suggested a direct formula for calculating the thickness, t_p, of an end plate for a four bolt unstiffened end plate:

$$t_p = \left\{ \frac{M_u/F_{py}}{((b_f/2)[(1/p_f) + (1/s)] + (p_f + s)(2/g))(h - t_f - p_f) + (b_f/2)[(h/p_f) + (1/2)]} \right\}^{1/2}$$

$$s = \sqrt{\frac{b_f g}{2}}$$

where F_{py} is the yield stress of the end-plate material and t_p is the thickness of the plate, b is the width, p_f is the distance from the beam flange to the bolt centerline, s is the distance to the last yield line, and the subscripts b and f refer to the plate and flange, respectively.

Once the plate thickness has been selected, the actual capacity of the connection can then be calculated as

$$M_u = \left\{ \left[\frac{b_f}{2}\left(\frac{1}{p_f} + \frac{1}{s}\right) + (p_f + s)\left(\frac{2}{g}\right) \right](h - t_f - p_f) + \frac{b_f}{2}\left(\frac{h}{p_f} + \frac{1}{2}\right) \right\} F_{py} t_p^2$$

Once this computation is made, it must be checked against the maximum capacity of the bolts. The latter is governed by prying action and can be computed based on the techniques discussed in Examples 4.3 and 4.4, or by the flowcharts from Murray shown as Figs. 4.16 through 4.18. The bolt forces must be checked separately

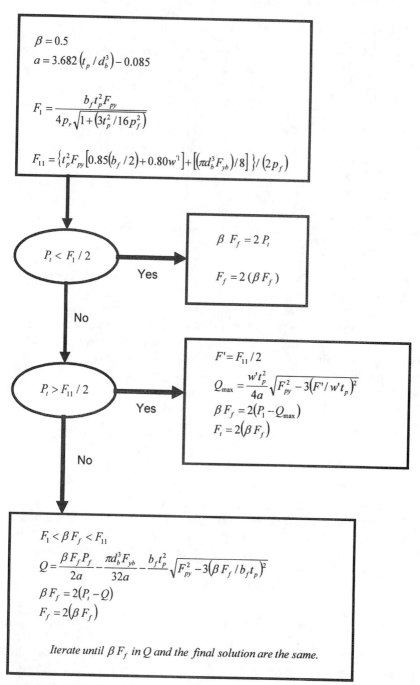

Figure 4.16 Flowchart for determining flange force for inner bolt (Murray and Meng, 1996).

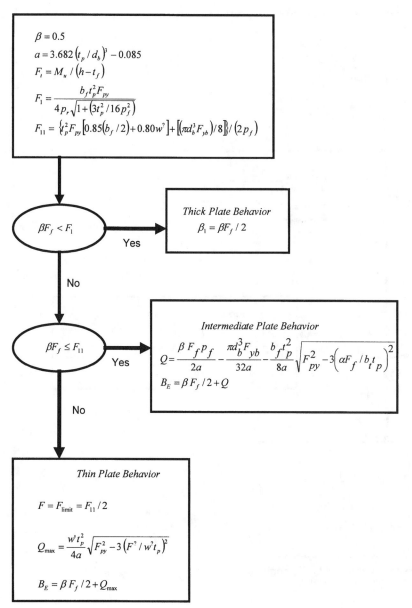

$$\beta = 0.5$$

$$a = 3.682\left(t_p / d_b\right)^3 - 0.085$$

$$F_t = M_u / \left(h - t_f\right)$$

$$F_1 = \frac{b_f t_p^2 F_{py}}{4 p_r \sqrt{1 + \left(3 t_p^2 / 16 p_f^2\right)}}$$

$$F_{11} = \left\{ t_p^2 F_{py}\left[0.85\left(b_f / 2\right) + 0.80 w'\right] + \left[\left(\pi d_b^3 F_{yb}\right) / 8\right]\right\} / \left(2 p_f\right)$$

$\beta F_f < F_1$ → **Yes** →

Thick Plate Behavior
$$\beta_1 = \beta F_f / 2$$

No

$\beta F_f \le F_{11}$ → **Yes** →

Intermediate Plate Behavior
$$Q = \frac{\beta F_f p_f}{2a} - \frac{\pi d_b^3 F_{yb}}{32a} - \frac{b_f t_p^2}{8a}\sqrt{F_{py}^2 - 3\left(\alpha F_f / b_f t_p\right)^2}$$
$$B_E = \beta F_f / 2 + Q$$

No

Thin Plate Behavior

$$F = F_{\text{limit}} = F_{11} / 2$$

$$Q_{\max} = \frac{w' t_p^2}{4a}\sqrt{F_{py}^2 - 3\left(F'' / w'' t_p\right)^2}$$

$$B_E = \beta F_f / 2 + Q_{\max}$$

Figure 4.17 Flowchart for determining inner bolt force (Murray and Watson, 1996).

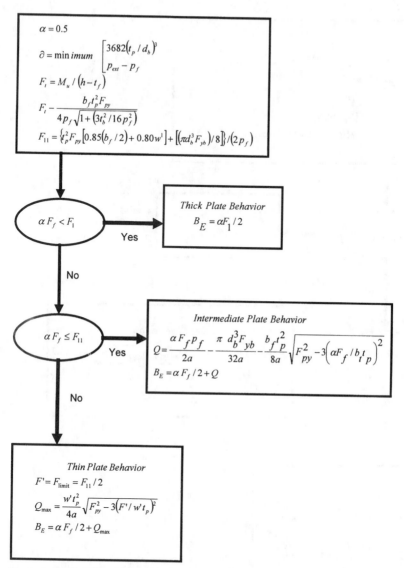

$$\alpha = 0.5$$

$$\partial = \min\,imum \begin{bmatrix} 3682(t_p / d_b)^3 \\ p_{ext} - p_f \end{bmatrix}$$

$$F_t = M_u / (h - t_f)$$

$$F_t - \frac{b_f t_p^2 F_{py}}{4 p_f \sqrt{1 + (3 t_b^2 / 16 p_f^2)}}$$

$$F_{11} = \left\{ t_p^2 F_{py} \left[0.85 (b_f / 2) + 0.80 w^1 \right] + \left[(\pi d_b^3 F_{yb}) / 8 \right] \right\} / (2 p_f)$$

$\alpha F_f < F_1$ — Yes →

Thick Plate Behavior
$$B_E = \alpha F_1 / 2$$

No

$\alpha F_f \le F_{11}$ — Yes →

Intermediate Plate Behavior
$$Q = \frac{\alpha F_f P_f}{2a} - \frac{\pi\, d_b^3 F_{yb}}{32a} - \frac{b_f t_p^2}{8a} \sqrt{F_{py}^2 - 3\left(\alpha F_f / b_f t_p\right)^2}$$
$$B_E = \alpha F_f / 2 + Q$$

No

Thin Plate Behavior
$$F' = F_{limit} = F_{11} / 2$$
$$Q_{max} = \frac{w' t_p^2}{4a} \sqrt{F_{py}^2 - 3\left(F' / w' t_p\right)^2}$$
$$B_E = \alpha F_f / 2 + Q_{max}$$

Figure 4.18 Flowchart for determining outer bolt force including prying action (Murray and Watson, 1996).

for the interior and exterior row of bolts and the prying action added to the most critical case. Note also that the equations are based on an assumed yield-line pattern. Different yield lines have been treated by other authors. The reader is referred to Murray and Kurete (1988), Murray (1990), Murray and Abel (1992), and Astaneh-Asl et al. (1991) for additional details.

Design Example 4.5. Determine the required end-plate thickness and bolt size for a four-bolt extended unstiffened moment end-plate connection. Use A572 grade 50 for both the beam and end plate, and A325 for the bolts. The factored design moment is 225 kip-ft. See Fig. 4.19 for details of the beam and end plate sizes.*

1. Calculate s and the required end-plate thickness t_p. Using the equations above and the dimensions in Fig. 4.18:

$$s = \frac{1}{2}\sqrt{b_t g} = \frac{1}{2}\sqrt{8(3.25)} = 2.55 \text{ in}$$

$$t_p = \sqrt{\frac{225\,(12)/50}{(8/2)[(1/1.625)+(1/2.55)]+[1.625+2.55)(2/3.25)](24-2.125)} + (8/2)[(1/2)+(24/1.625)]}$$

$$t_p = 0.51 \text{ in} \qquad \text{Use } t_p = \tfrac{5}{8} \text{ in}$$

2. Determine the critical moment, M_{crit}, as the smallest of the moment capacities of the end plate, M_{plate}, due to the formation of yield lines and failure of the bolts, M_{bolt}, due to prying action. The moment capacity governed by the end plate, M_{plate}, is calculated as follows:

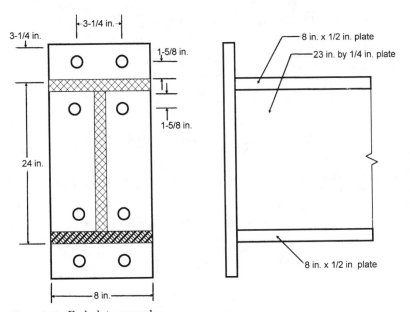

Figure 4.19 End-plate example.

*This example is from Murray and Abel (1992), with corrections in 1994 and 1995. It was kindly provided by Dr. T. M. Murray of the Virginia Polytechnic Institute.

$$M_{\text{plate}} = \frac{F_{py}t_p^2}{12}\left\{\left[\frac{b_f}{2}\left(\frac{1}{p_f} + \frac{1}{s}\right) + (p_f + s)\left(\frac{2}{g}\right)\right](h - p_t) + \frac{b_f}{2}\left(\frac{1}{2} + \frac{h}{p_f}\right)\right\}$$

$$= \frac{50(0.625)^2}{12}\left\{\left[\frac{8}{2}\left(\frac{1}{1.625} + \frac{1}{2.55}\right) + (1.625 + 2.55)\left(\frac{2}{3.25}\right)\right](24 - 2.125)\right.$$

$$\left. + \frac{8}{2}\left(\frac{1}{2} + \frac{24}{1.625}\right)\right\}$$

$$= 334.4 \text{ kips-ft}$$

To compute the capacity of the connection based on the bolts, a bolt trial size must be chosen. Assume ¾-in A325 bolts. The force yield capacity, P_t, of each pair of bolts, based on $F_y = 44$ ksi for the bolt material, is

$$P_t = 2 \times 44 \text{ ksi} \times 0.4418 \text{ in}^2 = 38.9 \text{ kips}$$

From the flowchart given in Fig. 4.16, determine the force in the inner bolts:

$$F_1 = \frac{b_f t_p^2 F_{py}}{4\,p_f\sqrt{1 + (3t_p^2/16\,p_f^2)}}$$

$$= \frac{8\,(0.625^2)50}{4(1.625)\sqrt{1 + [3(0.625^2)/(16 \times 1.625^2)]}} = 23.7 \text{ kips}$$

$$w^1 = \frac{b_f}{2} - (d_b + \tfrac{1}{16}) = \% - (0.75 + \tfrac{1}{16}) + 3.1875 \text{ in}$$

$$F_{11} = \frac{t_p^2\,F_{py}\,[0.85(b_f/2) + 0.80w] + [(\pi d_b^3 F_{yb})/8]}{2p_f}$$

$$= \frac{0.625^2\,(50)[0.85(8/2) + (0.80)3.1875] + \{[\pi 0.75^3(81)/8]\}}{2(1.625)} = 39.9 \text{ kips}$$

3. Determine the force in the flange, F_f. Since $P_t > F_{11}/2$, we have thin-plate behavior. Therefore:

$$a = 3.682\left(\frac{t_p}{d_b}\right)^3 - 0.085 = 3.682\left(\frac{0.625}{0.75}\right)^3 - 0.085 = 2.05 \text{ in}$$

$$F' = \frac{F_{11}}{2} = 20.0 \text{ kips}$$

$$Q_{\text{max}} = \frac{w't_p^2}{4a}\sqrt{F_{py}^2 - 3\left(\frac{F'}{w't_p}\right)^2} = \frac{3.1875\,(0.625)^2}{4\,(2.05)}$$

$$\times 50^2 - \frac{320.0}{3.1875 \times (0.625)^2}$$

$$= 7.12 \text{ kips}$$

$$F_t = \frac{M_u}{h - t_f} = \frac{225(12)}{24} - 0.5 = 114.9 \text{ kips}$$

$$\text{ß}F_f = 2 \, (P_t - Q_{max}) = 2(38.9 - 7.12) = 63.6 \text{ kips}$$

$$F_f = 2\text{ß}F_f = 2(63.6) = 127.2 \text{ kips}$$

$$M_{bolt} = F_f(h - t_f) = \frac{127.2}{12} \, (24 - 0.5) = 249.1 \text{ kip-ft}$$

$$M_{crit} = M_{bolt} = 249.1 \text{ kip-ft}$$

since

$$M_{plate} = 334.4 \text{ kip-ft} > M_{bolt}$$

4. Determine the inner end-plate behavior. From Fig. 4.17 and the values given previously for a, F_t, F_1, and F_{11}, $\text{ß}F_t = 0.5$ (114.9) = 57.5 kips $> F_{11} =$ 39.9 kips, inner end-plate behavior is thin-plate behavior. From the flowchart in Fig. 4.17, the inner bolt force is

$$B_f = \frac{\text{ß}F_t}{2} + Q_{max} = \frac{0.5 \, (114.9)}{2} + 7.12 = 35.8 \text{ kips}$$

5. Check the outer bolt force. From Fig. 4.18, the outer bolt force is equal to $\alpha F_f/2 = 28.7$ kips, so the inner bolts control at $B_f = 35.8$ kips.
6. Checking the bolt diameter:

$$d_b = \sqrt{\frac{2B_{max}}{\pi F_1}}$$

$$= \sqrt{\frac{2(35.8)}{\pi(44)}} = 0.72 \text{ in} \rightarrow \text{use } \tfrac{3}{4}\text{-in-diameter bolt as assumed}$$

While many models of end-plate behavior exist, there is relatively little work on the design of end plates for cyclic loads (Whittaker and Walpole, 1982; Tsai and Popov, 1988, 1990; Ghobarah et al., 1992; Nader and Astaneh-Asl, 1992). Nader and Astaneh-Asl (1992) reviewed the available data and proposed design provisions. They listed plastic yield-line formation in the end-plate and column flange bending as the most desirable failure modes.

For developing design provisions the end plate can be separated into two T stubs (Packer and Morris, 1977), which results in a very similar approach to design to that developed in the previous section. The design forces can be calculated from free-body diagrams such as those shown in Fig. 4.1a and 4.12. Replacing a with n and b with v in Fig. 4.12 to follow the nomenclature in (Astaneh-Asl, 1995) and using appropriate resistance ϕ_b and material overstrength factors α to satisfy capacity design criteria, equilibrium of forces between the force in the plate F_{ep}, and the force in the beam flange, F_{fb}, gives

$$\phi_b \, F_{ep} = \mu \phi_b F_{fb}$$

$$\phi_b \, \frac{(2)(M_v + M_{v''})}{v} = \mu \, \phi_b \, \frac{M_{pb}}{(d_b - t_{fb})}$$

$$\phi_b \, \frac{(2)(M_v + M_{v'})}{v} = \frac{M_{yc}}{(d_b - t_{fb})}$$

$$M_v = \frac{b_p t_p^2}{4} f_y \quad \text{and} \quad M_{v'} = \frac{[(b_p - (N/2)D')] \, t_p^2}{4} f_y$$

where α is the ratio of the connection yield moment to the plastic moment capacity of the beam, ϕ_b is the resistance factor for the bolts, F_{ep} is the force in the flange corresponding to yielding of the end plate, F_{fb} is the axial force in the beam flange corresponding to the plastic capacity of the beam, b_p is the width of the end plate, t_p is the thickness of the end plate, N is the number of bolts, and D' is the diameter of the bolt holes.

For end plates, it is recommended that two yield-line patterns be checked on the column flange (Fig. 4.15). One consists mostly of straight lines and the other incorporates curved ones (Packer and Morris, 1977; Astaneh-Asl and Nader, 1992).

$$F_{f1} = t_{fc}^2 f_y \left\{ \frac{n}{v} + \frac{n}{m} + (n - 0.5D')\left(\frac{1}{v} + \frac{1}{m}\right) + \pi + \pi \sec^2\left[\tan^{-1}\left(\frac{2}{\pi} \ln \frac{v}{m}\right)\right] \right\}$$

$$F_{f2} = t_{fc}^2 f_y \left[\left(\frac{1}{v} + \frac{1}{w}\right)(2m + 2n - D') + \frac{(2v + 2w + -D')}{m}\right]$$

$$w = \sqrt{[m(m + n - 0.5D')]}$$

To ensure that no out-of-plane bending occurs (Astaneh-Asl and Nader, 1992):

$$\phi_b F_{fb} = \phi_b \, \frac{M_{pb}}{(d - t_{fb})} \leq \left\{ \begin{matrix} \phi_b \, F_{f1} \\ \phi_b \, F_{f2} \end{matrix} \right.$$

$$\begin{cases} \text{if } \dfrac{b_p}{b_{fc}} = 1.0 & \text{then } t_{fc} \geq t_p \\[2ex] \text{if } \dfrac{b_p}{b_{fc}} = 0.5 & \text{then } t_{fc} \geq 0.65 \, t_p \end{cases}$$

where M_{pb} is the plastic capacity of the beam, t_{fb} is the thickness of the beam flange, d is the distance between flange centerlines, and b_{fc} is the width of the column flange. Interpolation is permitted between b_p/b_{fc} values of 1.0 and 0.5.

4.3.4 Flexible PR connections

The connections described in the previous sections all fall in the category of full-strength, partially restrained connections. With respect to stiffness, these connections have high initial stiffness and can probably be analyzed as rigid connections for service loads. There are a number of other common steel connections, primarily the top-and-seat angle, with and without stiffeners, that offer partial-strength, partial-restraint behavior. Design examples for this type of connection are available in the literature [see pp. 9-253 to 9-261 of the *Manual of Steel Construction*, LRFD (AISC, 1993), for example] and will not be covered here. In most cases these connections cannot provide sufficient lateral stiffness to resist large wind or earthquake loads unless all the connections in the structure are of this type and the effect of the slab is taken into account (Fig. 4.20) or the angles are stiffened (Fig. 4.21). For the design of this type of PR composite connections, shown in Fig. 4.20, the reader is referred to Chap. 23 of Chen (1997).

4.4   Considerations for analysis of PR frames*

The common practice for analysis of multistory frames assumes that joints are rigid and beam and columns intersect at their centerline. Using this method, there is no allowance for connection and panel zone flexibility, the spans of the beam and columns are overestimated, and the joints have no physical size and are reduced to a point. Since the PR behavior of most connections was recognized early, several modifications have been proposed to classical linear analysis techniques to account for connection flexibility. The first attempts involved modifying the slope-deflection method by adding the effect of linear rotational springs at beam ends (Batho and Rowan, 1934; Rathbun, 1936). Johnston and Mount (1941) gave a complete listing of coefficients to be used in the slope-deflection method including both the flexibility of the connections and the finite widths of the members. These methods were for hand calculations, and thus were limited to the analysis of relatively small structures. In an exception to this, Sourochnikoff (Sourochnikoff, 1950) used the beam-line method along with experimental results obtained by Rathbun (1936) to compute the nonlinear cyclic response of a one-story one-bay partially restrained frame. Monforton and Wu (1963) incorporated linear connection flexibility into the computerized direct stiffness method. This development permitted the analysis of large structures, but was still limited to linear analysis where the connections have constant stiffness. Lionberger and Weaver (1969) published the results from a program that per-

---

*This section is reproduced from Chap. 5 of Background Reports: "Metallurgy, Fracture Mechanics, Welding, Moment Connections and Frame System Behavior," FEMA-288, FEMA, Washington, DC, March 1997.

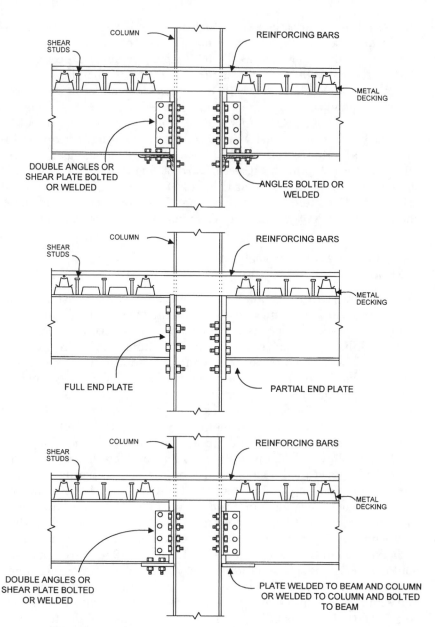

Figure 4.20 Flexible composite PR connections.

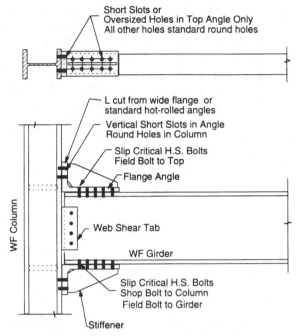

Short Slots or
Oversized Holes in Top Angle Only
All other holes standard round holes

L cut from wide flange or
standard hot-rolled angles

Vertical Short Slots in Angle
Round Holes in Column

Slip Critical H.S. Bolts
Field Bolt to Top

Flange Angle

WF Column

Web Shear Tab

WF Girder

Slip Critical H.S. Bolts
Shop Bolt to Column
Field Bolt to Girder

Stiffener

Figure 4.21 Stiffened seat connection (Astaneh-Asl, 1995).

formed fully dynamic lateral load analysis on plane frames. The connections in their program were modeled by a nondegrading bilinear
model, which included the sizes of the rigid panel zones. Moncarz and
Gerstle (1981) used a nondegrading trilinear model to analyze steel
partially restrained frames subjected to lateral load reversals. When
the first databases for connections were developed (Frye and Morris,
1975; Ang and Morris, 1984; Nethercot, 1985; Kishi and Chen, 1986),
nonlinear expressions for moment-rotation curves became widely
available. This led to the development of numerous computer programs that modelled the nonlinear behavior of PR connections. Shin
(1991) and Shin and Leon (1992) devised hysteresis rules for nonsym-
metrical composite connections with degradation of the unloading
stiffness based on the maximum attained rotation, and implemented
them in a dynamic nonlinear plane-frame analysis program.

The dynamic performance of frames incorporating partially
restrained composite connections has been studied by Astaneh-Asl,
Nader, and Harriot (1991), Nader and Astaneh-Asl (1992), and Leon
and Shin (1995). The numerical results obtained by Leon and Shin
(1995) using a modified trilinear degrading model showed excellent
agreement with test results from a two-story, two-bay, half-scale frame.
The analytical studies for PR frames showed good seismic performance

for ground motions expected in zones of low to moderate seismicity. In particular, they showed less problems with local buckling of members and equal or better energy dissipation capacity than rigid frames. In addition, these studies showed that the lateral drifts of PR frames were within ± 20% of those of companion rigid frames when four-, six-, and eight-story frames were subjected to the El Centro, Parkfield, and Pacoima ground motions. The results of these studies confirmed those of Nader and Astaneh-Asl (1992), and verified their shake table results. Further verification of the good performance can be found in the work of Osman et al. (1993) who presented the analysis results for eight-story frames with end-plate connections and flexible panel zones of various thickness.

4.5   References

Ackroyd, M. H., and Gerstle, K., "Behavior of Type 2 Steel Frames," *Journal of the Structural Division*, ASCE, vol. 108, no. ST7, July 1982, pp. 1541–1558.

AISC, *Specification for the Design, Fabrication, and Erection of Structural Steel for Buildings*, 5th ed., revised (June 1949), American Institute of Steel Construction, New York, 1947.

AISC, *Specification for the Design, Fabrication, and Erection of Structural Steel for Buildings*, 8th ed., American Institute of Steel Construction, Chicago, 1978.

AISC, *Load and Resistance Factor Design Specification for Structural Steel Buildings*, 1st ed., American Institute of Steel Construction, Chicago, 1986.

AISC, *Load and Resistance Factor Design*, 2d ed., vol. II, American Institute of Steel Construction, Chicago, 1998.

AISC, *Seismic Provisions for Structural Steel Building, Load and Resistance Factor Design*, American Institute of Steel Construction, Chicago, 1992 and 1997.

AISC, *Load and Resistance Factor Design Specification for Structural Steel Buildings*, 2d ed., American Institute of Steel Construction, Chicago, 1993.

Ang, K. M., and Morris, G. A., "Analysis of Three-Dimensional Frames with Flexible Beam-Column Connections," *Canadian Journal of Civil Engineering*, vol. 11, no. 2, 1984, pp. 245–254.

ASCE, *Effective Length and Notional Load Approaches for Assessing Frame Stability: Implications for American Steel Design*, 1st ed., American Society of Civil Engineers, New York, 1977a.

ASCE, *Minimum Design Loads for Buildings and Other Structures*, American Society of Civil Engineers, Washington, D.C., 1977b.

Astaneh-Asl, A., *Seismic Design of Bolted Steel Moment-Resisting Frames, Steel Tips*, Structural Steel Education Council, Moraga, CA, 1995, 82 pp.

Astaneh-Asl, A., Nader, M. N., and Harriott, J. D., "Seismic Behavior and Design Considerations in Semi-Rigid Frames," *Proceedings of the 1991 National Steel Construction Conference*, AISC, Chicago, 1991.

Batho, C., and Rowan, H. E., "The Analysis of the Moments in the Members of a Frame Having Rigid or Semi-Rigid Connections, under Vertical Loads," Second Report of the Steel Structures Research Committee, Department of Scientific and Industrial Research, His Majesty's Stationery Office, London, 1934, pp. 177–199.

Bjorhovde, R., Brozzetti, J., and Colson, A., eds., *Connections in Steel Structures: Behaviour, Strength and Design, Proceedings of the Workshop on Connections held at the Ecole Normale Superiere, Cachan, France*, May 1987, Elsevier Applied Science, London, 1988a.

Bjorhovde, R., Brozzetti, J., and Colson, A., eds., *Connections in Steel Structures: Behaviour, Strength and Design, Proceedings of the Workshop on Connections held at* the *Ecole Normale Superiere, Cachan, France,* May 1987, Elsevier Applied Sciences, London, 1988*b.*

Bjorhovde, R., Colson, A., and Zandonini, R., eds., *Connections in Steel Structures III: Behaviour, Strength and Design, Proceedings of the Workshop on Connections held at Trento, Italy,* May, 1995, Elsevier Applied Science, London, 1996.

Bjorhovde, R., Colson, A., Haaijer, G., and Stark, J. W. B., eds., *Connections in Steel Structures II: Behaviour, Strength and Design, Proceedings of the Workshop on Connections held at Pittsburg,* April 1991, AISC, Chicago, 1992.

Chen, W. F., *Handbook of Structural Engineering,* CRC Press, Boca Raton, FL, 1997.

Chen, W. F., Goto, Y., and Liew, J. Y. R., *Stability Design of Semi-Rigid Frames,* Wiley, New York, 1995.

Chen, W. F., and Lui, E., *Stability Design of Steel Frames,* CRC Press, Boca Raton, FL, 1991, 380 pp.

Chen, W. F., and Toma, S., *Advanced Analysis of Steel Frames,* CRC Press, Boca Raton, FL, 1994.

CTBUH, *Semi-Rigid Connections in Steel Frames,* B. Lorenz, B. Kato, and W. F. Chen, eds., Council on Tall Buildings and Urban Habitat, Bethlehem, PA, 1993.

Eurocode 3, "Design of Steel Structures, Part 1: General Rules and Rules for Buildings," ENV 1993-1-1:1992, Comite Europeen de Normalisation (CEN), Brussels, 1992.

FEMA, "Interim Guidelines: Evaluation, Repair, Modification, and Design of Welded Steel Moment Frame Structures," FEMA 267, FEMA, Washington, DC, 1995.

FEMA, "Steel Moment Frames: Background Reports." FEMA 288, FEMA, Washington, DC (see Part 5: Seismic Performance of Bolted and Riveted Connections), 1997*a.*

FEMA, "Interim Guidelines: Advisory No. 1" FEMA 267-A, FEMA, Washington, DC, 1997*b.*

Frye, M. J., and Morris, G. A., "Analysis of Flexibly Connected Steel Frames," *Canadian Journal of Civil Engineering,* vol. 3, no. 3, 1975, pp. 280–291.

Gerstle, K. H., "Flexibly Connected Steel Frames," in *Steel Frames Structures: Stability and Strength,* R. Narayanan, ed., Elsevier Applied Sciences, London, 1985, pp. 205–240.

Ghobarah, A., Korol, R. M., and Osman, A., "Cyclic Behavior of Extended End-Plate Joints," *Journal of Structural Engineering,* ASCE, vol. 118, no. 5, May 1992, pp. 1333–1353.

Ghobarah, A., Osman, A., and Korol, R. M., "Behavior of Extended End-Plate Connections under Cyclic Loading," *Engineering Structures,* vol. 12, no. 1, 1990, pp. 15–27.

Goverdhan, A. V., "A Collection of Experimental Moment Rotation Curves Evaluation of Predicting Equations for Semi-Rigid Connections," M.Sc. Thesis, Vanderbilt University, Nashville, TN, 1984.

Griffiths, J. D., "End-Plate Moment Connections—Their Use and Misuse," *Engineering Journal,* AISC, vol. 21, 1st quarter, 1984, pp. 32–34.

Kishi, N., and Chen, W. F., *Data Base of Steel Beam-to-Column Connections,* vols. 1 and 2, Structural Engineering Report No. CE-STR-86-26, School of Civil Engineering, Purdue University, West Lafayette, IN, 1986.

Kulak, G. L., Fisher, J. W., and Struik, J. H. A., *Guide to Design Criteria for Bolted and Riveted Joints,* 2d ed., Wiley, New York, 1987.

Johnston, B., and Mount, E. H., "Analysis of Building Frames with Semi-Rigid Connections," *Transactions of ASCE,* vol. 107, March 1941, pp. 993–1019.

Leon, R. T., "Composite Semi-Rigid Construction," *Engineering Journal,* AISC, vol. 31, 2d quarter, 1994, pp. 57–67.

Leon, R. T., "Composite Connections," Chap. 23 in *Handbook of Structural Engineering,* W. F. Chen, ed., CRC Press, Boca Raton, FL, 1997.

Leon, R. T., and Shin, J. K., "Performance of Semi-Rigid Frames," in *Restructuring America and Beyond,* M. Sanayei, ed., Proceedings of Structures Congress XIII, ASCE, New York, 1995, pp. 1020–1035.

Leon, R. T., Hoffman, J., and Staeger, T., "Design of Partially-Restrained Composite Connections," *AISC Design Guide 9,* American Institute of Steel Construction, Chicago, 1996.

Leon, R. T., and Zandonini, R., "Composite Connections," in *Steel Design: An International Guide,* P. Dowling, J. Harding, and R. Bjorhovde, eds., Elsevier Publishers, London, 1992, pp. 501–522.

Lionberger, S. R., and Weaver, W., "Dynamic Response of Frames with Nonrigid Connections," *Journal of the Engineering Mechanics Division*, ASCE, vol. 105, EMI, February 1969, pp. 95–114.

Moncarz, P. D., and Gerstle, K. H., "Steel Frames with Nonlinear Connections," *Journal of the Structural Division*, ASCE, vol. 107, no. ST8, August 1981, pp. 1427–1441.

Monforton, G. R., and Wu, T. S., "Matrix Analysis of Semi-Rigidly Connected Frames," *Journal of the Structural Division*, ASCE, vol. 89, no. ST6, June, 1963, pp. 13–42.

Murray, T. M., "Extended End-Plate Moment Connections," *Steel Design Guide Series 4*, Publ. No. D804, AISC, Chicago, 1990.

Murray, T. M., and Abel, M. S., "Analytical and Experimental Investigation of the Extended Unstiffened Moment End-Plate Connections with Four Bolts at the Beam Tension Flange," Report No. CE/VPI-ST 93/08, Virginia Polytechnic Institute, Blacksburg (revised 1994, 1995), 1992.

Murray, T. M., and Kukreti, A. R., "Design of 8-Bolt Stiffened End Plate Moment Connections," *Engineering Journal*, AISC, vol. 25, 2d quarter, 1988, pp. 45–53.

Murray, T. M., and Meng, R. L., "Seismic Loading of Moment End-Plate Connections: Some Preliminary Results," in *Connections in Steel Structures III: Behaviour, Strength and Design*, R. Bjorhovde et al., eds., Elsevier Applied Science, London, 1996.

Murray, T. M., and Watson, D. P., "Strength of Moment End-Plate Connections with Multiple Rows at the Beam Tension Flange," in *Connections in Steel Structures III: Behaviour, Strength and Design*, R. Bjorhovde et al., eds., Elsevier Applied Science, London, 1996.

Nader, M. N., and Astaneh-Asl, A., "Seismic Design Concepts for Semi-Rigid Frames," *Report No. EERC/92-06*, University of California, Berkeley, 1992.

Nethercot, D. A., "Steel Beam to Column Connections—A Review of Test Data and Their Applicability to the Evaluation of the Joint Behaviour of the Performance of Steel Frames," CIRIA, London, 1985.

Osman, A., Ghobarah, A., and Korol, R. M., "Seismic Performance of Moment Resisting Frames with Flexible Joints," *Engineering Structures*, vol. 15, no. 2, 1993, pp. 117–134.

Packer, J. A., and Morris, G. A., "A Limit Design Method for the Tension Region of Bolted Beam-Column Connections," *Journal of the Institution of Civil Engineers*, vol. 55, no. 10, 1987.

Rathbun, J. C., "Elastic Properties of Riveted Connections," *Transactions of ASCE*, vol. 101, 1936, pp. 524–563.

Salmon, C. G., and Johnson, J. E., *Steel Structures: Design and Behavior*, 3d ed., HarperCollins, New York, 1996.

Shin, K. J., "Seismic Response of Frames with Composite Semi-Rigid Connections," Ph.D. Thesis, The Graduate School, University of Minnesota, MN, 1991.

Shin, K. J., and Leon, R., "Seismic Response of Semi-Rigid Composite Frames," *Structures Congress 92: Compact Papers*, ASCE, 1992, pp. 645–648.

Sourochnikoff, B., "Wild Stresses in Semi-Rigid Connections of Steel Frameworks," *Transactions of ASCE*, vol. 115, 1950, pp. 382–402.

Thorton, W. A., "Prying Action—A General Treatment," *Engineering Journal*, AISC, vol. 22, 2d quarter, 1985, pp. 67–75.

Thorton, W. A., "Strength and Serviceability of Hanger Connections," *Engineering Journal*, AISC vol. 29, no. 4, 4th quarter, 1992, pp. 145–149.

Tsai, K.-C., and Popov, E. P., *Steel Beam-Column Joints in Seismic Moment Resisting Frames*, Report No. UCB/EERC-88/19, Earthquake Engineering Research Center, University of California at Berkeley, CA, 1988.

Tsai, K.-C., and Popov, E. P., "Cyclic Behavior of End-Plate Moment Connections," *Journal of Structural Engineering*, ASCE vol. 116, no. 11, November 1990, pp. 2917–2930.

Whittaker, D., and Walpole, W. R., "Bolted End-Plate Connections for Seismically Designed Steel Frames," Research Report no. 82-11, Department of Civil Engineering, University of Canterbury, New Zealand, 1982.

Chapter

5

Seismic Design of Connections

James O. Malley
Raymond S. Pugliesi
Shane Noel

Degenkolb Engineers
San Francisco, CA

5.1 Special Design Issues for Seismic Design

The structural design philosophy for most loading conditions, such as gravity loads due to everyday dead and live loads or expected wind loadings, is that the structural system, including the connections, resist the loads essentially elastically, with a safety factor to account for unexpected overloading within a certain range. The parallel philosophy for resisting earthquake-induced ground motions is in strik-

(Courtesy of The Steel Institute of New York.)

ing contrast to that for gravity or wind loading. This philosophy has evolved over the years since the inception of earthquake-resistant structural design early in the twentieth century, and is continuing to develop as engineers learn more about the performance of structures subjected to strong earthquakes. The present general philosophy for seismic design has been most succinctly stated in the *Bluebook* of the Structural Engineers of California (SEAOC, 1996) for a number of years. The document states this approach as the following:

> Structures designed in conformance with these Requirements should, in general, be able to:
>
> Resist a minor level of earthquake ground motion without damage.
>
> Resist a moderate level of earthquake ground motion without structural damage, but possibly experience some nonstructural damage.
>
> Resist a major level of earthquake ground motion—of an intensity equal to the strongest earthquake, either experienced or forecast, for the build-

ing site—without collapse, but possibly with some structural as well as nonstructural damage.

It is expected that structural damage, even in a major design level earthquake, will be limited to a repairable level for most structures that meet these Requirements. In some instances, damage may not be economically repairable. The level of damage depends upon a number of factors, including the intensity and duration of ground shaking, structure configuration type of lateral force resisting system, materials used in the construction and construction workmanship.

It is clear, then, that when subjected to a major earthquake, buildings designed to meet the design requirements of typical building codes, such as the Uniform Building Code (UBC, 1997), are expected to have damage to both structural and nonstructural elements. The intent of the building code under this scenario is to avoid collapse and loss of life. Because of the economic impact, structural design to resist major earthquake ground motions with little or no damage has been limited to special buildings, such as postdisaster critical structures (for example, hospitals, police, and fire stations) or structures that house potentially hazardous materials (for example, nuclear power plants).

The structural design for large seismic events must therefore explicitly consider the effects of response beyond the elastic range. A mechanism must be supplied within some elements of the structural system to accommodate the large displacement demand imposed by the earthquake ground motions. In typical applications, structural elements, such as walls, beams, braces, and to a lesser extent columns and connections, are designed to undergo local deformations well beyond the elastic limit of the material without significant loss of capacity. Provision of such large deformation capacity, known as *ductility,* is a fundamental tenet of seismic design. Note that new technologies (for example, base isolation and passive energy dissipation) have been developed to absorb the majority of the deformations and, therefore, protect the "main" structural elements from damage in a major earthquake. Such applications are gaining increasing application in areas of high seismicity. Addressing such systems is beyond the scope of this text, which will focus on the seismic design of steel connections in typical applications.

In most cases, good seismic design practice has incorporated an approach that would provide for the ductility to occur in the members rather than the connections. This is especially the case for steel frame structures, where the basic material has long been considered the most ductile of all materials used for building construction. The reasons for this approach include the following:

- The failure of a connection between two members could lead to separation of the two elements and precipitate a local collapse.

- The inelastic response of members is more easily defined and more reliably predicted.

- The inelastic action of steel members generally occurs at locations where the distribution of strain and stress does not induce constraint that could lead to a state of triaxial tension. Under certain circumstances, these connections induce significant constraint that inhibits material yielding.

- Local distributions of strain and stress in connections can become quite complicated, and be very different from simple models typically used in design.

- Connection failures in frame structures could jeopardize the stability of the system by reducing the buckling restraint provided to the building columns.

- The repair of connection damage may be more difficult and costly than replacing a yielded or buckled member.

- Possible yield length of member versus connection.

Building codes have incorporated this philosophy into their seismic design requirements for a number of years. The most common method employed to incorporate this approach has been to require that the connections be designed to resist the expected member strength of the connecting elements, or the maximum load that can be delivered to the connection by the system. This implies that a conscious effort has been made by the designer to preclude the connections from undergoing severe inelastic demands. As such, a strength-based design approach as employed by the latest codes [for example, 1997 UBC and 1997 National Earthquake Hazards Reduction Program (NEHRP) provisions] is a much more direct and fundamental procedure than allowable stress methods that were previously followed. Seismic design of steel structures using LRFD is clearly a more rational, consistent, and transparent approach. As such, the 1997 AISC *Seismic Provisions for Structural Steel Buildings* (AISC, 1997) is based primarily on an LRFD approach. Connection design procedures in this document are based on Chapter J, "AISC LRFD Specification for Structural Steel Buildings" (AISC, 1994). An ASD section (Part III) has been generated directly from the LRFD requirements in Part I to support allowable stress design procedures.

The 1997 AISC *Seismic Provisions for Structural Steel Buildings* (AISC, 1997) include a number of requirements that are intended to ensure that this philosophy can be realized in the actual seismic performance. For example, a set of additional load combinations [Eqs. (4.1) and (4.2)] have been included to account for the expected structural system overstrength, termed Ω_0. In addition, the provisions require that the expected (rather than the nominal) yield strength of the materials be considered in comparisons of relative strengths between various members and/or connections. This term, R_y, ranges from 1.1 to 1.5 depending on the material specification chosen, with the highest value for modern A36 rolled shape material.

Other design approaches intend for the connections themselves to absorb substantial energy and provide major contributions to the displacement ductility demand. Examples of such a system would include both fully restrained (FR) and partially restrained (PR) connections in moment-resisting frames. To properly incorporate these elements in seismic design requires a much greater level of attention than for standard connection design or for moment connections to be subjected only to typical static loads. In addition to typical strength design requirements, such connections should take factors such as the following into account:

- Toughness of joining elements in the connections, including any weldments

- High level of understanding of the distribution of stress and strain throughout the connection

- Elimination (or at least control) of stress concentrations

- Detailed consideration of the flow of forces and the expected path of yielding in the connection

- Good understanding of the properties of the materials being joined at the connection (for example, through-thickness, yield-to-tensile ratio)

- The nature of the connection demands being high-strain, low-cycle versus low-strain high-cycle fatigue typical of other structural applications such as bridges

- The dynamic nature of the response which induces strain rates well below impact levels

- The need for heightened quality control in the fabrication, erection, and inspection of the connection.

While these types of considerations are particularly critical for connections where inelastic response is anticipated, it also behooves the designer to take factors such as these into account for all connections of the seismic force resisting system.

In the AISC seismic provisions, all connections in the lateral force-resisting system are required to meet a number of basic design requirements, which go beyond those required of joints in typical steel connections. For bolted connections, the design of bolted joints require the following:

- All joints must use fully tensioned, high-strength bolts
- Bearing design values are allowed, within the limits of the lower nominal bearing strength given in the AISC *LRFD Manual,* $2.4dtF_u$
- Bolted joints are not to be used in combination with welds
- Design of bolted joints should be such that nonductile modes do not control the inelastic performance.

For welded joints, the requirements include:

- Provision of approved welding procedure specifications that meet American Welding Society (AWS) D1.1 and are within the parameters established by the filler-metal manufacturer
- All complete-joint-penetration welds must have a Charpy V-notch (CVN) toughness AWS classification or manufacturers certification of 20 ft-lb at $-20°F$
- The repair of discontinuities created by items such as tack welds, erection aids, air-arc gouging, and flame cutting, as required by the engineer of record.

5.2   Connection Design Requirements for Various Structural Systems

Proper system selection is a critical element in successful seismic design. Various systems, such as fully and partially restrained moment-resisting frames, concentrically braced frames, and eccentrically braced frames, are addressed in the AISC seismic provisions. These provisions have specific requirements for the different structural systems that address connection design.

For moment-frame systems, special moment frame (SMF) and intermediate moment frame (IMF) connections have specified values for both inelastic deformation and strength capacities, since it is expected that these connections will absorb substantial energy during the design earthquake. Deformation capacities are to be demonstrated by qualified cyclic testing of the selected connection type. At the minimum acceptable inelastic deformation level (0.03 rad for SMF, 0.02 rad for IMF), the provisions require that the nominal beam plastic moment, M_p, be reached unless local buckling or a reduced beam approach is followed, in which case the value is reduced to $0.8\ M_p$. The minimum beam shear connection capacity is defined as resisting a combination of full-factored dead load, a portion of the live and snow load (if any), and the shear that would be generated by the expected moment capacity (including R_y) of the beam due to seismic actions. Finally, the joint panel zone shear is required to have a capacity able to resist the actions generated by Eqs. (4.1) and (4.2). For ordinary moment frames (OMF), the strength requirement is similar and the deformation limit is reduced to 0.01 rad. No specific joint panel zone requirements are defined for OMF systems.

The design requirements for PR connections in SMF and IMF are similar to those required for FR connections as described previously. For OMF structures, a set of requirements are provided to ensure a minimum capacity level of 50% of that of the weaker connected member, and that connection flexibility is considered in the determination of the overall frame lateral drifts.

The connection design requirements of AISC seismic provisions are similar for both special concentrically braced frames (SCBF) and ordinary concentrically braced frames (OCBF). For OCBF, the connections that are part of the bracing system must meet the lesser of the following:

- The nominal axial tensile strength of the bracing member, including R_y.

- The maximum force that can be transferred to the brace by the remainder of the structural system. An example of how this provision could be invoked would be the uplift capacity of a system with spread footing foundations.

- The amplified force demands, as defined by Eqs. (4.1) and (4.2).

For SCBF, the connection strength must exceed the lesser of the first two elements in this list, fully ensuring that the connections are not the weak elements in the system.

For OCBF and SCBF, both the tensile and flexural strength must be considered in the design of the connections. The flexural strength of the connections in the direction of brace buckling is required to be greater than the nominal moment capacity of the brace, unless they are specifically designed to provide the expected inelastic rotations that can be generated in the postbuckling state. This type of detail typically includes a single gusset plate where there is adequate separation between the end of the brace and the connecting element so that the gusset plate can bend unrestrained, as developed from research at the University of Michigan (Astaneh, 1989). In addition, the potential for buckling of gusset plates that may be used in bracing connections must be addressed. Finally, bolted connections should be checked for local failure mechanisms, such as net tension and block shear rupture, to ensure that these potentially brittle modes are avoided.

The other major system included in seismic design provisions is the eccentrically braced frame (EBF). This system, which was systematically developed through years of research at the University of

California by Professor Egor Popov and his students, was the first to explicitly require that elements and connections within the system be designed so as to limit the inelastic response to special members known as "links." For example, in the 1997 AISC seismic provisions, the design of connections between links and brace elements must consider both the expected overstrength of the material and the strain hardening that is expected to occur in properly detailed link elements. The design of such connections must also be detailed such that the expected response of the link elements is not altered.

In a number of EBF configurations, the link beams are located at the end of a bay, adjacent to a supporting column. Since severe inelastic rotation demands are expected in link beams during major seismic events, there was concern that, without special precautions, link-to-column connections in these EBF configurations may be subject to the same type of connection fractures that numerous moment connections suffered in the Northridge earthquake. As a result, the 1997 AISC seismic provisions require that these connections be tested to demonstrate that they have adequate rotation capacity. Without testing to qualify the connection detail, the links are conditions that are required to be proportioned to yield in shear and the connections must be reinforced to preclude inelastic behavior at the face of the column.

5.3   Design of Special Moment-Frame Connections

5.3.1   Introduction

This section provides an overview of the requirements and concepts for the design of special moment frame (SMF) connections. The design basis presented is established based on the recommendations of the SAC Joint Venture "Interim Guidelines" (SAC, 1995) and the "Interim Guidelines Advisory No. 1" (SAC, 1997) as well as requirements given in the AISC *Seismic Provisions for Structural Steel Buildings* (1997). First, general concepts and objectives for design will be outlined, followed by specific connection types and design examples.

Figure 5.1 shows a typical unreinforced detail for an SMF beam-to-column connection. The beam-to-column connection must be capable of transferring both the beam shear and moment to the column. Historically, the assumption for design has been that the beam shear is transferred to the column by the beam web connection and the moment is transferred through the beam flanges. Recent studies by Lee et al., (1997) and others have demonstrated that this assumption is far different from the actual behavior. Common practice prior to the

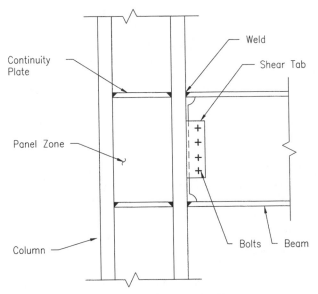

Figure 5.1   One-sided moment frame connection.

Northridge earthquake was to either bolt or weld the web to the column shear plate and to weld the beam flanges to the column flange
using a full-penetration groove weld. The panel zone (the column web
at the beam intersection) is subjected to a shear force due to these
moments applied by the beam.

In the design of SMF connections the engineer must set objectives
for both load and deformation capacities. Specifically, the load capacity requirement is based on the maximum attainable moment in the
beam. The connection to the column must be sufficiently strong to
develop the strength of the beam, thus reducing the risk of brittle
failure in the connection. Inelastic deformation capacity is required to
ensure ductility in predetermined locations under large deformation
demands.

Load capacities.   A common philosophy adopted after the Northridge
earthquake has been to design the connection at the column face to
remain nominally elastic, and force the inelastic deformation to occur
in the beam itself. The design strength of the connection between
beam and column is determined by using a "capacity design"
approach. The maximum probable moment and shear that the beam
is capable of achieving are determined based on the probable strength
of the beam. These maximums then become the design loads for the
connection. The connection to the column is then designed based on
nominal material properties.

The ability to estimate the maximum moment developed in the beam becomes quite important given the uncertainties regarding actual material behavior. The connection should be designed with the expectation of both beam overstrength and strain hardening in the plastic hinge region. A methodology for estimating the probable moment in the plastic hinge is presented in FEMA 267A (SAC, 1997). The approach taken here is based on the FEMA 267 methodology. A similar approach is presented in the AISC *Seismic Provisions for Structural Steel Buildings* (1997).

Beam overstrength should be accounted for by using the expected yield strength of the beam material. For example, the expected strength of A572, grade 50 steel is approximately 57 ksi, based on mill certificate test values. This value is based on tests of web material, however, which typically has a slightly higher strength than flange material. To convert this value to a flange yield stress, a factor of 0.95 is introduced. So for A572, grade 50 steel, the expected yield stress increase from the nominal is $(57/50) \times 0.95 = 1.08$, or approximately 1.1. (*Note:* For ASTM A36 steel this value is 1.5.)

The strain-hardening effect in the beam can be quantified by applying a factor of 1.1 to the expected flange yield stress. Recent connection testing (Yu et al., 1997) has shown that an increase by a factor of 1.1 is reasonable to account for strain hardening of the beam in the plastic hinge region. The resulting increase, including both overstrength and strain hardening, is $ß = 1.1 \times 1.1 = 1.21$, say 1.2. (*Note:* For ASTM A36 steel this value is 1.7.)

The location of the plastic hinge also must be accounted for. If the plastic hinge occurs at the face of the column ($x = 0$ in), the moment at the column face, M_f, will equal M_{pr}. However, it has been shown by numerous tests that the plastic hinge in a conventional SMF connection typically occurs away from the column face (or end of strengthened beam section), at a distance of approximately $x = d/3$ to $d/4$. Extrapolation over this distance to the column face results in an increased moment demand at the face of the column.

The moment demand at the column face is determined as follows (see Fig. 5.2):

1. Determine the maximum probable plastic moment of the beam, M_{pr}, including overstrength and strain hardening:

$$M_{pr} = ßM_p = ßZ_bF_y \qquad (5.1)$$

   where $ß = 1.2$ for A572 and A913 steel, and 1.7 for A36 steel, per FEMA 267A (SAC, 1997).

2. Extrapolate the moment to the column face from the assumed seismic inflection point at beam midspan to find the maximum beam moment, M_f:

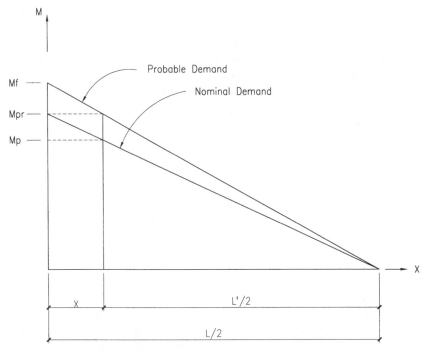

Figure 5.2 Beam seismic moment diagram.

$$M_f = M_{pr} + V_p x \tag{5.2}$$

where the shear

$$V_p = \frac{2\,M_{pr}}{L'} + \text{factored gravity loads at the hinge location}$$

$$M_f = M_{pr}\left(1 + \frac{2x}{L'} + \right) \text{factored gravity moment at column face} \tag{5.3}$$

3. The shear demand at the column face will be

$$V_f = \frac{2M}{L} + \text{factored gravity loads at the column face} \tag{5.4}$$

Thus, the nominal capacity of the connection at the column face must be designed to resist the load demands M_f and V_f.

Additionally, FEMA 267 (SAC, 1997) recommends that the calculated through thickness stress at the groove welds remain less than 0.9 times the nominal column yield stress. This stipulation not only guards against weld fracture, but column flange fracture as well. Research by SAC in 1998 will provide additional information related to the design for column flange through thickness stresses.

Deformation capacities. Obtaining large story drift ratios in an SMF is dependent on the inelastic rotation capacity of the connections. This inelastic rotation may occur by hinging of the beam or column, or by shear yielding in the panel zone, or by a combination of these two effects. As the strong column–weak beam (SCWB) is commonly preferred to weak column–strong beam (WCSB), the case of column hinging is not covered here. See Roeder et al. (1990) for further information on WCSB performance.

The story drift of a moment frame is closely related to the total joint rotation. This rotation is composed of both the elastic and inelastic deformations in the frame members (plastic hinges in the beam, shear yielding in the panel zone, etc.). Inelastic deformations in each component of the connection add cumulatively to the total plastic rotation of the connection. This parameter has become a valuable tool in determining the acceptability of connection designs. Connections that have exhibited adequate plastic rotation capacity in tests are generally thought to perform better in seismic events. Inelastic rotation demands may be estimated during the design using various nonlinear analysis techniques.

For special moment frame systems, AISC requires a minimum level of 0.03 rad of plastic rotation in a qualifying test. This may be obtained by a combination of yielding in the beam, panel zone, or column. The ability of a connection to withstand such deformation without significant loss of strength is heavily dependent on ductile detailing of the entire connection region.

5.3.2 Recent developments in connection design

Historically, moment connection design has relied on the previously described load-transfer assumptions and welded flange/bolted web connection details to allow the strength of the beam to fully develop prior to connection failure. Tests performed by Popov and Stephen (1970) and others indicated that this type of detail could be used for design, as the beam plastic moments were reached and, in some cases, significant amounts of ductility were observed. Indeed, it was believed that the typical steel SMF was well equipped to withstand large seismic force and deformation demands.

With the connection fractures caused by the Northridge earthquake came new questions related to this force transfer mechanism. Was the pre-Northridge connection detail fundamentally flawed? Can it be substantially improved by proper control over material and workmanship? Soon after the Northridge connection fractures were discovered, practitioners and researchers alike began to investigate these questions, and ultimately, to arrive at connection details that can be relied upon to deliver sufficient levels of force and deformation capacity.

Many successful testing programs have been performed that now provide guidance and direction for future work in this field. Many of the connection design concepts are still relatively new. Full-scale testing has become an extremely useful tool in helping to understand SMF connection behavior. At this stage, design by calculation rests heavily on the use of empirical data, much of which is only a few years old. While a great body of work has been produced since 1994, many questions regarding connection performance have yet to be answered.

It is with these concerns in mind that AISC (1997) and SAC (FEMA 267A, 1997) strongly suggest the use of connection designs which have been proven to consistently perform well in tests. Due to the variation of member sizes, material strengths, and other variables between projects, project-specific testing programs are encouraged. Alternatively, AISC provides specific acceptance criteria for using past test results of comparable connection designs (AISC, 1997).

The vast majority of research that ensued in this area stemmed from one of three primary philosophies: (1) a toughening scheme, (2) a strengthening scheme, and (3) a weakening scheme. Often, some or all of these schemes are used in combination.

5.3.3 Toughened connections

Design philosophy. To toughen the connection, significant attention is paid to the complete-penetration weld details between the beam and the column. Notch-tough electrodes are now typically specified (a common requirement is for Charpy V-notch values of 20 ft-lb at $-20°F$). In addition, backing bars are removed and replaced with reinforcing fillet welds in order to eliminate the notch effect at the root pass of the weld and to remove any weld flaws, which are more prevalent at the bottom flange where the beam web prevents continuous weld passes. An alternate approach to removal of the backing bar is simply to add a reinforcing fillet to secure the bar to the column flange. This is more commonly specified for the top flange weld. Research performed by Xue et al. (1996) supports this approach.

This scheme may be used either as a stand-alone design method or as a supplement to either of the second two schemes. The use of a notch-tough electrode and corrective measures for the backing bar notch effect are critical components to any connection design. In short, taking such measures to toughen the groove weld should be considered as a minimum amount of effort to ensure adequate ductile behavior of the connection and may not fully meet the SMF rotation requirements. Other recommendations include improved control in welding and inspection practices.

Another important aspect of the connection is the addition of column continuity plates. Although the use of continuity plates has been based on member geometry for some time, it is now recommended by SAC (FEMA, 1996) that "continuity plates be provided in all cases and that the thickness be at least equal to the thickness of the beam flange (not including cover plates) or one half the total effective flange thickness (flange plus cover plate)." Welding of continuity plates to column flanges should be performed with full-penetration groove welds, while the plate-to-column web weld may be a double-sided fillet. Notch-tough electrodes should be used in all cases, and care should be taken to avoid welding in the k region of the column.

5.3.4 Strengthened connections

Design philosophy. Another method of ensuring sufficient connection capacity is by strengthening the portion of the beam directly adjacent to the column, where the maximum moment occurs during seismic loading. The increased capacity near the column flange, M_f, forces the plastic hinge to form in the unstrengthened section of the beam (see Fig. 5.3).

The method used in this approach is to protect the previously vulnerable beam-flange complete-penetration welds with the addition of cover plates, rib plates, side plates, or haunches at the beam-to-column interface. The effective section modulus of the beam at the connection is increased, which decreases the bending stress at the

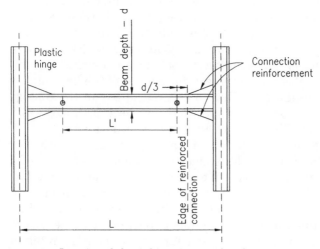

Figure 5.3 Location of plastic hinge in a one-bay frame.

extreme fiber of the section, as well as the total force resisted by the flange welds.

Strengthening these connections will invariably increase the stiffness of the frame. The effect this has on determining story drifts and building period must be considered in the design, but in most cases is a relatively minor effect.

Another consideration is the satisfaction of the AISC requirements for panel zone strength and the strong column–weak beam condition. The extrapolated moment, M_f, can now be well above the beam plastic moment, M_p, and must be considered. A minimum level of panel zone strength of at least 80% of M_f should be provided so that the panel zone is not too weak. Further research on this topic is being conducted.

Cover-plated connections. One popular method of strengthening the connection is by welding cover plates to the top and bottom beam flanges. Full-scale testing of cover-plated connections in recent years has been done by Engelhardt and Sabol (1995), Noel and Uang (1996), and others. In general, these tests have shown the ability of cover-plated connections to perform well in the inelastic range, although concern over the reliability of this detail has been raised in FEMA 267A (SAC, 1997).

Proper detailing is essential for obtaining ductile behavior from a cover-plated connection. Typically, cover plates are fillet-welded to the beam flanges and groove-welded to the column flange. A common detail is shown in Fig. 5.4. For ease of field erection, the bottom cover plate is oversized and the top plate undersized, to allow for downhand welding at each location. A variation to this technique uses oversized top and bottom cover plates, with the top plate shop-welded to the beam and the bottom plate field-welded. This allows the use of wider plates, while allowing downhand welding at both locations.

Note that only the long sides of the cover plates are welded to the beam flange. Welds loaded in the direction of their longitudinal axes perform significantly better in the inelastic region than those loaded in a perpendicular direction (AISC, 1994), hence cross-welds to the beam flanges at the end of the cover plates are not recommended.

Another detailing issue is the type of groove weld used at the cover-plate–to–column-flange connection. Two options are shown in Fig. 5.5 for this weld detail. Type I is the preferred detail. Although the type II detail uses less weld metal, the sharp angle of intersection between the cover-plate weld and the beam-flange weld creates a less desirable "notch" effect. From a fracture mechanics standpoint, the type II detail is more susceptible to horizontal crack propagation into the column flange. The designer must consider the amount of heat input and residual stresses in the joint region for either type detail. FEMA 267A (SAC, 1997) recommends a maximum total weld thick-

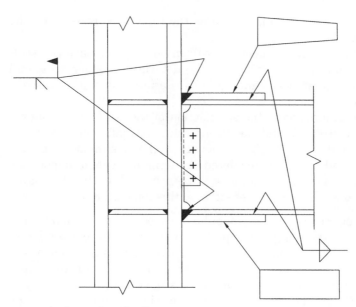

Figure 5.4 Cover-plate detail.

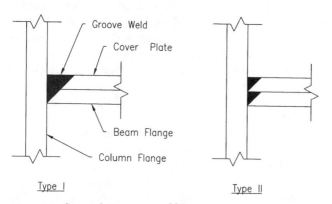

Type I   Type II

Figure 5.5 Cover-plate groove weld types.

ness of 2 times the beam flange thickness, or the thickness of the column flange, whichever is less. This is a means to conserve the amount of heat input to the welded joint region.

The thickness of the cover plate used is an essential variable to consider. The area of weld required between the strengthened beam section and the column face must be sufficient to resist the amplified beam moment, M_f. Once the required cross-sectional area of weld is obtained, it may be comprised of a combination of beam-flange weld

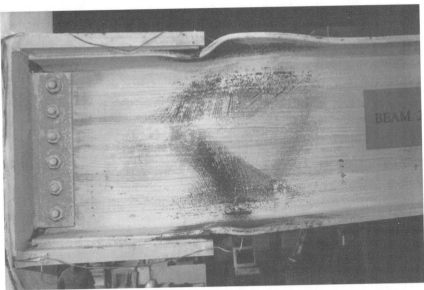

Figure 5.6   Cyclic performance of a flange-plated connection. (*Courtesy of Forell/Elsesser Engineers, Inc., San Francisco, CA.*)

and cover-plate weld or by plate weld alone if a thicker plate is used. The latter, known as a flange-plate connection, provides no direct connection of the beam flanges to the column flange, only to the cover plates. Full-scale tests of this type of connection are reported by Noel and Uang (1996) (see Fig. 5.6). If this option is chosen, care must be taken to avoid deformation incompatibility between the thin beam flange and relatively thick cover plate, resulting in premature fracture of the longitudinal cover-plate fillet welds.

Haunched connections.   Another method of strengthening the connection is by the addition of a haunch at the beam-flange–to–column-flange connection. The haunch is typically located on the bottom flange only, due to the presence of the floor slab on the top flange. The addition of a haunch to both flanges is a more expensive option, but has been shown to perform extremely well in tests. Haunches are typically made from triangular portions of structural tee sections or built-up plate, and stiffeners are provided in the column and beam webs (see Fig. 5.7).

Full-scale testing of bottom flange–welded haunch connections to date includes work done by Engelhardt et al., (1996), Popov and Stephen (1970), Uang and Bondad (1996), and Noel and Uang (1996). Whittaker, et al. (1995) report good performance of connections made with top and bottom flange–welded haunches. The bolted haunch has been studied by Ksai and Bleiman (1996).

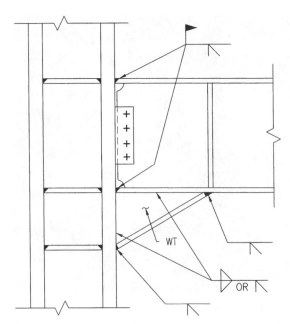

Figure 5.7 Bottom flange haunch connection.

Although recent work by Yu and Uang (1997) has brought into question the validity of the classical beam theory bending stress ($f = Mc/I$) in haunch design, a number of test specimens designed using this theory have performed very well (see Fig. 5.8). The geometry of the haunch should be such that: (1) the moment, M_f, is resisted while satisfying the $0.9F_{yc}$ through thickness requirement and (2) the haunch aspect ratio is sufficient to develop adequate force transfer from the beam flange. A moderate balance is required here as the longer the haunch is, the higher the demand moment at the column face, M_f, becomes. The design methodology presented by Yu and Uang (1997) recognizes a more realistic force transfer mechanism in the haunch connection. In this approach, the haunch flange is modeled as a strut which attracts vertical beam shear, hence reducing the beam moment, M_f, and the tensile stress at the beam flange welds.

Vertical rib-plate connections. A vertical rib plate serves a similar purpose as the cover plate and the haunch; strengthening the section by increasing the section modulus while distributing the beam-flange force over a larger area of the column flange (see Fig.5.11). The engineer may place a single rib plate at the center of the beam flange, but a more common practice is to position multiple ribs on each flange to direct the beam-flange force away from the center of the beam flange.

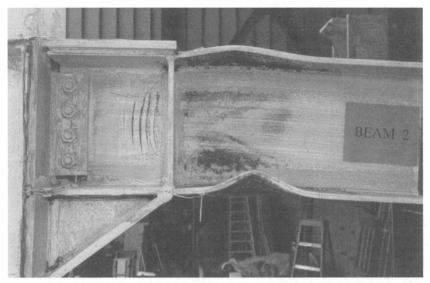

Figure 5.8 Yielding and buckling patterns of a beam subjected to cyclic loading. This connection incorporates both a bottom flange haunch and a top flange cover plate. (*Courtesy of Forell/Elsesser Engineers, Inc., San Francisco, CA.*)

By doing so, the stress concentration at the center of the beam-flange groove weld is somewhat alleviated.

Testing of rib-reinforced connections has been limited, but a few examples have shown that this method of strengthening can lead to ductile connection behavior (Engelhardt and Sabol, 1995; Anderson, 1995; and Tsai and Popov, 1988).

Example 5.1: Cover-Plate Connection Design In this example, we will modify a connection to include a cover-plate–strengthening scheme using the FEMA 267A (SAC, 1997) guidelines. See Fig. 5.9 for the predetermined frame geometry and member sizes. Type II complete-penetration groove welds will be used. For simplicity, gravity loading is neglected here.

$$L = 30 \text{ ft} = 360 \text{ in}$$

$$x = d = 35.85 \text{ in}$$

$$d_c = 18.67 \text{ in}$$

$$L' = L - d_c - 2x = 269 \text{ in}$$

$$h_c = 12 \text{ ft} = 144 \text{ in}$$

$$s_h = \tfrac{1}{4}d = 9 \text{ in}$$

$$F_{yc} = F_{yb} = 50 \text{ ksi}$$

$$F_{ye} = 54 \text{ ksi} \quad (\text{per FEMA 267A})$$

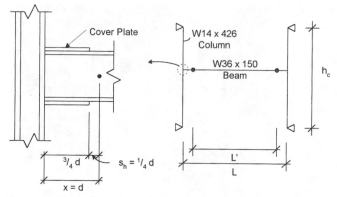

Figure 5.9 Frame and connection used in Example 5.1.

$$\phi = 0.9$$
$$\beta = 1.2$$
$$S_b = 504 \text{ in}^3$$
$$Z_b = 581 \text{ in}^3$$
$$Z_c = 869 \text{ in}^3$$

- Determine the probable moment demand at the column face:

$$M_{pr} = \beta\, Z_b F_y = 1.2(581 \text{ in}^3)(50 \text{ ksi})$$
$$= 34{,}860 \text{ kip-in}$$

$$V_p = \frac{2M_{pr}}{L'} = \frac{2(34{,}860 \text{ kip-in})}{269 \text{ in}}$$

$$= 259 \text{ kips}$$

$$M_f = M_{pr} + V_p x = 34{,}860 \text{ kip-in} + 259 \text{ kips (36 in)} = 44{,}190 \text{ kip-in}$$

- Approximate the cover-plate thickness:

$$Z_{req} = \frac{M_f}{F_{yb}} = \frac{44{,}190 \text{ kip-in}}{50 \text{ ksi}} = 883 \text{ in}^3$$

Try cover plates $1 \times 14 \times 27$ in long:

$$Z_{plate} = (1 \text{ in})(14 \text{ in})(35.85 \text{ in} + 1 \text{ in}) = 516 \text{ in}^3$$

$$Z_{new} = Z_b + Z_{plate} = 1097 \text{ in}^3 > Z_{req} = 883 \text{ in}^3, \text{ ok}$$

- Check the column-flange through-thickness stress at cover plates:

$$f_{tt} = \frac{M_f}{S_{new}}$$

$$S_{\text{new}} = S_b + S_{\text{plate}} = 504 + \frac{14(37.85^3 - 35.85^3)}{6(37.85)} = 1006 \text{ in}^3$$

$$f_{tt} = 43.9 \text{ ksi} \le \phi F_{ye} = 0.9(54 \text{ ksi}) = 49 \text{ ksi, ok}$$

∴ Use top and bottom cover plates, $1 \times 14 \times 27$ in

- Size the fillet welds (beam flange to plate) by estimating the force in the plate:

$$T_{\text{plate}} = f_{tt}A_{\text{plate}} = (43.9 \text{ ksi})(14\text{in}^2) = 615 \text{ kips}$$

To transfer T_{plate} from the beam flange via two 27-in-long fillet welds, the fillet welds are sized as follows. Force transferred through each of two fillets = 615 kips/2 = 308 kips:

$$L_{\text{weld}} = 27 \text{ in} \qquad \text{Force/inch} = \frac{308 \text{ kips}}{27 \text{ in}} = 11.4 \text{ kips/in}$$

∴ Use ⅝-in fillet weld $\phi R_n = 13.9$ kips/in > 11.4 kips/in, ok.

- Check the panel zone strength:

Strength:

$$R_v = 0.6\, F_y d_c t_p \left(1 + \frac{3b_{cf}t_{cf}^2}{d_b d_c t_p}\right) \tag{9.1}*$$

$$\phi = 0.75 \qquad P_u < 0.75 P_y$$

$$\phi R_v = 0.75(0.6)(50)(18.67)(1.875)\left(1 + 3\frac{(16.7)(3.05)^2}{(35.85)(18.67)(1.875)}\right)$$

$$= 1080 \text{ kips}$$

Demand:

$$V = V_{pz} - V_c$$

$$= \frac{M_c}{d + t_{\text{plate}}} - \frac{2M_c}{h_c} = M_c\left(\frac{1}{d + t_{\text{plate}}} - \frac{2}{h_c}\right)$$

$$M_c = M_{pr}\left(\frac{L}{L'}\right) = 34{,}860 \text{ kip-in } (1.33) = 46{,}364 \text{ kip-in}$$

$$V = 46{,}364 \left(\frac{1}{36.85} - \frac{2}{144}\right) = 614 \text{ kips} < \phi R_v = 1076 \text{ kips, ok}$$

∴ No doubler plate required.

- Check the SCWB requirements:

---

*Equation (9.1) is from AISC *Manual of Steel Construction,* LRFD, vol. II.

$$\frac{\Sigma M_{pc}}{\Sigma M_{pb}} \geq 1.0 \qquad\qquad (9.3)*$$

$$\Sigma M_{pc} = \Sigma Z_c\left(F_y - \frac{P_u}{A_g}\right) = 2(Z_c)(0.75\ F_y) = 65{,}175 \text{ kip-in}$$

$$\Sigma M_{pb} = M_c = 46{,}364 \text{ kip-in}$$

$$\frac{\Sigma M_{pc}}{\Sigma M_{pb}} = 1.4 > 1.0, \text{ ok}$$

∴ SCWB requirement is satisfied.

- Continuity plates:

Per FEMA 267 (SAC, 1996), add 1-in-thick continuity plates at the top and bottom flange level.

5.3.5   Weakened connections

Design philosophy.   Weakening the connection is achieved by removing a portion of the beam flange to create a reduced beam section, or RBS (see Fig. 5.10). The concept allows the designer to "force" a plastic hinge to occur in a specified location by creating a weak link, or fuse, in the moment capacity of the beam. Figure 5.11 shows the moment diagram of a beam under seismic loading. The geometry of the RBS must be such that the factored nominal moment capacity is not exceeded, and the through-thickness requirements are met, at the critical beam section adjacent to the column.

This method has potential benefits where the strengthening scheme had drawbacks. With a reduced M_p, the overall demand at the column face, M_f, must, by design, be less than the nominal plastic moment of the beam. Therefore, SCWB and panel zone strength requirements are easier to achieve.

The drawbacks for RBS come in the form of reduced stiffness. The reduction in overall lateral frame stiffness is typically quite small. On the other hand, the reduction in the flange area can significantly reduce the stiffness (and stability) of the beam flange, creating a greater propensity for lateral torsional buckling of the beam in the reduced section. The addition of lateral bracing is recommended for connections with RBS (Uang and Noel, 1996), and is required by the 1997 AISC seismic provisions.

Choice of RBS shape.   The shape, size, and location of the RBS all can significantly affect the connection performance. Various shapes

---

*Equation (9.3) is from AISC *Manual of Steel Construction*, LRFD, vol. II.

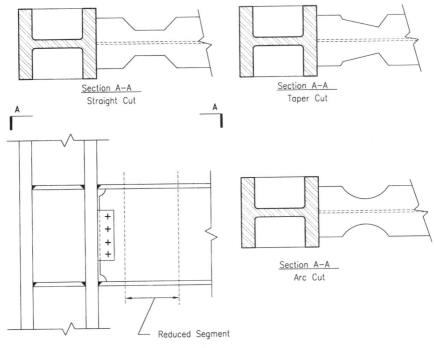

Section A-A
Straight Cut

Section A-A
Taper Cut

Section A-A
Arc Cut

Reduced Segment

Figure 5.0 Reduced beam section connection.

have been tested and used in new construction during the past several years. Test programs have been performed to investigate straight cut (Engelhardt et al., 1996), taper cut (Iwankiw and Carter, 1996; Uang and Noel, 1996), arc cut (Engelhardt et al., 1996) and drilled flanges.

Each RBS shape has benefits and shortcomings relative to each other. For instance, tapered cuts allow the section modulus of the beam to match the seismic moment gradient in the reduced region. This creates a reliable, uniform hinging location. However, stress concentrations at the reentrant corners of the flange cut may lead to undesired fracture at these locations as reported by Uang and Noel (1996) (see Fig. 5.12). Curved flange cuts avoid this problem, but do not give the benefit of uniform flange yielding, although test results indicate that plastification does distribute over the length of the reduced section. Using a combination of the two shapes, with curved reentrant corners on the taper cut section could also be a viable solution.

The lack of sharp reentrant corners and the ease of cutting has made the circular arc-cut reduction a viable option. In general, tests performed on arc-cut RBS connections have provided favorable results (Engelhardt and Sabol, 1996). The design methodology presented by FEMA (SAC, 1997) is applicable to various shapes of reduction cuts.

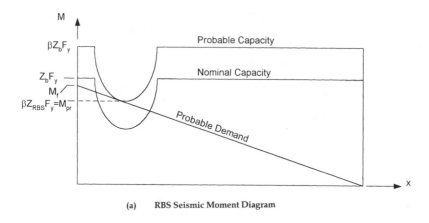

(a)   RBS Seismic Moment Diagram

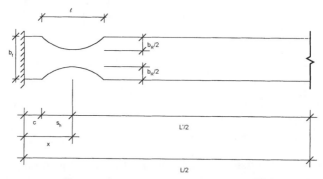

Figure 5.11 Reduced beam section moment diagram and flange geometry: (*a*) RBS seismic moment diagram and (*b*) RBS beam flange geometry (arc-cut section).

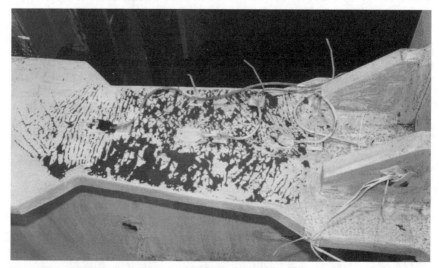

Figure 5.12 Yielding in the reduced section of a "taper-cut" beam flange subjected to cyclic loading. (*Courtesy of Ove Arup and Partners, Los Angeles, CA.*)

Geometry determination. Once a suitable RBS shape is obtained, sizing the cut becomes the next obstacle. Keeping in mind the FEMA 267A (SAC, 1997) requirements for connection capacity at the face of the column (see Sec. 5.3.1), as well as gravity loading demands at the location of the RBS, the size of reduction must be chosen appropriately.

Since member sizes in SMFs are typically governed by drift requirements, it is initially assumed that the reduced section will still work for strength under gravity loading. This load case must be checked after the geometry is chosen based on seismic loading.

Given a beam span L, depth d, and hinge location s_h, the reduction variables l, c, and b_R (see Fig. 5.11) define the seismic moment gradient and may be tailored to satisfy the requirements described previously. The majority of RBS connection tests have used relatively similar values for these essential variables. For instance, the length of reduction, l, has typically ranged between $0.75d$ and d. The distance of the RBS away from the column face, c, has typically been chosen as approximately $0.25d$, however, work by Engelhardt and Sabol (1997) justifies using a value of $0.75\ b_f$. These values have been shown to be effective in a number of testing programs.

The width of flange which is removed, b_R, determines the plastic modulus at the reduced section, $Z_{RBS} = Z_x - b_R t_f (d - t_f)$. This reduced modulus is then used to calculate the moment at the column face, M_f, using the method shown in Sec. 5.3.1. A practical upper bound on the value, b_R, has generally been 50% of the flange width, b_f. This limit is based on both stability and strength considerations. Excessive reduction can lead to premature lateral torsional buckling of the beam, which should be avoided. In the event that even a 50% flange reduction does not sufficiently reduce M_f. Supplemental strengthening may be considered in the area between the RBS and the column face. Reinforcing ribs at the column face have been shown to enhance the performance of RBS connections in tests (Uang and Noel, 1996).

Example 5.2   RBS Connection Design.   Using the unreinforced frame in Example 5.1, we will now design an RBS connection. The flange reduction will be an arc-cut shape. Again, we will use the guidelines of FEMA 267A (SAC, 1997) and gravity loads will be neglected (see Fig. 5.13).

$$L' = L - d_c - 2x = 302 \text{ in}$$

$$M_{pr} = \beta Z_{RBS} F_y = 1.2(Z_{RBS})(50 \text{ ksi})$$

$$= 60 Z_{RBS}$$

$$V_p = \frac{2M_{pr}}{L'} = \frac{2(60 \text{ ksi}) Z_{RBS}}{302 \text{ in}}$$

$$= 0.39 Z_{RBS}$$

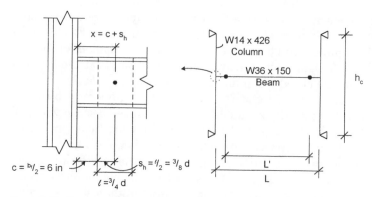

Figure 5.13 Frame and connection used in Example 5.2.

$$M_f = M_{pr} + V_p x = Z_{RBS}[60 + 0.39(19.5 \text{ in})]$$
$$= 67.6 \, Z_{RBS}$$

- Find the required flange reduction:

$$Z_{req} = \frac{M_f}{50 \text{ ksi}} = 1.35 \, Z_{RBS} \qquad Z_{req} = Z_b$$

$$Z_{RBS} \leq 0.74 \, Z_b = 430 \text{ in}^3$$

$$Z_{RBS} = Z_x - b_R t_f (d - t_f)$$

$$430 \text{ in}^3 = 581 \text{ in}^3 - b_R (1 \text{ in})(35.85 \text{ in} - 1 \text{ in})$$

$$b_R \geq 4.3 \text{ in}$$

$$\frac{b_R}{b_f} = \frac{6 \text{ in}}{12 \text{ in}} = 50\% \text{ reduction, ok}$$

∴ Try a 6-in flange reduction

$$Z_{RBS} = 372 \text{ in}^3$$

$$M_f = 67.6(372) = 25{,}147 \text{ kip-in}$$

- Check the through-thickness stress:

$$f_{tt} = \frac{M_f}{S_b} = \frac{25{,}147}{504} = 49 \text{ ksi} = \phi F_{ye} = 0.9(54 \text{ ksi}), \text{ ok}$$

∴ Use a 6-in beam-flange reduction (50%) (see Fig. 5.14).

- By comparison to Example 5.1, panel zone strength and SCWB requirements will be satisfied.
- Continuity plates:

Per FEMA 267 (SAC, 1996), add 1-in-thick continuity plates at the top and bottom flange level.

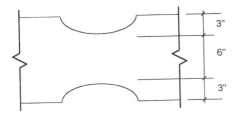

Figure 5.14 RBS flange reduction.

It should be noted that the preceding discussion presents some, but not all, of the connection design approaches that have been developed since the Northridge earthquake. In fact, a few approaches have been patented; these patented connections have not been addressed here.

5.4 Concentrically Braced Frames

5.4.1 Introduction

Concentric braced frames have found wide application in lateral force-resisting systems, typically having been chosen for their high elastic stiffness. This system is characterized by horizontal and vertical framing elements interconnected by diagonal brace members with axes that intersect. The primary lateral resistance is developed by internal axial forces in the framing members. The AISC seismic provisions make a distinction between ordinary concentrically braced frames (OCBF) and special concentrically braced frames (SCBF). SCBF frames are specifically detailed and typically sized to withstand the fully inelastic behavior of the lateral system. This section will describe the connection design for both types of concentric braced frames.

Figure 5.15 shows several types of braced frames. The K-braced frame shown is only allowed for use as an OCBF in roof structures and buildings under two stories when sized to remain essentially elastic. Such a configuration is not recommended in taller buildings because their inelastic behavior involves the hinging of columns which can jeopardize the gravity load-carrying capacity. The V-braced systems shown require the intersected beams to be specially designed when used in SCBF structures in order to ensure their stability once the bracing system begins to exhibit inelastic behavior. During a large earthquake, it is expected that the compression brace will buckle before the tension brace begins to yield. At the connection to the beam, there is an imbalance of forces from the braces that needs to be resolved by the beam member. As the lateral loading continues and both braces yield, the maximum force imparted to the beam will be the difference in the strengths of the buckling brace and the tension

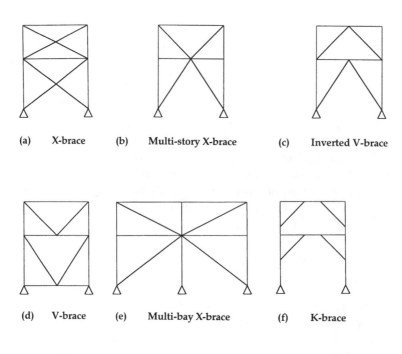

(a)   X-brace   (b)   Multi-story X-brace   (c)   Inverted V-brace

(d)   V-brace   (e)   Multi-bay X-brace   (f)   K-brace

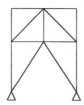

(g)   Inverted V-brace with Zipper Column

Figure 5.15 Concentric braced frame types: (*a*) X brace; (*b*) multistory X brace; (*c*) inverted V brace; (*d*) V brace; (*e*) multibay X brace; (*f*) K brace; and (*g*) inverted V brace with zipper column.

yielding brace. The direction in which this force acts depends on the bracing configuration. The brace connections and the beam need to be able to transfer these loads.

SCBF systems have different requirements than OCBF systems. The width-thickness ratio of the rectangular tube sections in SCBFs is limited to $110/\text{SQRT}(F_y)$. This is intended to minimize local buckling of the brace elements and results in larger wall or flange thickness. Since the connections are typically designed for the brace capacity in SCBF systems, the force level for the design of the connection will increase.

Due to the better behavior of the system, AISC allows more slender elements in SCBFs than in OCBFs. The slenderness of OCBFs are limited to $KL/r \leq 720/\sqrt{F_y}$, whereas SCBF braces are only limited to $KL/r \leq 1000/\sqrt{F_y}$. This seems to contradict testing which has shown that the hysterectic response during inelastic cyclic reversals improves as the slenderness of the compression member decreases. Locally, brace behavior is improved with stocky members, however, push-over analyses which analyze the entire system indicate that large reductions in the slenderness can cause the compression capacity to approach the tension capacity, which results in a soft story effect. This will occur if, once the compression braces of a story buckle, the tension members on the same story yield before compression members on other floors buckle. Since the buckling strength is close to the tension capacity, the postbuckling reduction in strength is often enough to yield the adjacent tension members. The addition of a "zipper column" as shown in Fig. 5.15, avoids this condition by distributing the forces throughout the height of the system as the members exceed their elastic limits.

5.4.2 Connection design and example

This section will present an example design of a connection in a special concentrically braced frame. Throughout the example, it will be noted how the criteria would differ for an ordinary concentrically braced frame. Figure 5.16 shows the brace configuration and Example 5.3 presents a spreadsheet outlining the entire connection design. The connection presented addresses the intersection of a tube steel brace with a beam-column connection. Similar approaches may be followed for other brace configurations and section types.

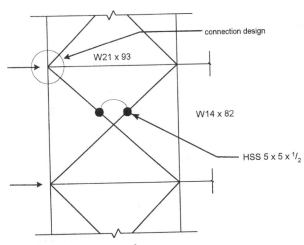

connection design

W21 x 93

W14 x 82

HSS 5 x 5 x $\frac{1}{2}$

Figure 5.16 Design example.

Force level. The design of the connection is dependent on the design forces during compression and tension. Designing the connection for the capacity of the member ensures the connection is not the yielding element in the system. The maximum force the connection will be subject to is the yielding of the brace member in tension defined as $R_y F_y A_g$.

The R_y factor accounts for the expected material overstrength and strain hardening of the member. Had this connection been designed for use in an OCBF, the design force level could have been reduced to the maximum expected force as defined in Eqs. (4.1) and (4.2) in the AISC *Seismic Provisions for Structural Steel Buildings* (1997).

The spreadsheet analysis begins by determining the section sizes, material, and geometry which will determine the brace's force magnitude and direction at the connection.

Force distribution. The force distribution from the brace to the beam and column can be calculated using the uniform force method described in AISC's vol. II, *Connections*. This method provides a rational procedure for determining the interface forces between the gusset plate and the horizontal and vertical elements at the connection. The axial and shear forces are distributed in the connection based on stiffness, while the required moment for equilibrium is assigned to the beam or column or to the beam and column equally.

An alternate method to determine the forces in the connection may use the fin truss approach originally proposed by Whitmore and modified by Astaneh (1989). This approach discretizes the gusset plate into radial elements and distributes the force based on axial stiffness and the angle of incidence. The procedure has been applied successfully on single-member connections, but appears overly conservative for multimember gusset-plate configurations when the forces are not independent of one another.

Example 5.3 defines the geometry of a rectangular plate where the $2t$ offset required to allow an unrestrained bending zone of the brace is provided between the points of support of the plate to the beam and column and the end of the brace. This configuration not only provides a simple plate geometry, but also eliminates the need for stiffeners on the plate. Had the plate utilized smaller leg dimensions, a tapered plate would be required, but the buckling line perpendicular to the brace would start from the bottom edge of the plate upward to a free edge of the plate. This free edge should be supported by stiffener plates to ensure that during buckling the plate remains stable and bends perpendicular to the brace. Should the buckling line migrate to the stiff supported points not perpendicular to the brace, such as the ends of the tapered plate without stiffeners, it is feared that the brace may effectively buckle about a different axis at each end

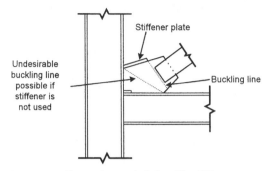

Figure 5.17 Tapered gusset plate with stiffeners.

imparting torsional forces into the brace. Figure 5.17 shows this condition.

Example 5.3 shows the dimensions calculated based on the geometry specified and the resulting load distribution using the uniform force method. Figures 5.18 and 5.19 show the geometric variables. The axial force on the beam and the shear force on the column can be significant from the gusset plate. In frames where the brace intersects the column from each side or the beam from top and bottom, large demands may overstress the section requiring either the size be increased or the webs be strengthened with doubler plates.

Brace–to–gusset-plate connection. The connection of the tube brace to the gusset plate uses four fillet welds along a slot to fit the gusset plate. Half of the force is transferred to the plate by each half of the tube section. The centroid of the half section is no longer at the centroid of the plate, but rather is offset from the face of the plate toward the remaining wall of the tube section. The eccentricity between this centroid and the weld to the plate creates bending along the length of the welds. Of course, equal and opposite bending exists on the other side of the plate resulting in no net bending on the plate provided the tube is slotted along its centerline. The welds must be sufficiently strong to resist this bending stress. Damage to similar connections was found after the Northridge earthquake where the lack of sufficient weld to resist this bending resulted in the sides of the tube peeling away from the gusset plate. Welds may be strengthened by either increasing their thickness or length. Although increasing the length is the most efficient locally, it may increase the gusset-plate size and connections to the beam and column depending on the configuration.

The connection of the brace to the gusset plate is also subject to block shear. For the tube steel brace, the plate may yield around the perimeter of the full tube section (two lines of shear and one of tension) or along each weld line in shear (four lines of shear). The tube

Example 5.3:   Design of Special Concentric Braced Frame Connection

Calculation of special concentric braced frame connection
to accomodate brace buckling behavior and conform to the
*AISC Seismic Provisions for Structural Steel Buildings
Uniform Force Method - LRFD 2nd ed. Vol. II*

(INPUT IS INDICATED BY BOLD ITALICS)

Global Input

General
    Title: *TS5x5x1/2 Brace*

Global Data

Location: (U)pper / (L)ower =   *L*

Brace Data
    Shape: Shape =   *TS5x5x.5*
    Orientation: (S)trong / (W)eak   *W*   relative to the out of plane direction
    K_ip: X-X eff. len. fact. =   *1.50*   brace effective length factor in the plane of the frame
    K_op: Y-Y eff. len. fact. =   *1.00*   brace effective length factor out of the plane of the fram
    Fybr: Yield =   *46* ksi
    Ry: Overstrength =   *1.1* kips   1.5 for A36, 1.3 for A572, 1.1 for other grades

    Put: Ry*Fy*Ag =   423 kips
    λ: KL/(r*pi)*sqrt(Fy/E) =   1.159   (Out-of-Plane Buckling Governs)
    Pcr: 0.658^(LAMBDA^2)*A*FY   219.18

Upper beam
    Shape: Shape =   *W21X93*
    TOS: Top of steel =   *10.00* ft
    Fyub: Yield =   *50.00* ksi

Lower beam
    Shape: Shape =   *W21X93*
    TOS: Top of steel =   *0.00* ft
    Fylb: Yield =   *50.00* ksi

    X: Bay width =   *10* ft   Horizontal length between work points

Column
    Shape: Shape =   *W14X82*
    Fycol: Yield =   *50.00* ksi
    Orientation: (S)trong/(W)eak =   *S*

Member Properties Summary

	Brace:				Upper beam:		
	section =	TS5x5x.5			section =	W21X93	
Abr:	area =	8.36	in^2	Abm:	area =	27.3	in^2
dbr:	depth =	5.0	in	dbm:	depth =	21.62	in
bfbr:	flange width =	5.0	in	bfbm:	flange width =	8.42	in
twbr:	wall thick. =	0.5	in	kbm:	k =	1.6875	in
tfbr:	wall thick. =	0.5	in	twbm:	web thick. =	0.58	in
rxx:	X-X rad. of gyr. =	1.8	in (strong axis)	tfbm:	flange thick. =	0.93	in
ryy:	Y-Y rad. of gyr. =	1.8	in (weak axis)	k1bm:	k1 =	0	in

	Lower beam:				Column:		
	section =	W21X93			section =	W14X82	
Abm:	area =	27.3	in^2	Acol:	area =	24.1	in^2
dbm:	depth =	21.62	in	dcol:	depth =	14.31	in
bfbm:	flange width =	8.42	in	bfcol:	flange width =	10.13	in
kbm:	k =	1.6875	in	kcol:	k =	1.625	in
twbm:	web thick. =	0.58	in	twcol:	web thick. =	0.51	in
tfbm:	flange thick. =	0.93	in	tfcol:	flange thick. =	0.855	in
k1bm:	k1 =	0	in	k1col:	k1 =	1	in

Connection Geometry

Lwpt:	Length from W.P. to end of gusset along brace =	**42**	**in.**
Ly:	Gusset dim. along col. from face-of-beam =	**22**	**in.**
Ly2:	Face of beam to start of column-gusset weld =	**5**	**in.**
Lx:	Gusset dim. along beam from face-of-column =	**21.07**	**in.**
Lx2:	Face of Column to start of beam-gusset weld =	**5**	**in.**
Lwld:	Weld length for brace flanges-to-gusset pl. =	**12**	**in.**
Lg:	Gusset dimension perpendicular to brace =	**9**	**in.**
t:	Gusset thickness =	**0.75**	**in.**
Fy_gus:	Gusset yield stress =	**36**	**ksi**
Fu_gus:	Gusset Ultimate stress =	**58**	**ksi**
θ:	Theta =	**0.785**	**rad.**
α:	Length to mid-gusset along bm. = (Lx-Lx2)/2+Lx2 =	**13.035**	**in.**
β:	Length to mid-gusset along col. = (Ly-Ly2)/2+Ly2 =	**13.5**	**in.**
eb:	dlb / 2 =	**10.81**	**in.**
ec:	dcol / 2 =	**7.155**	**in.**
L_unb:	Approximate unbraced length of brace =	**9.14214**	**ft.**

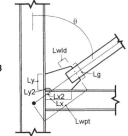

Figure 5.18

To specify a rectangular plate

Lx req'd =	25.73	in.
Ly req'd =	22.07	in.

Note: The brace length is calculated assuming that the same columns are used on each side of the bay and that the basic gusset geometry is the same at both ends of the brace. This length is used to calculate the brace compression capacity.

Bending Zone

t:	Gusset thickness =	0.75	**in.**
2t:	Current value for twice the gusset thickness =	1.5	**in.**
Lgb:	Bending at the beam-to-gusset connection =	2.40	**in. >= 2t OK**
Lgc:	Bending at the column-to-gusset connection =	1.74	**in. >= 2t OK**

Figure 5.19

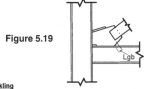

Compactness, Local Buckling

KL/r:	(Out-of-Plane Buckling Governs)	Maximum slenderness ratio =	91.4214 **< 1000/sqrt(Fy) OK**
b/t:		Compactness =	10 **< 110/sqrt(Fy) OK**

Figure 5.18   Gusset-plate connection geometry.
Figure 5.19   Gusset-plate distance requirements.

Moment Reduction I

	Current	Req'd for Zero Mom.	
Lx =	21.07	29.31	in.
Ly =	22.00	13.76	in.
α =	13.04	17.16	in.
β =	13.50	9.38	in.

Connection Stiffness
Any moment required for equilibrium will be carried by the stiffer of the beam-to-gusset and the column-to-gusset connection. The designer can choose roughly how to distribute the forces to the framing members.
Distribute forces:
    (E)qually to Beam and Column, Primarily to (B)eam, or Primarily to (C)olumn =   **E**   (Enter E, B, or C only)

Moment Reduction II

	Current	Req'd for Zero Mom.		
Lx =	21.07	25.05	in.	
Ly =	22.00	17.74	in.	
α =	13.04	15.02	in.	**Moment on interface**
β =	13.50	11.37	in.	**Moment on interface**

Load Distribution Manipulation
The beam and column shear capacities are checked for the additional shear being introduced into them.
Beams may also need to be checked for additional bending stresses.

Load Adjustments

				CONCLUSIONS
Vbm_r:	Reduce shear in beam by :	**0**	kips	
Hcol_2:	Other horizontal shear into column :	**0**	kips	
Vbm_2:	Gravity + other seismic shear into beam :	**0**	kips	

Final Interface Forces

Gusset-to-Column

Fvcol:	Phi*Aweb*Fv = 0.9x0.6 x fycol x dcol x twcol =	197.049	kips	
Hcol:	Horizontal force on column = Ec*Pu/r	96.50	kips	
Hcol_Total	Total column shear = Hcol + Hcol_2 =	96.50	kips	**OK < Fvcol**
Vcol:	Vertical force on column = Beta*Pu/r	153.32	kips	
Mc:	Moment from gusset on column = Huc*(Beta-Beta_bar)	-205.76	in.-kips	

Gusset-to-Beam

Fvbm:	Phi*Aweb*Fv = 0.9x0.6 x fybm x dbm x twbm =	338.569	kips	
Hbm:	Horizontal force on beam = Alpha*Pu/r	202.62	kips	
Vbm:	Vertical force on beam (incl. Vbm_r red.) = Eb*Pu/r - Vbm_r	145.80	kips	
Vbm_Total	Total beam shear = Vbm + Vbm_2 =	145.80	kips	**OK < Fvbm**
Mbm:	Moment from gusset on beam =	289.82	in.-kips	
fbm:	Bending stress on beam = Mb/S	1.51	ksi	**OK < Fybm**

Brace-to-Gusset Connection
This section designs the welded connection between the tube walls and the gusset plate. This section applies for TS sections only.
Additional and/or different checks will need to be performed for other shapes.

Brace Flange-to-Gusset Connection

	Electrode =	**E70**	
	Electrode correction =	1	[LRFD 2nd ed. p8-158]
FLFORCE:	Design force in one side of tube = (Pu / 2)=	211.5	kips
efl:	One side eccentricity =	1.88	in.
a:	efl / Lwld =	0.156	in.
c:	Interpolate from weld table values =	2.736	[LRFD 2nd ed. p8-163]

Eccentric weld table

a	c
0.150	2.750
0.200	2.640

GBlockTear:	Min of 0.75*(0.6*Fy*Agv+Fu*Ant) and 0.75*(0.6*Fu*Anv+Fy*Agt)	454.725	kips	**OK > Pu**
Gtear:	Gusset tearout cap. = 0.75x0.6 x Fu x t x Lwld x 2 =	470	kips	**OK > FLFORCE**
TStear:	Tube wall tearout = 0.9x0.6 x Fybr x tfbr x Lwld x 2 =	298	kips	**OK > FLFORCE**
	Required fillet weld size =	6.44	16ths in	

Design of Gusset Plate
Check the tension and buckling capacity of the gusset plate.

Design Data

	Slope (m)	Y intercept y = mx + b		Line Intersect.	Xint	Yint
Line A	-1.00	42.43		Lines A & B	29.00	13.43
Line B	3.37	-84.41		Lines A & C	27.88	14.55
Line C	3.73	-89.50		Lines A & D	9.61	32.82
Line D	0.00	32.78		Lines A & E	14.55	27.88
Line E	0.27	23.98		Lines A & F	19.39	23.04
Line F	1.00	3.66				

Tension

Weff:	Effective width of gusset for tension =	18.86	in.
Tcap:	Capacity of gusset in tension = (0.9*Fy x t x Weff) =	458.21	kips **OK > Pu**

Compression

K:	Gusset plate effective length factor =	0.80	
	Effective width of gusset for compression =	18.86	in. (per Whitmore's area)
L:	Average Length for Compression =	17.30	in. (used to calculate gusset
ry:	Radius of gyration (weak way) =	0.22	in. comp. capacity)
λ:	Lambda (gusset) = KL/r*pi*sqrt(FY/E)	0.72	
Fcr:	Fcr for gusset =	29.03	ksi
Pcap:	Compression capacity of gusset = 0.85*Ag*Fcr =	349.02	kips **OK > Pbuckle**

Design of Gusset-to-Beam Connection
Design of the gusset-to-beam coonection as a welded connection.

Gusset-to-beam interface forces

Hbm:	Horiz. force at the gusset-beam interface = Hb	202.62	kips
Vbm:	Vertical force at the gusset-beam interface = Vb	145.80	kips
Mbm:	Moment at the gusset-beam interface = Mb	289.82	in.-kips

Gusset Stresses

Fvcap:	Gusset shear capacity =0.9x0.6 x Fy =	19.44	ksi
fvult:	Shear stress on gusset = Hbm / (t x (Lx-Lx2)) =	16.81	ksi **OK < Fvcap**
fault:	Axial stress = Vbm / (t x (Lx-Lx2)) =	12.10	ksi
fbult:	Bending Stress = Mbm x 6 / (t x (Lx-Lx2)^2) =	8.98	ksi
	Combined normal stress = fault + fbult =	21.07	ksi **OK <Phi* Fy**

Weld Stresses

Average ultimate stress =	8.86	ksi/in.
Peak ultimate Stress =	10.11	ksi/in.
Peak stress / Average stress =	1.14	< 1.4 If this ratio is less than 1.4,
Implied orientation of stresses =	0.89747	rad size the weld for 1.4 times
Capacity increase for orientation =	1.35	the average ultimate stress.

Beam Web Stresses

Beam web stress =	19.67	ksi **OK < Fybm**

Weld Design Gusset-to-Beam

Electrode =	E70	
Electrode correction (LRFD 2nd ed. p8-158) =	1	
Required fillet weld size =	6.62	16ths in.

Design of Gusset-to-Column Connection
Design a fillet weld on each side of the gusset plate to the column.

Interface Forces

	Column Orientation =	Strong	
Hcol:	Horiz. force at the gusset-column interface = Hc	96.50	kips
Vcol:	Vertical force at the gusset-column interface = Vc	153.32	kips
Mcol:	Moment at the gusset-column interface = Mc	-205.76	in.-kips

Gusset Stresses

Fvcap:	Gusset shear capacity = 0.9x0.6 x Fy =	19.44	ksi
fvult:	Shear stress on gusset = Vcol / (t x (Ly-Ly2)) =	12.03	ksi	OK < Fvcap
fault:	Axial stress = Hcol / (t x (Ly-Ly2)) =	7.57	ksi
fbult:	Bending Stress = Mcol x 6 / (t x (Ly-Ly2)^2) =	5.70	ksi

	Combined normal stress = fault + fbult =	13.26	ksi	OK <Phi* Fy
	Implied orientation of stresses=	0.83436	rad
	Capacity increase for orientation=	1.32	

COLUMN STRONG WAY
Design welds between the gusset and the column flange.
Strong-Way Weld Stresses

	Average ultimate stress =	5.97	ksi/in.
	Peak ultimate Stress =	6.71	ksi/in.
	peak stress/average stress =	1.13	< 1.4	If this ratio is less than 1.4, size the weld for 1.4 times the average ultimate stress.

Column Web Stresses

	Column web yielding stress =	11.37	ksi	OK < Fy

Weld Design

	Electrode =	E70	
	Electrode correction (LRFD 2nd ed. p8-158) =	1	
	Required fillet weld size =	5.00	**16ths in.**

COLUMN WEAK WAY   Not Applicable
Design welds between the gusset and the column web, and plates between column flanges.
Weak-Way Weld Stresses

fvweb:	Unit shear weld to web = Vcol/(2 x (Ly-Ly2)) =	N.A.	kips/in.
Fvww:	Vert. shear cap. of col. web = 0.55 x Fy =	27.50	ksi
	Column web stress =	0.00	ksi	OK < Fvww

Weld Design Gusset-to-Column

	Electrode =	E70	
	Electrode correction (LRFD 2nd ed. p8-158) =	1	
	Required fillet weld size =	N.A.	**16ths in.**

Plate Design
Design plate to be welded to the gusset plate and the column flanges

PLfy:	Yield strength for the plate =	*36.00*	ksi
PLt:	Plate thickness =	*0.50*	in.
PLL:	Length of plate =	N.A.	in.
PLf:	Force in the plate =	N.A.	kips
Fvpl:	Ult. shear cap. of pl. = 0.55 x PLfy =	19.80	ksi
PLfv:	Shear stress in plate = (PLf / (2 x PLL x PLt)) =	N.A.	ksi	N.A.
PLfb:	Bending stress in plate =	N.A.	ksi	N.A.

Weld Design Plate-to-Column/Gusset

	Electrode =	E70	
	Electrode correction (LRFD 2nd ed. p8-158) =	1	
	Req'd fillet weld size of plate-to-gusset =	N.A.	**16ths in.**
	Req'd fillet weld of plate-to-column flange =	N.A.	**16ths in.**

Summary

USSET PL 'A'						WELDS (16th's)		
Thk (in)	Lcol (in)	Lbm (in)	Lbr (in)	Ld (in)	Lw (in)	A	B	C
0.75	22	21.07	12	42	9	5.00	6.62	6.44

PLATES					
Ref.	Thk (in)	Length (in)	idth (in)	Weld (16th)	Gap (in)
F	0.5	N.A.	12.6	N.A.	

section may also yield along the tube walls (four lines of shear). Other section types would have similar mechanisms.

Gusset-plate design. The gusset plate may either allow for out-of-plane rotation of buckling braces or may restrain the brace elastically. The design philosophy chosen will affect the slenderness ratio used for the brace. If the connection is not capable of restraining the rotation of the buckling brace, an effective length factor, K, of 1.0 is used. If, however, the connection can restrain the bending demands of the buckling brace a smaller value of K may be used. The connection must then be strong enough to restrain the bending capacity of the brace taken as $1.1R_yM_p$ about each axis. Although more efficient brace members may be utilized, more robust connections will be required which will at least partially offset the material savings.

An accepted design methodology for the gusset plate which allows member end rotation was researched by Goel and has been adopted by the AISC seismic provisions. The provisions require that the brace maintain a minimum distance of 2 times the thickness of the plate from the anticipated line about which the plate will yield flexurally as the brace buckles. This line is assumed to occur between points of restraint such as the end of the gusset-to-beam connection and gusset-to-column connection. Stiffener plates may also be used to support the plate. The design should also maintain this line perpendicular to the axis of the brace to ensure the brace will buckle perpendicular to the plane of the frame.

This example allows the buckling to occur in the out-of-plane direction while it is assumed that in-plane buckling is restrained and will not control the design. If rectangular sections with largely differing properties were chosen, the capacity in each direction would need to be investigated to determine which controls the design. An effective width of plate can be calculated using Whitmore's method presented in AISC and is checked for tension and compression. The tension capacity of the gusset is conservatively estimated at A_sF_y while the brace ultimate capacity is utilized. The gusset plate is checked for compression strength in an area where it is restrained by the beam, column, and/or stiffener plates on all sides but one: along the buckling line. The true effective buckling length is complicated at best, but conservatively may be estimated at 0.8. Alternately, 1.0 between hinge locations may also be used.

Gusset-plate-to-beam-and-column connection. The forces imparted from the gusset plate to the face of the beam and column are obtained from the analysis using the uniform force method. Unless specifically optimized otherwise, each connection will see axial, shear, and bending forces. The plate, as well as the welds, are designed to remain elastic under these forces. The capacity of the weld may be checked in a number of ways. It is conservative to calculate an effective eccentricity of the shear force to the weld and add it vectorially to the axial force resulting in an effective force with an eccentricity and angle to the weld. The AISC charts for eccentrically loaded weld groups may then be used to determine the weld capacity. AISC also requires that the connection be designed for the maximum of the peak stress and 1.4 times the average stress. The 40% increase in the average stress is recommended to provide ductility to the weld for force redistribution.

Beam-to-column connection. Connection of the beam to the column is designed to transfer the resulting axial, shear, and bending demands on the beam. Due to the connections' highly restrained configuration from the gusset plate(s), this connection must be very stiff to adequately resist the forces. Moment-frame type connections consisting of groove-welded flanges and either welded or bolted webs using slip-critical bolts are typically used. The web and flange connections are sized to develop their share of the forces at the joint. It is typically sufficient to use full-penetration-welded flanges and webs.

The last page of Example 5.3 summarizes the design and Fig. 5.20 shows the final detail of the connection.

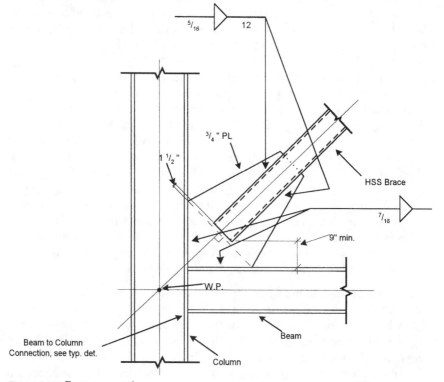

Figure 5.20 Brace connection.

5.5   Eccentrically Braced Frames

Eccentrically braced frames (EBFs) are braced frame systems which utilize a link beam created by the eccentric connection of the diagonal brace or braces. The system provides energy dissipation through inelastic deformation of the link. The link either yields in shear (short links) or in bending (long links), while the beams, columns, and braces in the system remain elastic.

The design of the connections in an EBF is very similar to that of the SCBF. The methodology used in Example 5.3 could be used to design a brace-to-beam or brace-to-column connection in an EBF with the following exceptions. First, where the SCBF was designed based on the capacity of the brace, in an EBF the expected capacity of the link is used to size the brace and beam connections. Second, since the brace is not intended to yield, providing the $2t$ buckling line is not necessary. Finally, the eccentricity of the brace to the beam creates large bending demands in the link which are resisted by the beam outside of the link and by the brace member. Although braces have traditionally been considered pinned, in an EBF a brace can attract significant bending due to their fixed connections which must be accounted for in the design of the brace and its connection to the beam and/or column. The additional bending on the gusset plate may be superimposed with the force distribution obtained from the uniform force method.

5.6   Commentary on the 1994 Northridge Earthquake Damage

The widespread damage to welded moment-resisting connections from the 1994 Northridge earthquake spurred a great deal of interest in better understanding and improving the performance of steel-frame structures in future earthquakes. Entities such as the American Institute of Steel Construction, various agencies of the federal government, and a number of private entities sponsored research and other activities toward this end. All of these efforts have contributed to a vast increase in the knowledge base through new innovations and the development of new details and approaches that will undoubtedly result in better design and construction, and ultimately, better performance of steel-frame structures in future earthquakes.

The most wide-reaching of these efforts to date has been by the SAC Joint Venture, with funding sponsored by the Federal Emergency Management Agency (FEMA). The initial phase of the SAC project resulted in the 1995 publication of FEMA 267, "Interim Guidelines: Evaluation, Repair, Modification and Design of Welded Steel Moment Frame Structures" (SAC, 1995). These interim guidelines provide substantive recommendations for the design of moment-resisting steel-frame structures, based on a broad consensus of the engineering community of the knowledge available at that time. Guidance is provided for the design of moment-resisting connections in order to avoid the type of fractures exhibited by the so-called pre-Northridge connection. A series of possible connection details are presented for consideration by design engineers. Recommendations for proper project control are also included. This document, which forms the basis for many of the improvements in the 1997 AISC seismic provisions, is an outstanding reference for design engineers performing the seismic design of any steel-framing system. It should also be noted that FEMA has funded a larger second phase of the SAC project to develop long-term standards of practice for the seismic design of moment-resisting steel frames. In 1997, FEMA 267A (SAC, 1997) was published to update FEMA 267 (SAC, 1995) with more recent information.

While the improvements made since the Northridge earthquake and those that are expected to occur with the completion of the second phase of the SAC project will undoubtedly result in better seismic performance of steel-frame buildings, it should be recognized that this field will continue to evolve with time. It is imperative, then, that engineers involved with the seismic design of steel structures stay abreast of the most recent code developments and, to as great an extent as possible, to ongoing research in this field.

But it must also be recognized that building codes and research will never be able to completely address all of the issues that engineers will address in the design of actual structures. Since no two building structures are identical, in either design or construction, a substantial amount of engineering judgment will always be required in the application of seismic design principles. Engineers must recognize that implementing the basics of good seismic design through proper system selection, configuration, member design, connection detailing, etc., will always be required in order to achieve the anticipated structural performance.

5.7 References

AISC, *Load and Resistance Factor Design Specification for Structural Steel Buildings,* American Institute of Steel Construction, Chicago, 1993.

AISC, *Load and Resistance Factor Design,* vol. II: *Connections,* American Institute of Steel Construction, Chicago, 1994.

AISC, *Seismic Provisions for Structural Steel Buildings,* American Institute of Steel Construction, Chicago, 1997.

Anderson, J. C., "Test Results for Repaired Specimen NSF#1," Report to AISC Steel Advisory Committee, June 1995.

Astaneh, Abolhassan, "Simple Methods for Design of Steel Gusset Plates," *Structures Congress Proceedings,* ASCE, San Francisco, CA, 1989, pp. 345–354.

Engelhardt, M. D., and Sabol. T. A., "Testing of Welded Steel Moment Connections in Response to the Northridge Earthquake," progress report to the AISC advisory committee on special moment-resisting frame research, October 1994.

Engelhardt, M. D., and Sabol, T. A., "Lessons Learned from the Northridge Earthquake: Steel Moment Frame Performance," *Proceedings, New Directions in Seismic Design,* Steering Committee of Symposium on a New Direction in Seismic Design, Tokyo, October 1–12, 1995.

Engelhardt, M. D., and Sabol, T. A., "Reinforcing of Steel Moment Connections with Cover Plates: Benefits and Limitations," *Engineering Structures, The Journal of Earthquake, Wind, and Ocean Engineering,* vol. 20, nos. 4–6, April–June 1998, p. 510.

Engelhardt, M. D., Winneberger, T., Zekany, A. J., and Potyraj, T. J., "The Dogbone Connection: Part II," M.S.C., August 1996.

Ivankiw, R. N., and Carter, C. J., "The Dogbone: A New Idea to Chew On," *Modern Steel Construction,* vol. 36, no. 4, 1996, pp. 18–23.

Ksai, K., and Bleiman, D., "Bolted Brackets for Repair of Damaged Steel Moment Frame Connections," 7th U.S.–Japan Workshop on the Improvement of Structural Design and Construction Practices: Lessons Learned from Northridge and Kobe, Kobe, Japan, January 1996.

Noel, S., and Uang, C. M., "Cyclic Testing of Steel Moment Connections for the San Francisco Civic Center Complex," test report to HSH Design/Build, Structural Systems Research Project, Division of Structural Engineering Report No. TR-96/07, University of California, San Diego, 1996.

Popov, E. P., and Stephen, R. M., "Cyclic Loading of Full-Size Steel Connections," Earthquake Engineering Research Center Report UCB/EERC-70-3, University of California, Berkeley, 1970.

Richard, R. M., "Analysis of Large Bracing Connection Designs for Heavy Construction," *National Steel Construction Conference Proceedings,* AISC, Chicago, 1986, pp. 31.1–31.24.

Roeder et al., "Seismic Performance of Weak Column-Strong Beam Steel Moment Frames," *Proceedings of the Fourth U.S. National Conference of Earthquake Engineering,* vol. 2, Palm Springs, CA, May 20–24, 1990, Earthquake Engineering Research Institute, El Cerrito, CA.

SAC, "Interim Guidelines: Evaluation, Repair, Modification and Design of Welded Steel Moment Frame Structures," Program to Reduce the Earthquake Hazards of Steel Moment Frame Structures, Federal Emergency Management Agency, Report FEMA 267/SAC-95-02, SAC Joint Venture, Sacramento, CA, 1995.

SAC, "Interim Guidelines Advisory No. 1: Supplement to FEMA 267," Program to Reduce the Earthquake Hazards of Steel Moment Frame Structures, Federal Emergency Management Agency, Report FEMA 267A/SAC-96-03, SAC Joint Venture, Sacramento, CA, 1997.

Tsai, K. C., and Popov, E. P., "Steel Beam-Column Joints in Seismic Moment-Resisting Frames," Earthquake Engineering Research Center Report UCB/EERC-88/19, University of California, Berkeley, 1988.

Uang, C. M., and Bondad, D. M., "Dynamic Testing of Full-Scale Steel Moment Connections," *Proceedings of 11th World Conference on Earthquake Engineering,* Acapulco, CD-ROM, Paper 407, Permagon Press, New York, 1996.

Uang, C. M., and Noel, S., "Cyclic Testing of Rib-Reinforced Steel Moment Connection with Reduced Beam Flanges," test report to Ove Arup & Partners, structural systems research project, Division of Structural Engineering Report No. TR-95/04, University of California, San Diego, 1995.

Uang, C. M., and Noel, S., "Cyclic Testing of Strong- and Weak-Axis Steel Moment Connection with Reduced Beam Flanges," final report to the City of Hope, Division of Structural Engineering Report No. TR-96/01, University of California, San Diego, 1996.

Whittaker, A., Bertero, V., and Gilani, A., "Testing of Full-Scale Steel Beam-Column Assemblies," SAC Phase I Report, SAC Joint Venture. Sacramento, CA, 1995.

Xue, M., et al., "Achieving Ductile Behavior of Moment Connections—Part II," *Modern Steel Construction,* vol. 36, no. 6, 1996, pp. 38–42.

Yu, Q. S., Noel, S., and Uang, C. M., "Experimental and Analytical Studies on Seismic Rehabilitation of Pre-Northridge Steel Moment Connections: RBS and Haunch Approaches," Division of Structural Engineering Report No. SSRP-97/08, University of California, San Diego, 1997.

Structural Steel
Details

David R. Williams
Williams Engineering Associates
Virginia Beach, VA

(Courtesy of The Steel Institute of New York.)

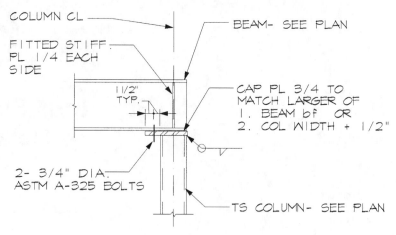

COLUMN CL

FITTED STIFF.
PL 1/4 EACH
SIDE

1 1/2"
TYP.

BEAM- SEE PLAN

CAP PL 3/4 TO
MATCH LARGER OF
1. BEAM bf OR
2. COL WIDTH + 1/2"

2- 3/4" DIA.
ASTM A-325 BOLTS

TS COLUMN- SEE PLAN

BEAM FRAMES OVER TS COLUMN

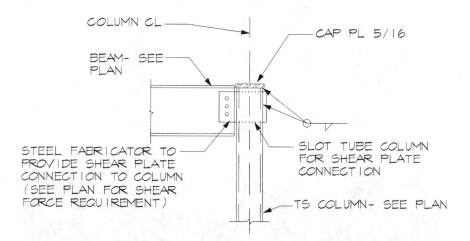

COLUMN CL

BEAM- SEE
PLAN

CAP PL 5/16

STEEL FABRICATOR TO
PROVIDE SHEAR PLATE
CONNECTION TO COLUMN
(SEE PLAN FOR SHEAR
FORCE REQUIREMENT)

SLOT TUBE COLUMN
FOR SHEAR PLATE
CONNECTION

TS COLUMN- SEE PLAN

BEAM FRAMES INTO TS COLUMN

BEAM SHEAR CONNECTION DETAILS
NOT TO SCALE   (DETAIL T5-CSH1)

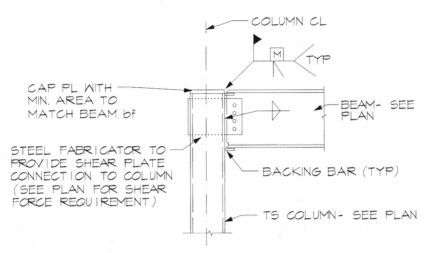

COLUMN CL

M TYP

CAP PL WITH
MIN. AREA TO
MATCH BEAM bf

BEAM- SEE
PLAN

STEEL FABRICATOR TO
PROVIDE SHEAR PLATE
CONNECTION TO COLUMN
(SEE PLAN FOR SHEAR
FORCE REQUIREMENT)

BACKING BAR (TYP)

TS COLUMN- SEE PLAN

BEAM FRAMES ONE SIDE OF TS COLUMN

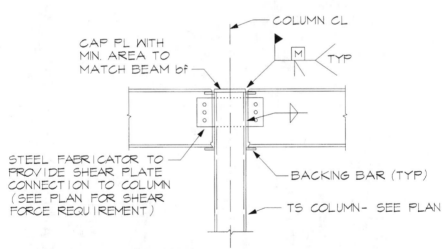

COLUMN CL

CAP PL WITH
MIN. AREA TO
MATCH BEAM bf

M TYP

STEEL FABRICATOR TO
PROVIDE SHEAR PLATE
CONNECTION TO COLUMN
(SEE PLAN FOR SHEAR
FORCE REQUIREMENT)

BACKING BAR (TYP)

TS COLUMN- SEE PLAN

BEAM FRAMES BOTH SIDES OF TS COLUMN

BEAM MOMENT CONNECTION DETAILS
NOT TO SCALE   (DETAIL T5-CM01)

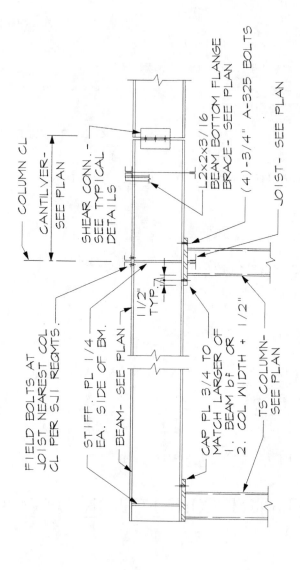

COLUMN CL

CANTILVER- SEE PLAN

SHEAR CONN. - SEE TYPICAL DETAILS

L2×2×3/16 BEAM BOTTOM FLANGE BRACE- SEE PLAN

(4)-3/4" A-325 BOLTS

JOIST- SEE PLAN

FIELD BOLTS AT JOIST NEAREST COL CL PER SJI REQMTS.

STIFF. PL 1/4 EA. SIDE OF BM.

BEAM- SEE PLAN

1 1/2" TYP.

CAP PL 3/4 TO MATCH LARGER OF 1. BEAM bf OR 2. COL WIDTH + 1/2"

TS COLUMN- SEE PLAN

END TS COLUMN

INTERIOR TS COLUMN

TYPICAL CANTILEVER ROOF BEAM DETAILS

NOT TO SCALE

(DETAIL T5—CBM1)

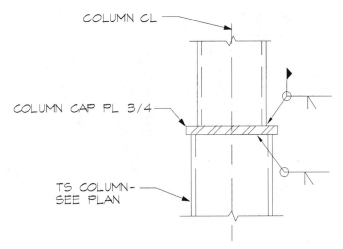

COLUMN CL

COLUMN CAP PL 3/4

TS COLUMN-
SEE PLAN

TS COLUMN SPLICE, FILLET WELDED

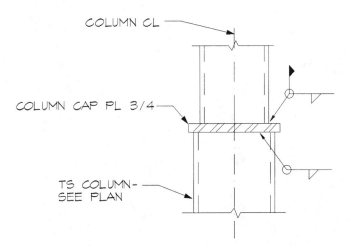

COLUMN CL

COLUMN CAP PL 3/4

TS COLUMN-
SEE PLAN

TS COLUMN SPLICE, PENETRATION WELDED

TYP. TS COLUMN SPLICE DETAILS

NOT TO SCALE   (DETAIL T5-COL1)

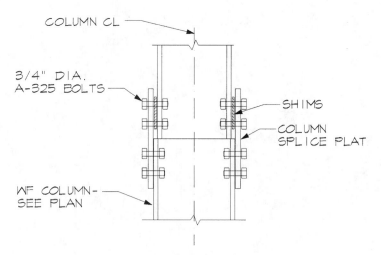

COLUMN CL

3/4" DIA.
A-325 BOLTS

SHIMS

COLUMN
SPLICE PLAT

WF COLUMN-
SEE PLAN

WF COLUMN SPLICE, BOLTED

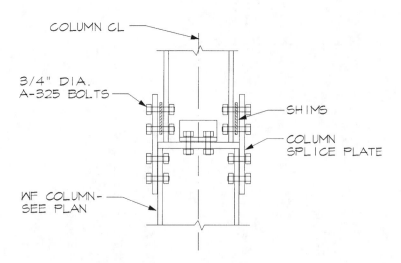

COLUMN CL

3/4" DIA.
A-325 BOLTS

SHIMS

COLUMN
SPLICE PLATE

WF COLUMN-
SEE PLAN

WF COLUMN SPLICE, BUTT PLATE

TYP. WF COLUMN SPLICE DETAILS

NOT TO SCALE   (DETAIL T5-COL2)

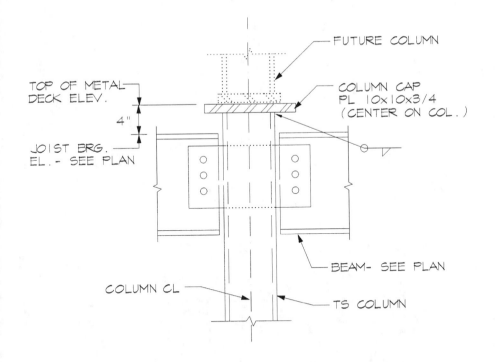

TOP OF METAL
DECK ELEV.

4"

JOIST BRG.
EL. - SEE PLAN

FUTURE COLUMN

COLUMN CAP
PL 10x10x3/4
(CENTER ON COL.)

BEAM- SEE PLAN

COLUMN CL

TS COLUMN

TYP. FUTURE COLUMN CONNECTION
NOT TO SCALE   (DETAIL T5-COL3)

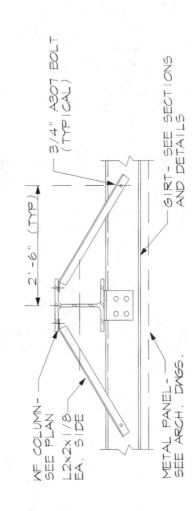

NOTES:

1. SEE WALL ELEVATIONS DETAILS FOR LOCATIONS.

2. DETAIL AT PERIMETER COLUMNS IS SHOWN. DETAIL AT CORNER COLUMNS IS SIMILAR.

3/4" A307 BOLT (TYPICAL)

2'-6" (TYP.)

GIRT- SEE SECTIONS AND DETAILS

WF COLUMN- SEE PLAN

L2x2x1/8 EA. SIDE

METAL PANEL- SEE ARCH. DWGS.

TYPICAL COLUMN BRACE DETAIL

NOT TO SCALE

(DETAIL T5-BRCE1)

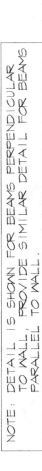

NOTE: DETAIL IS SHOWN FOR BEAMS PERPENDICULAR TO WALL, PROVIDE SIMILAR DETAIL FOR BEAMS PARALLEL TO WALL.

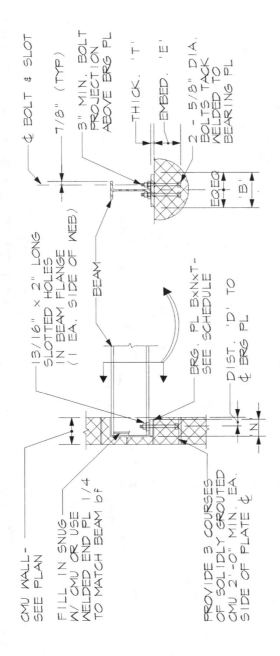

¢ BOLT & SLOT

7/8" (TYP)

3" MIN. BOLT PROJECTION ABOVE BRG PL

THICK. 'T'

EMBED. 'E'

2 - 5/8" DIA. BOLTS TACK WELDED TO BEARING PL

13/16" x 2" LONG SLOTTED HOLES IN BEAM FLANGE (1 EA. SIDE OF WEB)

BEAM

BRG. PL BxNxT- SEE SCHEDULE

DIST. 'D' TO ¢ BRG PL

CMU WALL - SEE PLAN

FILL IN SNUG W/ CMU OR USE WELDED END PL 1/4 TO MATCH BEAM bf

PROVIDE 3 COURSES OF SOLIDLY GROUTED CMU 2'-0" MIN. EA. SIDE OF PLATE ¢

'FIXED' CONNECTION

TYPICAL BEAM BEARING ON MASONRY DETAILS
(DETAIL T5-CMU1)

NOT TO SCALE

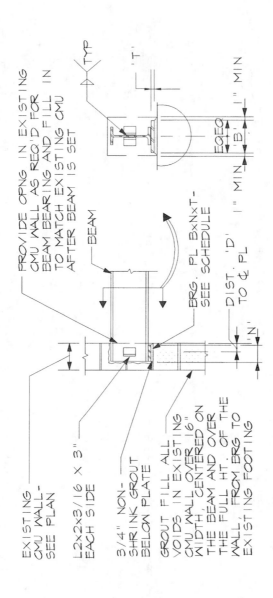

EXISTING CMU WALL- SEE PLAN

PROVIDE OPNG IN EXISTING CMU WALL AS REQ'D FOR BEAM BEARING AND FILL IN TO MATCH EXISTING CMU AFTER BEAM IS SET

L2x2x3/16 x 3" EACH SIDE

BEAM

3/4" NON- SHRINK GROUT BELOW PLATE

BRG. PL BxNxT- SEE SCHEDULE

DIST. 'D'   1" MIN
TO ₵ PL

'T'

EQEQ
'B'   1" MIN

'N'

GROUT FILL ALL VOIDS IN EXISTING CMU WALL OVER 16" WIDTH, CENTERED ON THE BEAM AND OVER THE FULL HT. OF THE WALL, FROM BRG TO EXISTING FOOTING

TYP

TYPICAL BEAM BEARING ON EXISTING MASONRY DETAILS
(DETAIL T5-CMU3)

NOT TO SCALE

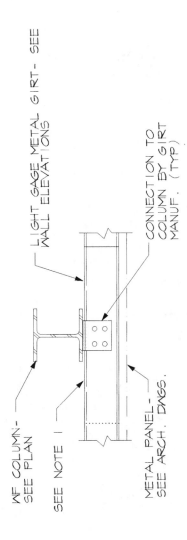

NOTES:
1. LAP GIRTS AT COLUMN AS REQUIRED BY THE MANUFACTURER.
2. PROVIDE GIRT WEB STIFFENERS AS REQUIRED BY THE MANUF.

LIGHT GAGE METAL GIRT- SEE WALL ELEVATIONS

CONNECTION TO COLUMN BY GIRT MANUF. (TYP)

WF COLUMN- SEE PLAN

SEE NOTE 1

METAL PANEL- SEE ARCH. DWGS.

TYPICAL LIGHT-GAGE METAL GIRT CONNECTION DETAILS
(DETAIL T5-GIRT1)

NOT TO SCALE

339

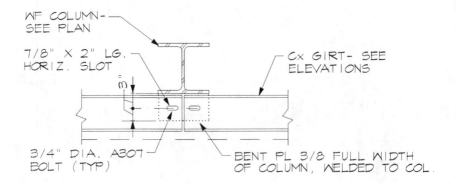

WF COLUMN—
SEE PLAN

7/8" X 2" LG.
HORIZ. SLOT

3"

Cx GIRT- SEE
ELEVATIONS

3/4" DIA. A307
BOLT (TYP)

BENT PL 3/8 FULL WIDTH
OF COLUMN, WELDED TO COL.

AT TYP. GIRTS

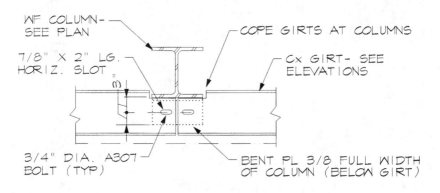

WF COLUMN—
SEE PLAN

COPE GIRTS AT COLUMNS

7/8" X 2" LG.
HORIZ. SLOT

3"

Cx GIRT- SEE
ELEVATIONS

3/4" DIA. A307
BOLT (TYP)

BENT PL 3/8 FULL WIDTH
OF COLUMN (BELOW GIRT)

AT COPED GIRTS

TYP. GIRT CONNECTION DETAILS
NOT TO SCALE   (DETAIL T5–GIRT4)

NOTE: COORD. SIZE AND LOCATION OF ROOF
OPENINGS WITH ACTUAL EQUIPMENT SELECTED.

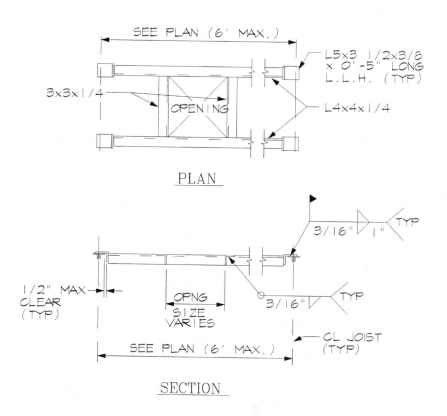

SEE PLAN (6' MAX.)

L5x3 1/2x3/8
x 0'-5" LONG
L.L.H. (TYP)

3x3x1/4

OPENING

L4x4x1/4

PLAN

3/16" 1" TYP

1/2" MAX
CLEAR
(TYP)

OPNG
SIZE
VARIES

3/16" TYP

CL JOIST
(TYP)

SEE PLAN (6' MAX.)

SECTION

TYP. ROOF OPENING DETAIL

NOT TO SCALE   (DETAIL T5–R01)

CENTER STUDS ON
THE BEAM UNLESS
OTHERWISE REQUIRED
(SEE NOTE 3 BELOW)

8 x T
MAX.
SPACING

DECK THICKNESS
(T)- SEE PLAN

5/8" DIA. PUDDLE WELD
WHERE NO STUD IS REQUIRED

COMPOSITE BEAM-
SEE SCHEDULE

DECK IS PERPENDICULAR OR SKEWED TO BEAM

NOTES:

1. THE MIN. NUMBER OF STUDS FOR EACH BEAM IS SHOWN IN
THE COMPOSITE BEAM SCHEDULE.

2. SPACE STUDS AS EVENLY AS POSSIBLE IN AVAILABLE
DECK FLUTES. WHERE STUD SPACING EXCEEDS THE MAX.
SPACING ALLOWED, PROVIDE ADDITIONAL STUDS TO
SATSIFY THE SPACING REQUIREMENTS.

3. WHERE THE NUMBER OF STUDS EXCEEDS THE NUMBER OF
FLUTES, PROVIDE TWO STUDS IN EVERY OTHER FLUTE,
STARTING AT EACH END OF THE BEAM. THE TRANSVERSE
SPACING BETWEEN TWO STUDS IN A SINGLE FLUTE SHALL
BE 4 x STUD DIAMETERS (MIN.).

4. SEE THE COMPOSITE BEAM SCHEDULE FOR ADDITIONAL
REQUIREMENTS. TURN THE NATURAL BEAM CAMBER UP.

TYP. COMPOSITE BEAM ELEVATION
NOT TO SCALE   (DETAIL T5-COMP1)

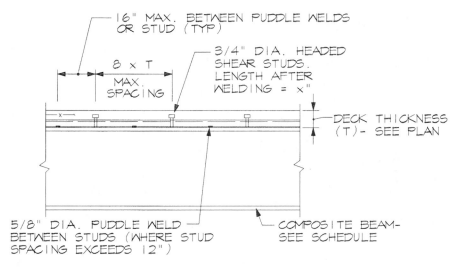

16" MAX. BETWEEN PUDDLE WELDS
OR STUD (TYP)

8 x T
MAX.
SPACING

3/4" DIA. HEADED
SHEAR STUDS.
LENGTH AFTER
WELDING = x"

DECK THICKNESS
(T)- SEE PLAN

5/8" DIA. PUDDLE WELD
BETWEEN STUDS (WHERE STUD
SPACING EXCEEDS 12")

COMPOSITE BEAM-
SEE SCHEDULE

DECK IS PARALLEL TO BEAM

COMPOSITE BEAM SCHEDULE

MARK	BEAM	MAX. END REACTION	MIDSPAN CAMBER	NUMBER OF STUDS	REMARKS
CB-1	W10x15	11 KIPS	—	12	INTERIOR
CB-2	W10x15	8 KIPS	—	8	INTERIOR
CB-3	W10x26	13 KIPS	—	12	INTERIOR
CB-4	W10x15	8 KIPS	—	3	CANTILEVER
CB-5	W10x26	13 KIPS	—	5	CANTILEVER
CB-6	W10x12	7 KIPS	—	3	CANTILEVER
CB-7	W18x76	86 k (@RT)	0.6"	22	GIRDER
CB-8	W16x40	36 k (@RT)	—	16	GIRDER

NOTE: NATURAL CAMBER OF BEAM SHALL BE TURNED UP.

TYP. COMPOSITE BEAM ELEVATION
NOT TO SCALE   (DETAIL T5-COMP2)

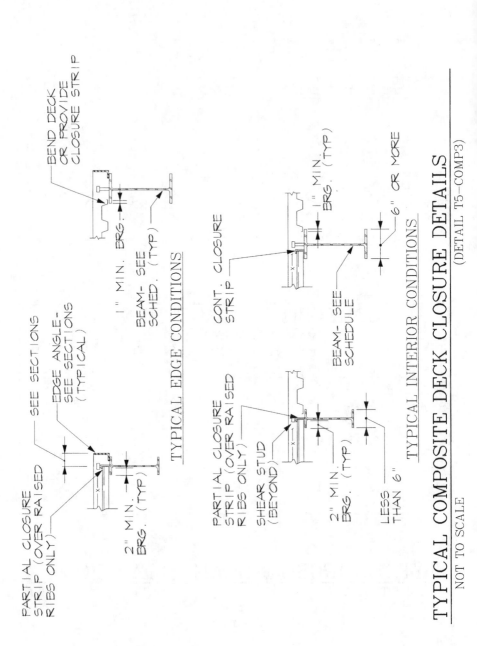

PARTIAL CLOSURE STRIP (OVER RAISED RIBS ONLY)

SEE SECTIONS

EDGE ANGLE - SEE SECTIONS (TYPICAL)

2" MIN. BRG. (TYP)

BEND DECK OR PROVIDE CLOSURE STRIP

1" MIN. BRG.

BEAM- SEE SCHED. (TYP)

TYPICAL EDGE CONDITIONS

PARTIAL CLOSURE STRIP (OVER RAISED RIBS ONLY)

SHEAR STUD (BEYOND)

2" MIN. BRG. (TYP)

LESS THAN 6"

CONT. CLOSURE STRIP

1" MIN. BRG. (TYP)

BEAM- SEE SCHEDULE

6" OR MORE

TYPICAL INTERIOR CONDITIONS

TYPICAL COMPOSITE DECK CLOSURE DETAILS

(DETAIL T5-COMP3)

NOT TO SCALE

344

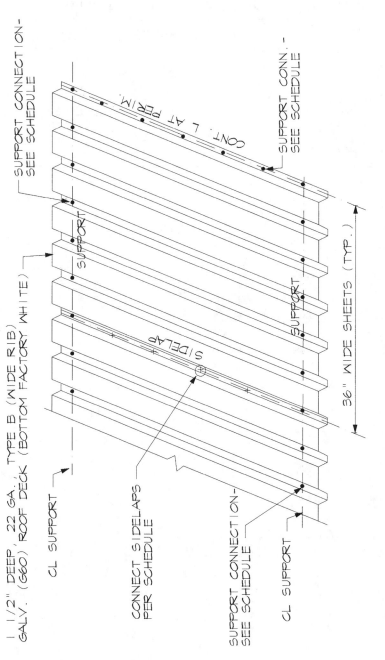

1 1/2" DEEP, 22 GA., TYPE B (WIDE RIB)
GALV. (G60) ROOF DECK (BOTTOM FACTORY WHITE)

SUPPORT CONNECTION -
SEE SCHEDULE

SUPPORT CONN. -
SEE SCHEDULE

CONT. L AT PERIM.

CL SUPPORT

SUPPORT

SUPPORT

SUPPORT

36" WIDE SHEETS (TYP.)

SIDELAP

CONNECT SIDELAPS
PER SCHEDULE

SUPPORT CONNECTION -
SEE SCHEDULE

CL SUPPORT

TYPICAL ROOF DECK ATTACHMENT DETAIL

NOT TO SCALE

(DETAIL T5—RDIA2)

345

ROOF DECK CONNECTION SCHEDULE	
ZONE	DECK CONNECTION
ZONE 1 (DIAPH. CAPACITY= 470#/LF)	WELDS IN 36/7 PATTERN (6)-#10 TEKS @ SIDELAPS
ZONE 2 (DIAPH. CAPACITY= 290#/LF)	WELDS IN 36/7 PATTERN (3)-#10 TEKS @ SIDELAPS
ZONE 3 (DIAPH. CAPACITY= 290#/LF)	WELDS IN 36/4 PATTERN (3)-#10 TEKS @ SIDELAPS

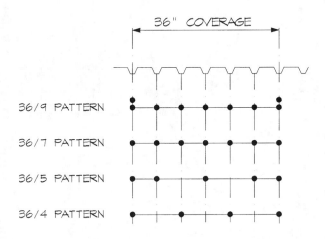

36" COVERAGE

36/9 PATTERN

36/7 PATTERN

36/5 PATTERN

36/4 PATTERN

ROOF DECK ATTACHMENT PATTERNS
NOT TO SCALE   (DETAIL T5-RDIA4)

NOTES:
1. AT JOIST NEAREST (OR ON) COLUMN CL
    CONNECT JOISTS WITH FIELD BOLTS PER
    S.J.I. REQUIREMENTS.
2. OFFSET JOISTS IF BEAM FLANGE IS
    LESS THAN 5" AND PROVIDE MINIMUM OF
    2 1/2" JOIST BEARING.

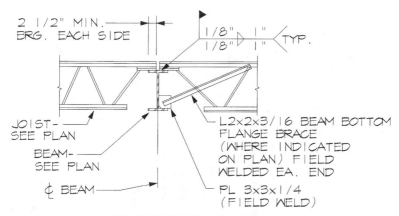

2 1/2" MIN.
BRG. EACH SIDE

1/8" ▷ 1"
1/8" ▷ 1" ⟨TYP.

JOIST-
SEE PLAN

BEAM-
SEE PLAN

₵ BEAM

L2x2x3/16 BEAM BOTTOM
FLANGE BRACE
(WHERE INDICATED
ON PLAN) FIELD
WELDED EA. END

PL 3x3x1/4
(FIELD WELD)

AT INTERIOR BEAMS

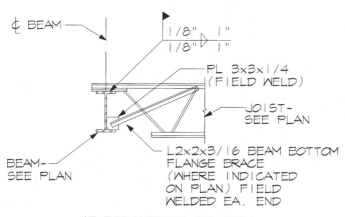

₵ BEAM

1/8" ▷ 1"
1/8" ▷ 1"

PL 3x3x1/4
(FIELD WELD)

JOIST-
SEE PLAN

BEAM-
SEE PLAN

L2x2x3/16 BEAM BOTTOM
FLANGE BRACE
(WHERE INDICATED
ON PLAN) FIELD
WELDED EA. END

AT PERIMETER BEAMS

TYP. K-JOIST CONNECTION DETAILS
NOT TO SCALE   (DETAIL T5-J1)

NOTES:
1. AT JOIST NEAREST (OR ON) COLUMN CL,
    CONNECT JOISTS WITH FIELD BOLTS PER
    S.J.I. REQUIREMENTS.
2. OFFSET JOISTS IF BEAM FLANGE IS
    LESS THAN 8" AND PROVIDE MINIMUM OF
    4" JOIST BEARING.

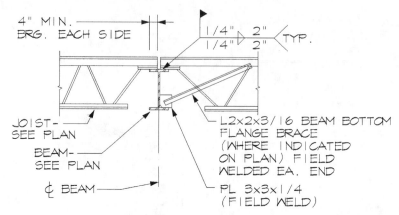

4" MIN.
BRG. EACH SIDE

1/4" ▷ 2"
1/4" ▷ 2" ⟨TYP.

JOIST-
SEE PLAN

BEAM-
SEE PLAN

₵ BEAM

L2x2x3/16 BEAM BOTTOM
FLANGE BRACE
(WHERE INDICATED
ON PLAN) FIELD
WELDED EA. END

PL 3x3x1/4
(FIELD WELD)

AT INTERIOR BEAMS

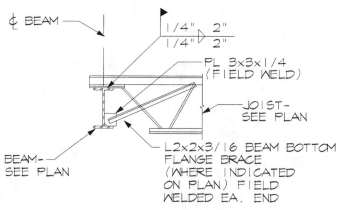

₵ BEAM

1/4" ▷ 2"
1/4" ▷ 2"

PL 3x3x1/4
(FIELD WELD)

JOIST-
SEE PLAN

L2x2x3/16 BEAM BOTTOM
FLANGE BRACE
(WHERE INDICATED
ON PLAN) FIELD
WELDED EA. END

BEAM-
SEE PLAN

AT PERIMETER BEAMS

TYP. LH–JOIST CONNECTION DETAILS

NOT TO SCALE   (DETAIL T5–J2)

NOTES:
1. CONCENTRATED LOAD LOCATED AT JOIST PANEL POINT LOCATION - NO ADDITIONAL ANGLES REQUIRED.
2. CONCENTRATED LOAD (100 LBS. OR HEAVIER) NOT LOCATED AT JOIST PANEL POINT LOCATION - PROVIDE ⌐L1×1×1/8 TO PANEL POINT AS SHOWN.

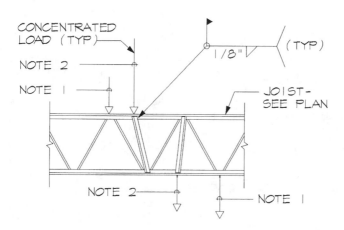

CONCENTRATED LOAD (TYP)

NOTE 2

NOTE 1

1/8" (TYP)

JOIST- SEE PLAN

NOTE 2

NOTE 1

TYP. CONCENTRATED LOAD DETAIL
NOT TO SCALE   (DETAIL T5–J5)

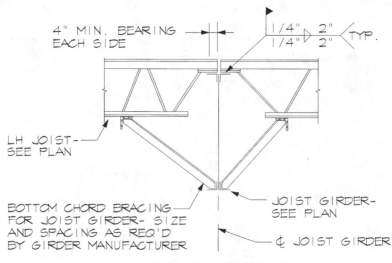

4" MIN. BEARING
EACH SIDE

1/4" 2"
1/4" 2" TYP.

LH JOIST-
SEE PLAN

BOTTOM CHORD BRACING
FOR JOIST GIRDER- SIZE
AND SPACING AS REQ'D
BY GIRDER MANUFACTURER

JOIST GIRDER-
SEE PLAN

₵ JOIST GIRDER

LH–SERIES JOISTS

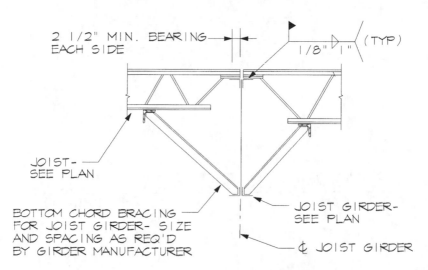

2 1/2" MIN. BEARING
EACH SIDE

1/8" 1" (TYP)

JOIST-
SEE PLAN

BOTTOM CHORD BRACING
FOR JOIST GIRDER- SIZE
AND SPACING AS REQ'D
BY GIRDER MANUFACTURER

JOIST GIRDER-
SEE PLAN

₵ JOIST GIRDER

K– SERIES JOISTS

TYP. JOIST TO GIRDER CONNECTION
NOT TO SCALE   (DETAIL T5–JG1)

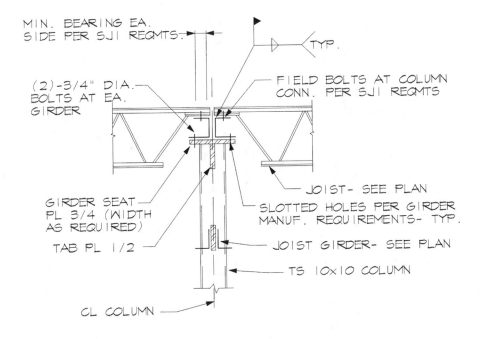

MIN. BEARING EA.
SIDE PER SJI REQMTS.

TYP.

(2)-3/4" DIA.
BOLTS AT EA.
GIRDER

FIELD BOLTS AT COLUMN
CONN. PER SJI REQMTS

JOIST- SEE PLAN

GIRDER SEAT
PL 3/4 (WIDTH
AS REQUIRED)

SLOTTED HOLES PER GIRDER
MANUF. REQUIREMENTS- TYP.

TAB PL 1/2

JOIST GIRDER- SEE PLAN

TS 10x10 COLUMN

CL COLUMN

TYP. DETAIL AT INTERIOR COLUMN CL
NOT TO SCALE   (DETAIL T5-JG2)

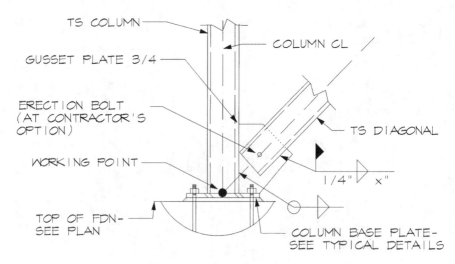

TS COLUMN

COLUMN CL

GUSSET PLATE 3/4

ERECTION BOLT
(AT CONTRACTOR'S
OPTION)

TS DIAGONAL

WORKING POINT

1/4" ▷ x"

TOP OF FDN-
SEE PLAN

COLUMN BASE PLATE-
SEE TYPICAL DETAILS

DETAIL A, AT TS COLUMN

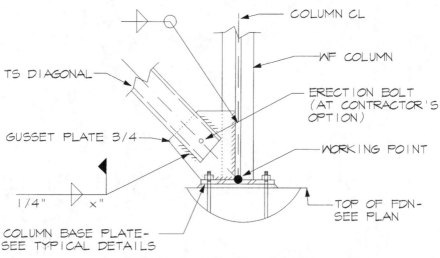

COLUMN CL

WF COLUMN

TS DIAGONAL

ERECTION BOLT
(AT CONTRACTOR'S
OPTION)

GUSSET PLATE 3/4

WORKING POINT

1/4" ▷ x"

TOP OF FDN-
SEE PLAN

COLUMN BASE PLATE-
SEE TYPICAL DETAILS

DETAIL A, AT WF COLUMN

BRACED BAY DETAILS

NOT TO SCALE   (DETAIL T5-BAYD1)

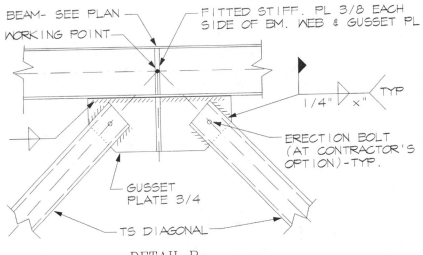

BEAM- SEE PLAN

WORKING POINT

FITTED STIFF. PL 3/8 EACH
SIDE OF BM. WEB & GUSSET PL

1/4" × " TYP

ERECTION BOLT
(AT CONTRACTOR'S
OPTION)-TYP.

GUSSET
PLATE 3/4

TS DIAGONAL

DETAIL B

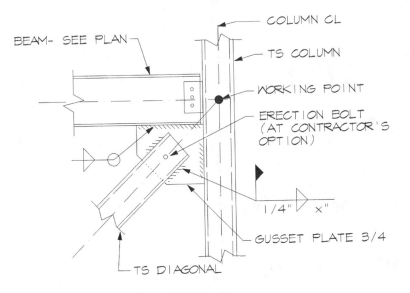

COLUMN CL

BEAM- SEE PLAN

TS COLUMN

WORKING POINT

ERECTION BOLT
(AT CONTRACTOR'S
OPTION)

1/4" × "

GUSSET PLATE 3/4

TS DIAGONAL

DETAIL C

BRACED BAY DETAILS
NOT TO SCALE   (DETAIL T5-BAYD2)

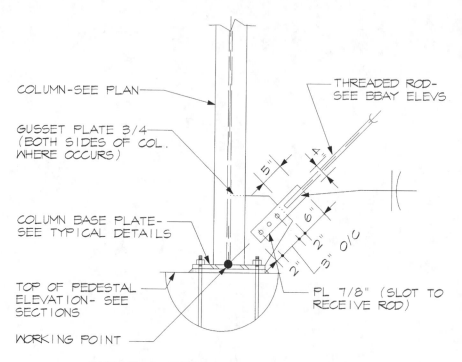

NOTE: PROVIDE BOLTED
CONNECTION BETWEEN
PL 7/8" AND THE 3/4"
GUSSET PER THE FORCES
INDICATED ON THE BRACED
BAY ELEVATIONS

COLUMN-SEE PLAN

THREADED ROD-
SEE BBAY ELEVS

GUSSET PLATE 3/4
(BOTH SIDES OF COL.
WHERE OCCURS)

COLUMN BASE PLATE-
SEE TYPICAL DETAILS

TOP OF PEDESTAL
ELEVATION- SEE
SECTIONS

PL 7/8" (SLOT TO
RECEIVE ROD)

WORKING POINT

DETAIL A, USING THREADED RODS

BRACED BAY DETAILS
NOT TO SCALE   (DETAIL T5-BAYD4)

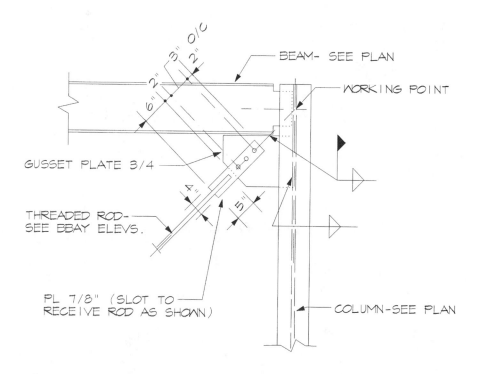

BEAM- SEE PLAN

WORKING POINT

GUSSET PLATE 3/4

THREADED ROD-
SEE BBAY ELEVS.

PL 7/8" (SLOT TO
RECEIVE ROD AS SHOWN)

COLUMN-SEE PLAN

DETAIL C, USING THREADED RODS

BRACED BAY DETAILS
NOT TO SCALE   (DETAIL T5-BAYD5)

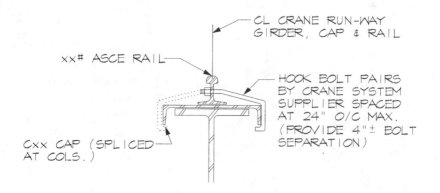

CL CRANE RUN-WAY
GIRDER, CAP & RAIL

xx# ASCE RAIL

HOOK BOLT PAIRS
BY CRANE SYSTEM
SUPPLIER SPACED
AT 24" O/C MAX.
(PROVIDE 4"± BOLT
SEPARATION)

Cxx CAP (SPLICED
AT COLS.)

TYPICAL CRANE RAIL CONNECTION
NOT TO SCALE

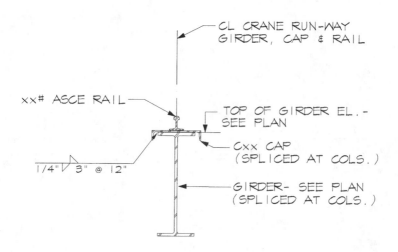

CL CRANE RUN-WAY
GIRDER, CAP & RAIL

xx# ASCE RAIL

TOP OF GIRDER EL. -
SEE PLAN

Cxx CAP
(SPLICED AT COLS.)

1/4" 3" @ 12"

GIRDER- SEE PLAN
(SPLICED AT COLS.)

TYPICAL RUN WAY GIRDER DETAIL
NOT TO SCALE   (DETAIL T5-CRAN1)

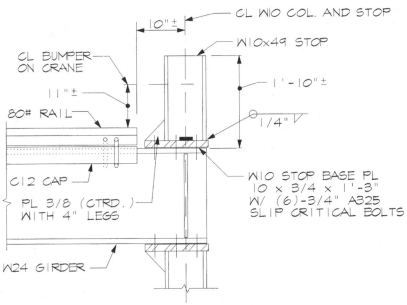

CL WIO COL. AND STOP

WIOx49 STOP

10" ±

CL BUMPER
ON CRANE

1'-10"±

11" ±

1/4"

80# RAIL

C12 CAP

WIO STOP BASE PL
10 x 3/4 x 1'-3"
W/ (6)-3/4" A325
SLIP CRITICAL BOLTS

PL 3/8 (CTRD.)
WITH 4" LEGS

W24 GIRDER

HIGH CAPACITY CRANE (30 TON MAX.)

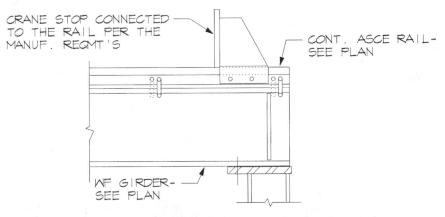

CRANE STOP CONNECTED
TO THE RAIL PER THE
MANUF. REQMT'S

CONT. ASCE RAIL-
SEE PLAN

WF GIRDER-
SEE PLAN

LOW CAPACITY CRANE (5 TON MAX.)

TYPICAL CRANE STOP DETAILS
NOT TO SCALE   (DETAIL T5–CRAN4)

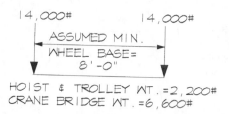

14,000#   14,000#

ASSUMED MIN.
WHEEL BASE=
8'-0"

HOIST & TROLLEY WT.=2,200#
CRANE BRIDGE WT.=6,600#

ASSUMED CRANE WHEEL LOADINGS
(2 TON CRANE, 2 WHEELS PER END TRUCK)

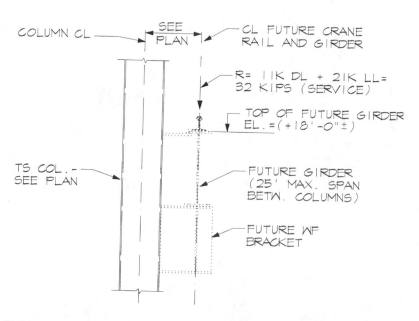

COLUMN CL ——

SEE
PLAN

—— CL FUTURE CRANE
RAIL AND GIRDER

—R= 11K DL + 21K LL=
32 KIPS (SERVICE)

— TOP OF FUTURE GIRDER
EL.=(+18'-0"±)

TS COL. -
SEE PLAN

—FUTURE GIRDER
(25' MAX. SPAN
BETW. COLUMNS)

—FUTURE WF
BRACKET

ASSUMED FUTURE CRANE CONDITION
NOT TO SCALE   (DETAIL T5-CRAN8)

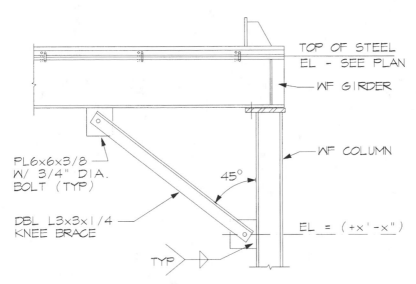

TOP OF STEEL
EL - SEE PLAN

WF GIRDER

WF COLUMN

PL6x6x3/8
W/ 3/4" DIA.
BOLT (TYP)

45°

DBL L3x3x1/4
KNEE BRACE

EL = (+x'-x")

TYP

TYPICAL KNEE BRACE DETAIL

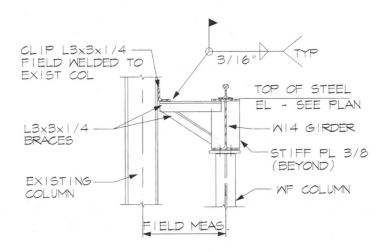

CLIP L3x3x1/4
FIELD WELDED TO
EXIST COL

3/16"   TYP

TOP OF STEEL
EL - SEE PLAN

L3x3x1/4
BRACES

W14 GIRDER

STIFF PL 3/8
(BEYOND)

EXISTING
COLUMN

WF COLUMN

FIELD MEAS.

TYPICAL GIRDER BRACE DETAIL

MISC. CRANE FRAMING DETAILS

NOT TO SCALE   (DETAIL T5-CRAN9)

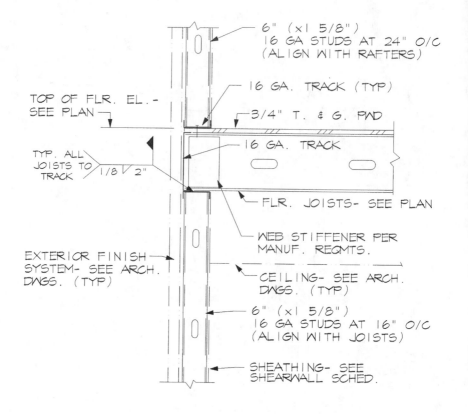

6" (x1 5/8")
16 GA STUDS AT 24" O/C
(ALIGN WITH RAFTERS)

16 GA. TRACK (TYP)

TOP OF FLR. EL.-
SEE PLAN

3/4" T. & G. PWD

16 GA. TRACK

TYP. ALL
JOISTS TO   1/8 ⌵ 2"
TRACK

FLR. JOISTS- SEE PLAN

WEB STIFFENER PER
MANUF. REQMTS.

EXTERIOR FINISH
SYSTEM- SEE ARCH.
DWGS. (TYP)

CEILING- SEE ARCH.
DWGS. (TYP)

6" (x1 5/8")
16 GA STUDS AT 16" O/C
(ALIGN WITH JOISTS)

SHEATHING- SEE
SHEARWALL SCHED.

TYPICAL FLOOR JOIST BEARING DETAIL

NOT TO SCALE   (DETAIL T5–CF1)

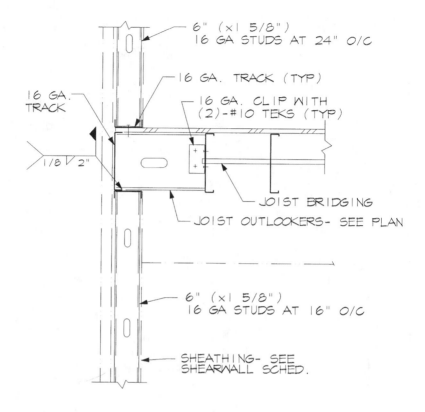

6" (×1 5/8")
16 GA STUDS AT 24" O/C

16 GA. TRACK (TYP)

16 GA.
TRACK

16 GA. CLIP WITH
(2)-#10 TEKS (TYP)

1/8 V 2"

JOIST BRIDGING

JOIST OUTLOOKERS- SEE PLAN

6" (×1 5/8")
16 GA STUDS AT 16" O/C

SHEATHING- SEE
SHEARWALL SCHED.

TYP. FLOOR JOIST NONBEARING DETAIL
NOT TO SCALE   (DETAIL T5-CF2)

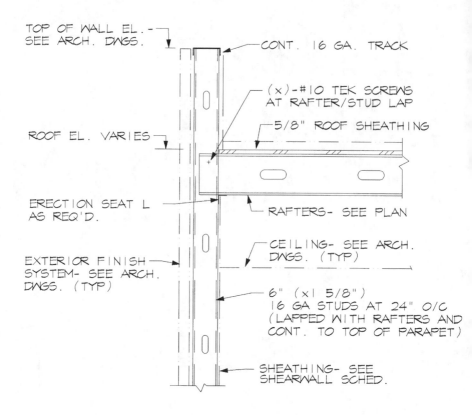

TOP OF WALL EL. -
SEE ARCH. DWGS.

CONT. 16 GA. TRACK

(x)-#10 TEK SCREWS
AT RAFTER/STUD LAP

5/8" ROOF SHEATHING

ROOF EL. VARIES

ERECTION SEAT L
AS REQ'D.

RAFTERS- SEE PLAN

CEILING- SEE ARCH.
DWGS. (TYP)

EXTERIOR FINISH
SYSTEM- SEE ARCH.
DWGS. (TYP)

6" (x1 5/8")
16 GA STUDS AT 24" O/C
(LAPPED WITH RAFTERS AND
CONT. TO TOP OF PARAPET)

SHEATHING- SEE
SHEARWALL SCHED.

TYPICAL ROOF JOIST BEARING DETAIL
NOT TO SCALE   (DETAIL T5-CF3)

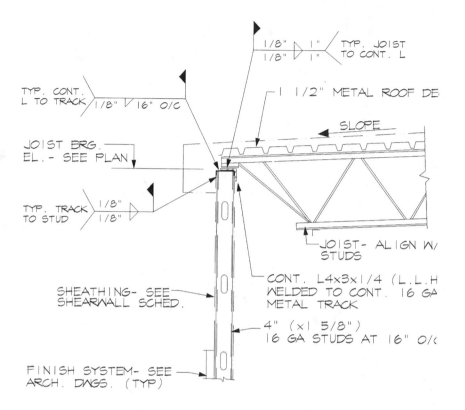

1/8" ◁ 1"
1/8" ◁ 1"
TYP. JOIST
TO CONT. L

1 1/2" METAL ROOF DE

TYP. CONT.
L TO TRACK / 1/8" ▽ 16" O/C

SLOPE

JOIST BRG.
EL. - SEE PLAN

TYP. TRACK \ 1/8"
TO STUD / 1/8"

JOIST- ALIGN W/
STUDS

CONT. L4x3x1/4 (L.L.H
WELDED TO CONT. 16 GA
METAL TRACK

SHEATHING- SEE
SHEARWALL SCHED.

4" (x1 5/8")
16 GA STUDS AT 16" O/C

FINISH SYSTEM- SEE
ARCH. DWGS. (TYP)

TYP. BAR JOIST ROOF BEARING DETAI
NOT TO SCALE
(DETAIL T5-CF7)

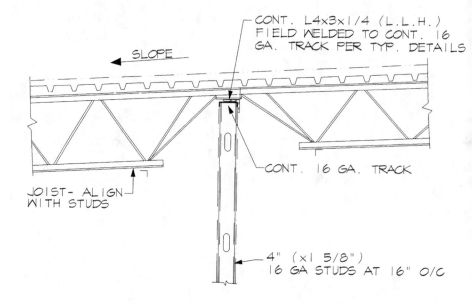

SLOPE

CONT. L4x3x1/4 (L.L.H.)
FIELD WELDED TO CONT. 16
GA. TRACK PER TYP. DETAILS

CONT. 16 GA. TRACK

JOIST- ALIGN
WITH STUDS

4" (x1 5/8")
16 GA STUDS AT 16" O/C

TYP. INTERIOR JOIST BEARING DETAIL
NOT TO SCALE   (DETAIL T5–CF8)

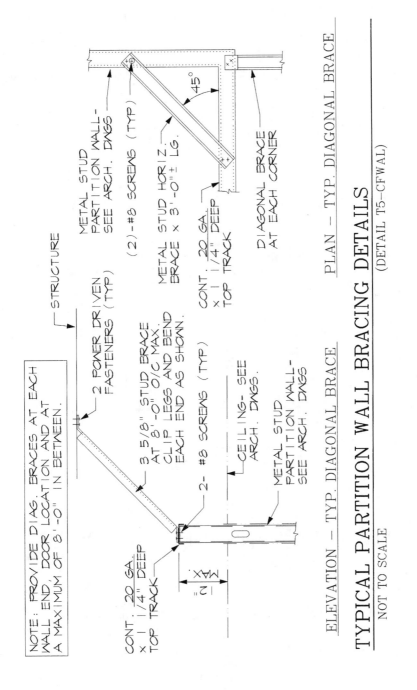

NOTE: PROVIDE DIAG. BRACES AT EACH WALL END, DOOR LOCATION AND AT A MAXIMUM OF 8'-0" IN BETWEEN.

STRUCTURE

METAL STUD PARTITION WALL- SEE ARCH. DWGS

2 POWER DRIVEN FASTENERS (TYP)

(2)-#8 SCREWS (TYP)

45°

METAL STUD HORIZ. BRACE × 3'-0"± LG.

3 5/8" STUD BRACE AT 8'-0" O/C MAX. CLIP LEGS AND BEND EACH END AS SHOWN.

CONT. 20 GA. × 1 1/4" DEEP TOP TRACK

DIAGONAL BRACE AT EACH CORNER

PLAN – TYP. DIAGONAL BRACE

CONT. 20 GA. × 1 1/4" DEEP TOP TRACK

2- #8 SCREWS (TYP)

CEILING- SEE ARCH. DWGS.

METAL STUD PARTITION WALL- SEE ARCH. DWGS

MAX. 2 1"

ELEVATION – TYP. DIAGONAL BRACE

TYPICAL PARTITION WALL BRACING DETAILS
(DETAIL T5-CFWAL)

NOT TO SCALE

COLD–FORMED METAL HEADER SCHEDULE

MARK	SECTION	DESCRIPTION	GAUGE	JAMB STUDS	
				JACK	FULL–HT.
H-1	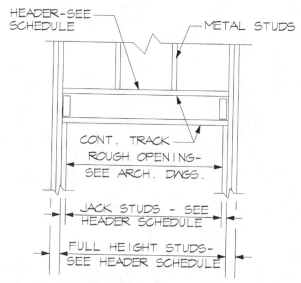	(2)- 6"(x1 5/8")	16 GA.	SINGLE	SINGLE
H-2		(2)- 8"(x1 5/8")	16 GA.	SINGLE	DOUBLE

HEADER-SEE SCHEDULE

METAL STUDS

CONT. TRACK

ROUGH OPENING- SEE ARCH. DWGS.

JACK STUDS - SEE HEADER SCHEDULE

FULL HEIGHT STUDS- SEE HEADER SCHEDULE

TYPICAL FRAMED OPENING DETAIL
NOT TO SCALE   (DETAIL T5–CFHDR)

6.8 Reference

Williams, David R., *Structural Details Manual*, McGraw-Hill, 1998.

7

Connection Design
for Special Structures

Lawrence A. Kloiber
Lejeune Steel
Minneapolis, MN

7.1 Introduction

Design drawings are intended to convey the engineer of record's (EOR) concept of the structure to the builder. As in any communication, there are always ample opportunities for misinterpretations or even a failure to communicate important information. The chance of a communication failure increases when constraints, such as time or budget, impact the drawing preparation and when the structural system involves unique, complex, or heavy members. Typically most of the engineer's design efforts involve laying out the structural system, structural analysis, and designing the members. Connections are often a last-minute addition to the drawing. They are usually communicated by use of schedules and standard details or, in the case of unique connections, a representative detail. In a complex structure it

(Courtesy of The Steel Institute of New York.)

is almost impossible for the designer's details to show the variations required to accommodate all of the loads, member sizes, and geometry required for special connections.

Traditionally, the structural engineer establishes the strength and stiffness requirements for all connections on the design drawings along with the preferred method of force transfer. The fabricator's engineer/detailer then develops connections that comply with these guidelines. The scope of this work may vary from only establishing detail dimensions to selecting the type of connection and sizing the connection material. When designing connections for special structures it is often necessary to develop connections that involve nonstandard details or at least a modification of standard details. It is important to clearly show load-transfer requirements and to work with the fabricator to design connections that can be economically fabricated and erected and still meet all structural requirements.

Unique connections for structures, such as long-span truss connections, space-frame connections, heavy plate connections, and splices in group IV and V shapes, present special problems. Standard connections have been refined over the years and the problems are known. Every time you develop connections for new systems you have to be on the alert for unforeseen problems. Long-span truss connections must carry large forces while allowing for mill and fabrication tolerances and still provide easy assembly in the field. Heavy connections may have material and structural compatibility problems. Space-frame connections have access and dimensional tolerance problems. All of these may involve economic and constructibility issues that require input from the fabricator and erector.

The completion of structural design drawings marks the close of only part of the total connection design process when designing special structures. Connections often need to be modified for reasons of constructibility and economy during the detailing phase. With special connections the need for modification may even arise during the fabrication or erection phases of construction when constructibility problems are discovered. Special connections as mentioned previously have not been subjected to the trials of repeated use and unforeseen problems sometimes occur during construction.

Converting these design drawings into a structure requires a partnership between the EOR and the fabricator. Each has a role, the EOR as the designer and the fabricator as the builder. While they may assist each other, they remain solely responsible for the separate duties. The fabricator may size connections and propose changes in details and material but this is done as a builder not a designer. The engineer may help with construction by providing dimensional information or checking construction loads for the erector but the fabricator/erector remains responsible for the fit and constructibility of the structure. This design-build approach to developing connections works well for special structures.

The EOR may elect to show representative connections, the type of load transfer that is needed, along with the required connection design forces rather than attempting to dimension each connection. The fabricator then sizes the connection material based on these requirements and provides all of the detail dimensioning. The EOR can then review and verify that the connections are adequate for his or her design. This method of developing connection details utilizes the knowledge and experience of both the EOR and the fabricator in the most efficient way.

The detailing phase should start with a predetail conference with the EOR, fabricator, and where necessary the erector and general contractor or construction manager attending. Preliminary sketches and schedules of some connections may be submitted at this time. This initial meeting is an effort by the entire construction team to understand the structural concept, verify whether all the needed information is shown, and determine if there are any obvious constructibility problems.

Connection design does not even stop with the approval of shop drawings by the EOR. The beginning of shop fabrication presents additional challenges. Material ordered for the project may not conform to specifications, fabrication errors may occur, and unforeseen constructibility problems might be discovered. The fabricator must evaluate each problem to determine if a modification or repair is necessary. Even though shop supervision or quality control personnel may identify the problem, it is important that the fabricator's engineer review and document any modification. If it is determined that the connection, even after repair or modification, will not meet the original standards, the proposed action must be submitted to the EOR to determine if the connection as fabricated will be fit for its intended purpose.

The erection of the steel frame serves as a check of the fabricator's efforts to detail and fabricate connections that fit perfectly. If the erector cannot put the bolt in the hole, it may be necessary to modify the connection. Most minor fit-up problems can be resolved with reaming, slotting, or shimming. Larger-dimensional errors or other constructibility problems may require the fabricator to develop a new connection detail that requires the approval of the engineer. Again, it may not be feasible to provide a connection that meets the original design standard and the EOR will be called on to make a fitness for purpose determination.

It is very important that the EOR be made aware of, and carefully review, any connection modifications during the entire detailing, fabrication, and erection of special structures. The load transfer is often so complex that only the EOR can evaluate the effect of any modification on the service and strength of the structure.

7.2   Lateral Load Systems

Bracing systems usually involve some of the most complex shop details, require the most labor to fabricate, and are the members most likely to have field fit-up problems. These members, however, are often shown with the least detail on the design drawing. Typical bracing elevations in addition to members sizes and the location of the work point should show the complete load path with all of the forces. The connection designer must be able to determine how the loads accumulate and are transferred from the origin of the force to the foundation in order to design all of the connections for the appropriate forces. This includes knowing diaphragm shears and chord forces, collector forces, and pass-through forces at bracing connections. The designer's failure to provide a complete load path may result in critical connections not being able to deliver the design forces to the bracing system.

Diaphragm chords and drag struts often serve as gravity load members in addition to being part of the bracing system. It is important to design the connections for these members for both shear and axial load. Wide-flange beams with heavy framing angles can transfer axial force of the amount found in drag struts for most bracing systems. See Chap. 2 of this book for more detail on how to design this type of connection and the limits on capacity while still maintaining a flexible-type connection. When these struts are joist or joist girders, special connection details are required. Joist and joist girder end connections typically are able to transmit only a few kips because of the eccentricity between the seated end connection and the axis of the top chord. A field-welded tie consisting of a plate or pair of angles near the neutral axis of the top chord angles is the preferred method of passing axial forces across these types of joints. Drag struts with very large transfer forces occur when it becomes necessary to transfer the entire horizontal force of a brace to a brace in a nearby bay. Members with large axial forces will usually require heavy connections that will be rigid. Consideration should be given to designing these members with fully restrained connections.

The use of concentric work points at bracing joints makes the analysis of the frame and the design of members easier but may subject the connection material to eccentric loads or result in awkward, uneconomical details. This usually occurs when bracing slopes are extreme or member sizes vary substantially in size. It is important when designing bracing to determine if the work points chosen will result in efficient use of connection material. The connection of the diagonal bracing member to the strut beam and column can be efficiently designed using procedures such as the uniform force method found in Chap. 2 (see Fig. 7.1).

Shear-wall systems are simpler to detail and normally involve knowing only drag strut forces or diaphragm forces. These force transfer systems must be clearly shown and detailed. When the structure depends on shear walls, precast panels, or horizontal diaphragms for lateral stability, it is important that the general contractor and erector know this. The erector, by standard practice, provides erection bracing only for lateral loads on the bare frame. The construction sequence may require the general contractor or erector to provide additional bracing because the permanent lateral load system is not complete.

Moment frame systems are often very conservatively shown with notes calling for connections with a strength equal to the full section. While this may be required in seismic zones, lateral load systems are often designed for wind and the members are often sized for stiffness. There can be a substantial savings if the connections are sized for the actual design forces rather than the member strength. It is important when designing connections for wind frames to know the size of the

Figure 7.1   Large gusset plate due to concentric work point for brace.

moment in each direction. Typically, the maximum tension at the bottom flange is substantially less than at the top flange due to the combination of wind and gravity moments. When designing moment connections and checking column stiffener requirements, the use of these reduced tension loads may provide simpler, more economical connections.

7.3 Long-Span Trusses

Long-span trusses can be divided into three general types based on the methods of fabrication and erection. Trusses up to approximately 16 ft deep and 100 ft in length can be shop-fabricated and shipped to the field in one piece. When these trusses are over 100 ft, they can be shop-assembled in sections and shipped to the field in sections for assembly and erection. Trusses over 16 ft deep are generally fabricated and shipped as individual members for assembly and erection in the field. The first two types usually have standard connections that are discussed in other chapters. The third type, because of the size and loads carried, has special connection design requirements.

  Deep long-span trusses typically use wide-flange shapes with all of the flanges in the same plane as the truss. If all of the members are approximately the same depth, connections can be made using gusset plates that lap both sides at the panel joints. When designing web members, it is important to look at the actual depth of the chord member rather than the nominal depth. For example, when using a W 14 × 311 for a chord where the actual depth is more than 17 in, it would be better to use a W 16 × 67 than a W 14 × 61 for a web member. The W 14 × 61 would require the use of 2⅛-in fills on each side. Truss panels should be approximately square for the most efficient layout of gusset-plate connections when using Pratt-type configurations. When using Warren-style panel, it may help to increase the slope of the web member to make a more compact joint while reducing the force and length of the compression diagonal. Chords should, where possible, splice at panel points in order to use the gusset plates and bolts already there as part of the splice material. This also makes it possible to provide for camber or roof slope by allowing a change in alignment at a braced point. The gusset plates on a Pratt-type truss will be extended on the diagonal side to allow for bolt placement in this member. For this reason, the gusset plates should be first sized to accommodate the web member connections and the chord splice placed near the center of the plate rather than at the actual panel point intersection. The plate size is then checked for chord splice requirements.

This type of truss typically uses high-strength bolts for all connections. Traditionally, these have been bearing-type connections in standard holes. This requires either very precise computer numerical controlled (CNC) drilling or full shop assembly with reaming or drilling from the solid. Even with current CNC equipment it is very difficult with heavy members to obtain the tolerances needed for reliable field fit up when using standard 1/16-in oversize holes. Shop assembly and reaming or drilling from solid is very expensive and because of the overall truss size it may not be possible for some fabricators to do this. There has been a trend in recent years to use oversize holes and slip-critical bolts to allow tolerances that are readily achievable by most drill lines. While this increases the number of bolts and gusset-plate sizes, this can be offset by using larger A490 bolts. Bolt material costs are approximately proportional to the strength provided. For example, while the cost of a 1-in-diameter A490 bolt is more than a ⅞-in A325, the number of bolts required is substantially less. While the cost of the bolt material required does not change as the size and grade increase, the cost of plate, hole making, and installation costs decrease so larger-diameter higher-strength bolts are usually cost-effective.

Bolt size selection is also dependent upon the magnitude of force to be transmitted, the net section requirements of the members, and the tightening methods to be used. Generally for the loads and member sizes used in this type of truss, a 1-in-diameter A490 bolt is an efficient choice. The AISC specification provisions that use yield on the gross section and fracture on the net section generally make it possible to use oversize holes for most members without increasing section size. The exception to this may be chord splices where double gage lines are sometimes used. In this case it may be necessary to use two or three rows of bolts at a single gage as lead-in bolts (see Fig. 7.2). It is also important to check for shear lag using the net section provisions of the AISC specification when connecting only to the flanges of members. If possible, all bolts should be designed for single shear. This is especially true at splices that change slope since any splice plate on the inside will have to accommodate the change in slope by skewing the holes in a relatively narrow width plate. It may, however, be necessary to use bolts in double shear at tension chord splices to limit splice length. Compression chord splices should generally be designed as finished to bear type joints with bolts sized for half of the design force. Since these bolts will be slip-critical in oversize holes, it may require the use of mild steel shims in the joint to achieve the detailed chord dimension. Oversize holes should generally be detailed in all plies of material. This will allow the use of full-size drift pins to fair up the hole and make it easier to align the truss. Slip-critical bolts require special procedures to properly tension the bolts and must be inspected to ensure the required tension is achieved. While a

Figure 7.2   Large truss gusset plate with lead-in bolts at splice.

slip into bearing is a service failure and not a collapse, it is important to establish a quality program that will ensure the work meets the design requirements.

Since most trusses will be assembled in large sections on the ground, it is important to design the major gusset-plate connections so all of the bolts, except the splices between sections, can be tensioned and inspected on the ground where they are easier to install and inspect. Secondary framing connections should be made with plates shop-welded to the gusset plates rather than using some of the truss connection bolts for both connections (see Fig. 7.3).

A trial joint should be assembled, tensioned, and inspected with the fabricator, erector, bolt supplier, independent testing laboratory, and the engineer of record present. Written procedures for both bolt installation and inspection for the project should be developed and agreed upon by all parties (see Fig. 7.4).

7.4   Space-Frame Structures

The space truss form is often selected for either architectural appearance or because of depth limitations. Since one-way long-span trusses are easier to fabricate and erect, they will almost always be more economical than space trusses even though they will weigh more.

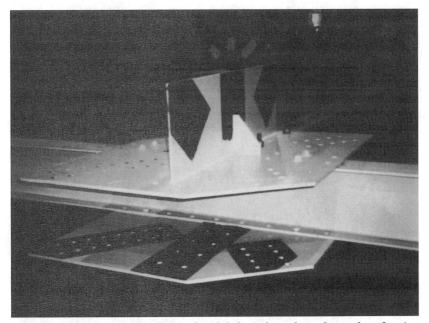

Figure 7.3 Truss gusset plate designed with bolts independent of secondary framing connections.

Figure 7.4   Trial joint assembly to establish bolt tensioning and inspection procedures—NWA Hangar, S.E. McClier and Computerized Structural Design.

Space-frame structures have connections that must transfer forces on all three axes. They have access, dimensional tolerance, and through-thickness strength problems. Because of the complexity of these joints, it is important to try and develop some type of universal connector that can be reliably fabricated or to design the structure with large shop-welded assemblies that can be connected in the field. There are patented space frames that use special steel connectors. These are typically lighter structures with somewhat limited configurations which are not covered here.

Connectors for field assembly of structural steel space frames can be designed using a through plate for the major chord force and intersecting plates for members in the other planes. These intersecting plates are generally complete-joint-penetration–welded to the primary plate. When the geometry is especially complex, it may be necessary to use a center connection piece, usually a round member, to provide access to weld the joint. In either case, the through-thickness strains due to welding make it advisable to use a low sulfur steel with a good through-thickness ductility. This material is expensive and not readily available so it is important to standardize on as few plate thicknesses as possible and use this only where needed. The welding procedure and filler metal should be evaluated to determine if it is

Figure 7.5 Space-frame connector—Carver-Hawkeye Arena, S.E. Geiger—Berger.

Figure 7.6 Space-frame connector in welding fixture—Carver-Hawkeye Arena.

adequate for both the design and fabrication requirements (see Figs. 7.5 and 7.6). The attachment of the truss members to these connectors in the field can either be by welding or bolting. Shop-welding and field-bolting may provide better quality control but this system generally requires two connections. Field-welding typically requires only one connection and generally provides more fit-up tolerance. If field-welding is used, it is important to try to use primarily fillet welds and, if possible, limit the out-of-position welding. Because some type of erection connection or shoring will be required until the structural weld is made, space trusses should, where possible, be designed so they can be ground-assembled. Their inherent stiffness allows them to be hoisted or jacked into final position after full assembly (see Fig. 7.7). This is very important for economy, quality, and safety.

Space trusses can also be designed so they can be shop-welded into panels of a size that can be shipped, thereby reducing the number of field connections. Shipping limitations will normally limit these panels to about 15 ft deep and about the same width. This size will allow the shop to rotate the panel and position it for efficient welding (see Fig. 7.8).

Hollow structural sections (HSS) are often used for truss members because of their appearance and axial-load capacity. Connections of direct-welded HSS require special design procedures. The connection

Figure 7.7 Space-frame module 42 by 126 ft being lifted into place—Carver-Hawkeye Arena.

Figure 7.8 Space-frame module 15 by 60 ft rotated for shop welding—Minneapolis Convention Center, S.E. Skilling, Ward, Magnusson and Barkshire.

Figure 7.9 Solid 6-inch-square reinforcement for HSS joint—Minneapolis Convention Center.

limit state can be various modes of wall failure in addition to weld rupture due to stress concentrations. These stress concentrations are caused by the difference in the relative flexibility of the chord wall when compared to the axial stiffness of the web member. The chord wall thickness required for connections is an important factor when designing members. It may be necessary to increase wall thickness or insert a heavy section at the branch to transmit the design forces. (See Fig. 7.9.)

Welds for HSS-to-HSS connections should be sized to ensure adequate ductility to prevent rupture at design loads. This can be easily accomplished using the effective length concepts given in the AISC *Hollow Structural Sections Connections Manual*[4] or the AWS/*Structural Welding Code* (ANSI/AWS D1.1-98).[5] A more conservative procedure would be to use a weld with an effective throat 1.1 times the wall thickness of the web or branch member. This is intended to make sure the wall of the web member will yield and redistribute stress before the weld ruptures. The ratio is based on E70XX electrodes and A500 grade B material. Direct-welded HSS connections of the T, Y, and K type should, where possible, utilize fillet or partial-penetration welds. Unbacked complete-joint-penetration welds that must be made from one side require special welder certification and are very difficult to make and inspect. Butt splices in HSS may require complete-joint-penetration welds. This type of weld should be made using steel backing to allow the use of standard weld procedures and welder certifications. For more detailed information on these connections, see Refs. 1, 3, 4, and 5.

7.5 Examples of Connections for Special Structures

Examples of connections developed for special structures can be helpful to illustrate the types of problems that are encountered and some idea of how connections can be adapted to meet special requirements.

The first project is a 42-story office building that uses a perimeter moment frame coupled with a braced core as the lateral load-resisting system. When a free-body diagram of the connection forces for the brace members in the core was prepared prior to developing connections, it was discovered that the axial loads for the horizontal struts given in the connection schedule were substantially less than the horizontal component of the brace diagonal. The EOR reviewed the lateral load analysis and discovered that when the structure was modeled, a stiffness factor was assigned to the floor diaphragm to provide for the interaction between the moment frame and the braced core. In the model the floor was carrying part of the brace force. While the brace loads may actually be transmitted in this manner, the EOR decided to follow conventional practice and size the steel for the full brace force rather than rely on this type of composite action. All of the horizontal struts were resized and connections were then developed for the full horizontal component of the diagonal force. Diagonal braces were wide-flange sections using claw angle-type connections with 1-in-diameter A325 SC bolts in oversize holes.

The second project is a sports arena using a skewed chord space truss supported on eight columns with the roof located at the bottom chord of the trusses (see Fig. 7.10). Each type of connection was clearly shown on the design drawing along with the forces to be used to determine the number of bolts and welds required for each connection. While reviewing the forces given for the bottom chord, it was noted that the bottom chord members had been modeled as axially loaded pin-ended members with vertical end reactions due to the roof dead and live loads on the bottom chord. A check of the actual connection which consisted of a plate on each side of the web that was welded between the flanges of W 27 sections indicated the connection was rigid. After reviewing this compatibility concern, the EOR decided to size the connections for the axial forces and vertical end reactions given using N-type values for all of the bolts. A check of the connec-

Figure 7.10   Bottom chord connection for space truss—Carver-Hawkeye Arena.

tions using X-type values for the bolts indicated adequate reserve strength for possible end moments.

The exposed top chord, diagonals, and connectors were all made of ASTM A588 material left unpainted so it could weather. While the fabricator was detailing these connections, the EOR became aware of a study[2] that indicated, under certain conditions where moisture had access to the inside of a joint, the expansive pressure of the continuing corrosion could overstress the bolts and lead to failure. Connection details were modified to make sure the recommendations on minimum plate thickness and maximum bolt spacing were complied with. Special restrictive fabrication tolerances were established for connection material flatness in order to ensure the connection bolts would be able to clamp the full surface together. The fabricator, by using techniques such as prebending plate prior to welding and using heat-straightening after welding, was able to eliminate almost all distortion due to shop welds (see Fig. 7.11). The high-strength bolts were able to pull the plates together so there were no gaps in the connections (see Fig. 7.12).

The third project is a 57-story office building that uses a unique lateral load system. The wind in the longitudinal direction is resisted by five-story bands of vierendeel trusses that span 97 ft between con-

Figure 7.11   Heat straightening of connection plates after welding—Carver-Hawkeye Arena.

Figure 7.12   Exposed top chord connection showing fit-up after bolting—Carver-Hawkeye Arena.

Figure 7.13 Vierendeel framing system—Norwest Financial Center, S.E. CBM Inc.

crete super columns. The vierendeel trusses were designed as horizontal tree girders with verticals spliced at midheight between floors
(see Fig. 7.13).

These splice connections were first designed with partial-joint-penetration field welds. The shop connections of these W 24 verticals to
the horizontal girder were complete-joint-penetration welds. The combination of weld shrinkage due to these shop welds along with the
distortion of the girder due to welding and the rolling tolerance of the
vertical section made it almost impossible to achieve the proper fit-up
of the field-welded joint without a lot of expensive remedial work.
Since the field splice was at the inflection point of the vertical, there
were only axial loads and shears to be transmitted through the connection. It was decided to use an end-plate–type connection with slip-
critical bolts in oversize holes to accommodate the fabrication and
rolling tolerances. In addition, the members were detailed short and a
$\frac{3}{8}$-in shim pack was provided to bring each joint to the proper elevation. The modification of this connection was one of the keys to the
early completion of the erection of this structure (see Figs. 7.14 and
7.15).

The fourth project is a 37-story mixed-use structure that uses a
megatruss bracing system for wind loads. The bracing truss has
nodes at five-story intervals and uses wide-flange members for stiff-

Figure 7.14 Shimming of Field splice of vierendeel verticals—Norwest Financial Center.

Figure 7.15 Splice of vierendeel verticals showing alignment—Norwest Financial Center.

ness. The connections of the truss at the nodes were designed as partial-joint-penetration groove-welded butt splices. Because of past experience with poor fit-up the EOR specified that joint fit-up had to comply with AWS D1.1 prequalified joint requirements with no build-out permitted.

The combination of mill, fabrication, and erection tolerances would have made it impossible to achieve this type of fit on these heavy W 14 members. It was decided to add a field splice in all of the diagonals midway between nodes using a lap-plate–type splice. This allowed the erector to position the lower half tight to the node and then jack the upper half tight to the upper node. The brace members had all been sized for axial stiffness and the design forces were typically less than half of the member capacity. The lap plates were designed and fillet welded for the actual brace force (see Figs. 7.16 and 7.17).

The fifth project is an exhibition hall consisting of three lamella domes 210 ft in diameter surrounded by a 60-ft-wide delta-type space truss made of hollow structural sections (see Fig. 7.18). Each dome is supported by a series of sloping pipe struts from four columns. The domes vertically support the inside of the space truss and the space truss laterally constrains the domes (see Fig. 7.19). The total struc-

Figure 7.16   Bracing node connection showing fit-up—Plaza Seven, S.E. CBM Inc.

Figure 7.17   Adjustable midheight splice of bracing diagonal—Plaza Seven.

Figure 7.18   Lamella dome and delta space frame—Minneapolis Convention Center.

Figure 7.19   Dome and space-frame column and pipe supports—Minneapolis Convention Center.

Figure 7.20 Top-chord field splice of delta space frame—Minneapolis Convention Center.

ture is approximately 900 ft long without an expansion joint. The EOR laid out the space truss so modular units could be shop-fabricated in units 15 ft wide and 60 ft long. The top and bottom chords of these units were offset and oriented in the 60-ft direction to minimize splicing (see Fig. 7.20). Each unit had two top chords and one bottom chord. This resulted in double top chords at the splice between units. These chords were connected by flare V-groove field welds at the panel points. The bottom chords were detailed with a short connector stub to which a section of cross-chord was butt-welded in the field.

The diagonals of the delta truss intersect the bottom chords at 45° to the vertical and the chords. These members were typically 6-in-square HSS and would have overlapped at the panel point. To avoid this the EOR detailed a connector consisting of intersecting vertical plates on top of each chord. Initially it was planned to provide complete-joint-penetration welds for these diagonal connections. However, when weld procedures were developed, it became apparent that the restricted access to these joints would make both welding and inspection very difficult. The connection was redesigned using partial-joint-penetration and fillet welds sized for the actual loads in the members with allowances as required for uneven load distribution (see Figs. 7.21 and 7.22).

All of the butt splices in the chord were detailed as complete-joint-penetration welds using internal steel backing so a standard V-groove weld could be used (see Fig. 7.23).

Figure 7.21   Space-frame bottom-chord connection showing fit-up—Minneapolis Convention Center.

Figure 7.22   Space-frame bottom-chord showing weld—Minneapolis Convention Center.

The connection of the sloping pipes to bottom ring of the dome consisted of a series of radial plates that were complete-joint-penetration–welded to a 6-in-thick connector plate on the ring (see Fig. 7.24). The EOR was concerned about possible brittle fracture of these heavy welded plate connections and specified material ductility using stan-

Figure 7.23   Space-frame bottom-chord splice connection—Minneapolis Convention Center.

Figure 7.24   Gusset-plate connections for pipe struts to dome—Minneapolis Convention Center.

dard Charpy V-notch testing. When orders were placed, the material supplier informed the fabricator that the standard longitudinal Charpy test would not measure the through-thickness properties needed to accommodate the weld strains. The design and construction team consulted with metallurgists and fracture mechanics experts to develop a specification and testing procedure that would ensure adequate through-thickness ductility. The testing procedure called for through-thickness samples to be taken near the center of the plate. A minimum through-thickness reduction in area of 20%, along with a minimum Charpy value of 15 ft-lb @70°F in all three axes, was specified. While the through-thickness Charpy test is not a reliable indicator of ductility, it was decided to do this test as a general comparison with the properties in the other two directions. The producer supplied a low-sulfur, vacuum-degassed, and normalized material with inclusion shape control. All material was 100% ultrasonically inspected at the mill. There were no through-thickness problems due to welding strains. Since this project was built, several mills have developed proprietary low-sulfur materials with excellent through-thickness properties and ASTM now has a specification, A770, for through-thickness testing.

The lamella domes were designed using wide-flange shapes shop-welded into diamond patterns (see Fig. 7.25). Since the fabricator was

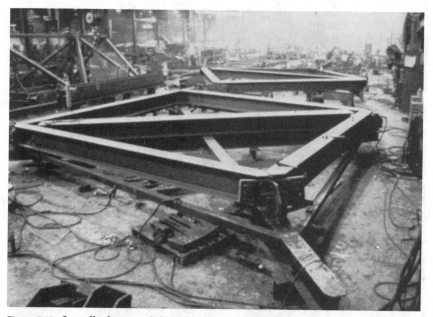

Figure 7.25   Lamella dome module in fabrication—Minneapolis Convention Center.

Figure 7.26   Start of ground assembly of dome—Minneapolis Convention Center.

nearly even, the 24-ft-wide diamonds at bottom ring were shop-fabricated and delivered to the site. A bolted web splice was provided between diamonds and for the ring beams to the diamonds. This provided both an erection splice and was adequate for any out-of-plane loads. The flanges of these members were complete-joint-penetration–welded.

The entire space frame project was ground-assembled. The space trusses were assembled in units 60 by 75 ft and hoisted by crane onto shoring towers and perimeter columns. The dome was assembled ring by ring on the ground using shores as required (see Fig. 7.26). The dome assembly, including the deck, was then jacked into place (see Fig. 7.27). When the slotted pipe supports were slipped over and welded to the gusset plates on the heavy weldments described here, a new concern arose. The misalignment of the gusset plates due to the angular distortion caused by the one-sided groove welds along with the erection tolerances of the structure resulted in some bowing of the connection plates. The EOR reviewed the forces and added stiffeners, where required, to prevent buckling due to any misalignment of the plates in compression.

The sixth project is a multiuse sports and events center. The roof framing consists of 26-ft-deep trusses spanning 206 ft that are framed at one end to a jack truss spanning 185 ft. All of the truss members

Figure 7.27 Jacking rods in position for lifting dome—Minneapolis Convention Center.

are W 14 sections oriented with the flanges in the vertical plane and connected with lap-type gusset plates on each flange. All of the typical truss connections used slip-critical high-strength bolts in oversized holes. Chords were spliced at panel points and W 16 sections were used for web members as recommended previously.

The connection of the 206-ft trusses to the jack truss presented a special problem because of the large reaction that had to be carried by the framing angles (see Fig. 7.28). Originally it was planned to intersect the work points of the end connection at the center of the jack truss top chord. A free body diagram of the connection, however, showed the bolts would have to develop a moment of 465 ft-kips in addition to carrying a shear of 450 kips. Even if it was possible to get enough bolts in the plate, the angles could not develop the moment. By moving the work point to the face of the truss it was possible to eliminate the bending in the outstanding leg of the angle and to reduce the eccentricity to the bolt group in the other leg to 4.5 in. The eccentric reaction on the jack truss was easily balanced by adding a 14-kip axial connection at the bottom chord. The bending stress in the jack truss vertical was checked and found to be acceptable.

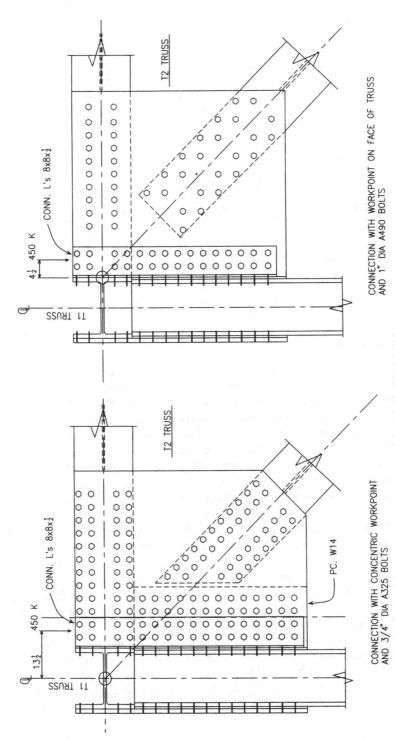

CONN. L's 8x8x½

450 K

4½

CONN. L's 8x8x½

450 K

13½

CONNECTION WITH WORKPOINT ON FACE OF TRUSS
AND 1" DIA A490 BOLTS

CONNECTION WITH CONCENTRIC WORKPOINT
AND 3/4" DIA A325 BOLTS

PC. W14

T1 TRUSS

T1 TRUSS

T2 TRUSS

T2 TRUSS

Figure 7.28 Connection roof truss to jack truss—Mankato Civic Center, S.E. MBJ.

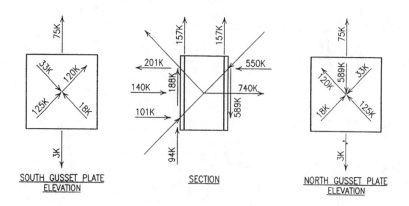

SOUTH GUSSET PLATE   SECTION   NORTH GUSSET PLATE
    ELEVATION   ELEVATION

CONNECTION COULD BE SIMPLIFIED BY MOVING WORK POINT TO THE FACE OF
GUSSET PLATE TO REDUCE ECCENTICITY IN TRUSS CONNECTION

Figure 7.29   Free body diagram of roof truss to box truss—NWA Hangar.

7.6   Conclusion

It is helpful when designing special connections to start with a free body diagram of the connection. The free body should usually be cut at the connection face and all forces shown (see Fig. 7.29). While it may be necessary to use advanced techniques, such as finite element analysis or yield-line theory to evaluate stiffness of elements, it is important to first try to establish the best force path. Care should be taken to make sure this is a complete path. Both sides of the connection must be able to transmit the force. An evaluation should be made of any connection eccentricity. It may be better to design the member for the eccentricity instead of the connection.

A check should also be made of the flexibility of the connection if it was modeled as a pin in the analysis. It may be possible to ignore connection fixity in axially loaded members, such as trusses, as long as the members are modeled with flexible connections and all loads are applied at panel points. Preliminary connection design should be done prior to final member selection. It is impossible to effectively size members without taking into account connection requirements.

Because of the constructibility and economic concerns, the design of special connections will almost always require input from the fabricator and erector, so the EOR must either obtain this input in advance or be prepared to evaluate proposed means and methods modifications during the construction stage.

7.7 References

1. Packer, Jeffrey A., and Henderson, J. E., *Hollow Structural Section Connections and Trusses—a Design Guide,* Canadian Institute of Steel Construction, Toronto, 1997.
2. Brockenbrough, R. L., "Considerations in the Design of Bolted Joints for Weathering Steel," *AISC Engineering Journal,* vol. 20, no. 1, 1st quarter, 1983, p. 40.
3. AISC, *Specification for the Design of Steel Hollow Structural Sections,* American Institute of Steel Construction, Chicago, 1997.
4. AISC, *Hollow Structural Sections Connections Manual,* American Institute of Steel Construction, Chicago, 1997.
5. AWS, *Structural Welding Code, Steel,* ANSI/AWS D1.1-98, American Welding Society, Miami, 1998.

8

Inspection and Quality Control

Robert E. Shaw, Jr.
Steel Structures Technology Center, Inc.
Novi, MI

(Courtesy of The Steel Institute of New York.)

8.1   Fastener Quality Control and Testing

The quality of the fastener components begins with the manufacturer of the steel. Steel is purchased by bolt, nut, and washer manufacturers to rigid chemical specifications so that, after manufacture and, if needed, heat treatment, the desired mechanical properties will be achieved. The quality of the steel is verified through the use of mill test reports provided by the steel mill and reviewed by the fastener manufacturer.

Treatment of the as-provided steel is sometimes necessary to prepare the steel for the manufacturing operations. This includes annealing, or softening, of the steel to make it more workable in the machinery. For bolts, drawing is needed to make the steel the exact diameter required.

The manufacturer will make several thousand to several hundred thousand components in each production lot, depending upon the type of product and the manufacturing facility. A *production lot* is defined as those items from the same lot of steel produced in the same equipment and heat-treated, lubricated, and tested at the same time. Testing is performed during production to verify that dimensional tolerances are met. Random sampling may be performed for physical testing, but generally this is left until the completion of the production run. Physical testing is performed following the completion of heat treatment, if performed, and following any galvanizing application.

The type of testing required depends upon the type of product being manufactured. Test requirements are specified in the applicable ASTM specification. Bolts are tested for strength and ductility. Additional tests are required for A490 bolts and galvanized assemblies. Special tests, called *rotational-capacity tests,* are performed when the bolts are to be used in bridge construction. Hardness and stripping tests are performed on nuts, and hardness tests are performed on washers.

Bolt strength is tested in a tensile testing machine. A wedge of either 6 or 10° is placed under the head of the bolt, then the bolt is pulled to failure. The failure must take place in the threads of the bolt, between the nut and the head. Failures directly underneath the bolt head, in the shank, or by stripping of the threads are unacceptable. The elongation of the bolt is also measured as tensile loading is applied. The bolt must satisfy the requirements for minimum proof load, which is established as 70% of the minimum required tensile strength for A325 bolts and 80% of the minimum required tensile strength for A490 bolts. The proof load establishes that the bolt will not yield prematurely at a low stress level, and therefore not provide the tension desired when installed using established techniques. When bolts are too short to fit into a tensile testing machine, they are tested using hardness test methods to establish minimum and maximum strength levels. All A490 bolts receive additional magnetic particle testing to detect microcracking.

Bolts that are galvanized must be supplied as an assembly, with the washers and nuts that are to be used with the bolts. A rotational-capacity test is performed to verify that the effect of galvanizing and the overtapping of the nut did not adversely affect the assembly performance. The test involves deliberately overtightening the assembly in a test fixture, ensuring the bolt and nut have adequate strength, then verifying that the threads of the bolt and nut resist stripping.

For bridge work, the testing includes checking the torque requirements for tightening the assembly, with a maximum torque value used to confirm the effectiveness of the nut lubrication.

Nuts are tested for stripping strength on a threaded mandrel, with a block attempting to push the nut off the end of the mandrel. Hardness tests are also performed to verify proper nut strength.

Fastener components are physically tested using statistical sampling techniques, as prescribed by the applicable American Society for Testing and Materials (ASTM) specifications. Zero defects in strength and proof-load requirements are permitted. Should the item fail a strength, proof-load, or hardness test, if used, then the entire production lot is rejected. Generally, lots rejected on the basis of strength are heat-treated again, and then retested.

Bolts may have small cracks, called *bursts*, in the head of the bolt. Bursts are acceptable provided they meet the limits for the type bolt, as prescribed in the applicable ASTM specification. A325 bolts may have bursts, provided they do not enter the chamfer circle on the top of the bolt head, do not enter the washer face on the underside of the bolt head, or reduce the dimension across the points of the head below American National Standards Institute (ANSI) values. For inch-series A490 bolts, head bursts are limited in width to 0.010 in, plus 0.025 times the bolt diameter, in inches. For metric A490 bolts, head bursts are limited in the same manner as A325 bolts. Physical sampling for head bursts in bolts involves taking a specified sample, based upon lot size. Approximately 5% of the sampled bolts may have rejectable head bursts. These bolts may be sold, but may not be used by the purchaser. A second visual inspection is required by the purchaser to reject those with unacceptable bursts.

8.2 Bolt Preinstallation Testing

Bolt inspection is a multistep operation that begins before the bolt installation starts. The following steps should be included in a bolt inspection program:

1. Check the materials and certifications. Verify that the materials supplied comply with the project specifications. Review the product certification papers for ASTM and project compliance. Check that the certifications match with the supplied product.

2. Check for proper storage conditions. The various fastener components should be kept separate by lot until time for installation. Preassembly of bolts, nuts, and washers prior to use by the installer is satisfactory as long as lot control is maintained. Proper storage also includes protection from the elements, maintaining adequate lubrica-

tion and keeping the materials free from dirt, sand, grit, and other foreign materials.

3. Check the assembly of bolt, nut, and washer (if used) in a bolt calibration device for material quality, verifying that it is capable of achieving the required pretension without breaking, thread stripping, or excessive installation effort. For bridge work, the American Association of State Highway and Transportation Officials (AASHTO) requires that two assemblies of each bolt lot, nut lot, and washer lot, as supplied and to be assembled and installed in the shop or field, be tested with a special rotational-capacity test. This test verifies material strength and ductility, thread stripping resistance, and the efficiency of the lubrication.

4. Check the validity of the installation technique for that group of fasteners. Perform the installation technique in a bolt calibration device, or with a "calibrated" direct tension indicator (dti) if the bolt is too short to fit into the calibrator. Verify that at least the minimum required tension, plus 5%, is achieved using the specified technique. For the calibrated wrench method, observe the calibration of the wrenches before the start of the work each day.

5. Verify the knowledge of the installation crew. The previous two steps should be performed by the installation crew. By observing these tests, the crew demonstrates to the inspector their knowledge of the proper technique.

6. Check the faying surface condition of the fabricated steel for proper conditions prior to erection.

7. Check the size and quality of the bolt holes in the fabricated steel.

8.3 Bolt Installation and Testing

The first step in installing bolts is the proper snugging of the joint. If the joint is not properly snugged, no pretensioning method will work correctly.

The majority of bolts in buildings must be tightened only to the "snug-tight" condition. Certain shear-bearing joints must be pretensioned, as well as slip-critical joints and direct tension joints.

The definition of *snug tight* is stated in the American Institute of Steel Construction (AISC) LRFD *Specification* in Section J3.1: "The snug-tight condition is defined as the tightness attained by either a few impacts of an impact wrench or the full effort of a worker with an ordinary spud wrench that brings the connected plies into firm contact."

If the joint is not in solid contact, the pretensioning method employed may fail to achieve the proper pretension for the bolts in the joint. Pretensioning the first bolt in the group will only serve to further draw down the gap between the steel elements. The installer erroneously assumes the first bolt is tight. The next bolt tightened further draws down any remaining gap, and the initial bolt becomes looser still. This can become a compounding series in some joints.

8.3.1 Turn-of-the-nut installation method

The turn-of-the-nut method has been around since the 1950s. The current "turns" table has been in use since 1978.

The principle behind the turn-of-the-nut method is the controlled elongation of the bolt. Because of the pitch of the threads, turning the nut a prescribed rotation elongates the bolt a certain amount. The elongation has a direct correlation to the bolt tension. As bolts become larger in diameter, the number of threads per inch decreases accordingly; therefore, the same number of turns will provide at least the required amount of pretension for a given length-to-diameter ratio.

Table 8.1 provides the required turns for given bolt length-to-diameter ratios, as provided in the Research Council on Structural Connection (RCSC) Specification Table 5. No such table has been prepared for metric bolts.

As an example, with flat surfaces and bolts less than or equal to four diameters in length, say a $\frac{3}{4}$- by 3-in bolt, a one-third turn must be provided. A $\frac{3}{4}$- by 5-in bolt would receive a one-half turn. A $\frac{3}{4}$- by $6\frac{1}{2}$-in bolt would receive a two-thirds turn.

TABLE 8.1   RCSC Table-5, Rurn-of-Nut Rotation

Bolt length (underside of head to end of bolt)	Disposition of outer face of bolted parts		
	Both faces normal to bolt axis	One face normal to bolt axis and other sloped not more than 1:20 (beveled washer not used)	Both faces sloped not more than 1:20 from normal to bolt axis (beveled washers not used)
Up to and including 4 diameters	1/3 turn	1/2 turn	2/3 turn
Over 4 diameters but not exceeding 8 diameters	1/2 turn	2/3 turn	5/6 turn
Over 8 diameters but not exceeding 12 diameters	2/3 turn	5/6 turn	1 turn

For bolts over 12 diameters in length, too much variation exists to provide tabular values. It is required that the installer use a bolt tension calibrator, such as a Skidmore-Wilhelm, to determine the number of turns required to provide the required bolt pretension.

The sloping surfaces provisions apply when there is a slope to the surface beneath the bolt head or nut. This slope must not exceed 1:20, or about 3°. Extra rotation is needed to overcome the loss caused by the bending at the head or nut; therefore, a one-sixth turn is added for each sloping surface. If the slope exceeds 1:20, a beveled washer must be used equal to the slope.

If the sloping surfaces are caused by the $16\frac{2}{3}\%$ (10°) bevel used for channel and S-section flanges, then a standard $16\frac{2}{3}\%$ beveled washer must be used. The required turns increase for the sloping surface is not required, because the beveled washer has returned the head or nut to the parallel condition.

There is a tolerance to the amount of applied rotation. For turns of one-half or less, the nut may be under- or overrotated by no more than 30°. For turns of two-thirds or more, the nut may be under- or overrotated by 45°. If the nut does not receive sufficient rotation, the desired pretension may not be achieved. The potential risk from overrotation is that the bolt may be stretched to the point of breaking, or to the point where nut stripping may occur.

The installation sequence should start with snugging the joint. Following inspection for snug, the installation crew is permitted to matchmark the end of the bolt and a corner of the nut. The crew applies the required turns from RCSC Table 5, and the joint is inspected to verify the applied turns by checking the matchmark rotation.

The installation crew may also use the "watch the wrench chuck" method for turn-of-the-nut, electing not to matchmark. The inspector must monitor the crew's efforts to verify that the proper technique is routinely applied during the pretensioning.

8.3.2 Calibrated wrench installation method

The calibrated wrench method uses a special type impact wrench to tighten the bolts. Rather than impact until the wrench operator releases the trigger, the wrench is adjusted to automatically stop impacting when a certain resistance is felt by the wrench. The wrench is adjusted so that it stops impacting when the bolt has achieved at least the required tension, as determined using a bolt calibration device, but does not overrotate the fastener beyond the turn-of-the-nut tables.

Pneumatic calibrated impact wrenches depend upon an internal cam unit for control. When the desired resistance, actually the torque, is reached, the cam unit shifts and the wrench stalls out. If the air pressure or air volume is inadequate, however, the control mechanism will not function properly and will continue to impact the fastener, although at a slower, weaker level. For this reason, the calibrated impact wrench must be calibrated with a given air-supply condition. The wrench should be calibrated using the same compressor and pressure settings, air hose, and air-hose length that will be used on the work. If an additional wrench is to be driven off the compressor, the wrench calibration should be checked with both wrenches in operation simultaneously as well as individually. If a significant length of hose from compressor to wrench is either added or removed, then the wrench should be recalibrated.

Calibration of the wrench is required every day, before installation begins, with three fastener assemblies of each diameter, length, grade, and lot. An assembly, by specification, would be comprised of a bolt from a specific production lot with a nut from a specific production lot. Washers representative of those being used in the work must be included in the test, but lot control for washers is not mandated except under AASHTO requirements for bridge work.

If there is a significant difference in the quality of fastener lubrication, then the wrench must be calibrated for the varying lubrication conditions. A well-oiled bolt, washer, and nut assembly will require considerably less torque than one that is nearly dry or one that exhibits some indications of rust. Hence, if the wrench is calibrated using well-oiled bolts, then used on a poorly lubricated bolt, the resultant bolt pretension will be less. The same concerns apply if the bolt, nut, or washer surfaces contain dirt, grit, or sand.

Efficient calibration becomes key to the use of the calibrated wrench method of installation. Some projects have used a separate calibrated wrench for each given diameter, length, and grade being installed, changing wrenches when a different bolt group is being installed. Every wrench is calibrated each morning. It is possible to simplify operations by calibrating the wrench to properly install a wider range of bolts. This would be accomplished by setting the wrench to install at a tension well above the required pretension, yet not high enough to exceed the turns tables. This same wrench setting could be tested on another length of bolts of the same diameter and grade, and then another. Perhaps one wrench setting could be used throughout the day for all bolts of the same diameter and grade being installed, or perhaps only two wrenches would be needed instead of several.

Snugging the joint can be done with either the calibrated wrench (actually in the uncalibrated condition, releasing the trigger when snug is achieved), with an impact wrench, or with a hand wrench for lighter framing. After snugging, the joint should be inspected to verify snug, then approved for pretensioning of the bolts. After the wrenches are calibrated, pretensioning can begin. The wrench operator should tighten the bolts using a systematic pattern, observing the chuck rotation as tightening proceeds. If the rotation of the nut exceeds the turns table for turn-of-the-nut, the wrench calibration should be rechecked. After tightening all the bolts in the pattern, the operator should return to "touch up" each bolt in the pattern. Only the calibrated wrench method calls for such "touching up."

8.3.3   Direct tension indicator installation method

The *direct tension indicator,* or dti, is a special load-cell device used as proof that the required tension has been provided in the assembly. The manufacturing and testing of the dti itself is governed by ASTM F959. The effectiveness of the dti, however, is also dependent upon the techniques used in installing the fastener.

The dti has protrusions (bumps) formed into the device that will be compressed when the bolt is pretensioned. The gap remaining between the dti face and the fastener element against which it is placed should not close below a specified gap until after the fastener has reached the required fastener tension. By use of a feeler gage to verify that the gaps have suitably closed, one can verify that the bolt has been pretensioned.

The preferred position of the dti is with the bumps facing outward directly underneath the bolt head. During installation, the bolt head must be held from turning to prevent abrasion of the bumps, thereby rendering the dti measurement invalid. For building applications, this condition would call for a 0.015-in (0.38-mm) feeler gage to be refused entry in half or more of the gaps of the dti. For bridges, a 0.005-in (0.13-mm) feeler gage is used.

If the bolt head will be turned, a hardened F436 washer must be placed between the bolt head and dti. If the dti is placed at the nut end of the assembly, an F436 washer must be used between the dti and nut, whether or not the nut is allowed to turn. In each of these cases, a 0.005-in (0.13-mm) feeler gage is used to check the gaps.

The dti bumps must face outward away from the steel to keep the dti from cupping outward, opening the gaps larger and voiding the measurement technique. Standard F436 washers are needed behind the dti when the dti is used over an outer ply containing an oversized or slotted hole. This prevents the dti from cupping into the hole and voiding the gap measurement technique.

The joint is first snugged using a systematic technique, then inspected. The installer should check that at least half the gaps do not refuse the feeler gage. This is done to ensure that the bolt did not reach its required tension during snugging, then subsequently loosen when adjacent bolts were snugged. Since the dti is inelastic, it will not rebound to increase the gap when the preload is released. Therefore, a bolt "overtightened" during snugging, then subsequently loosened, will still appear to be properly tensioned by having an adequate number of refusals for the postinstallation check. After the entire joint has been verified as snug, the installation crew can then proceed to further tighten each bolt until at least half the dti bumps refuse entry of the feeler gage.

8.3.4   Alternate design fastener (twist-off bolt) installation method

The *twist-off bolt* is a specially designed bolt that has a spline at the end that is used by the installation tool. The spline is designed to shear off from the torque generated by the wrench. This torque is the result of wrench efforts to turn the nut in the clockwise (tightening) direction, as well as counteracting efforts to turn the bolt shank in the counterclockwise direction. When the twist-off bolt's head grabs the steel, the bolt shank itself will not turn, and all rotation takes place with the nut.

The spline has been designed so that it will not shear off until the bolt is above the required tension, but will not be too strong to cause overtightening of the bolt beyond reasonable limits. When the spline shears off, the wrench no longer functions.

The twist-off bolt is completely dependent upon the torque-tension relationship, which can vary greatly depending upon the quality and type of lubrication of the assembly. The manufacturers of twist-off bolts generally produce a very consistent and durable lubricant that resists water, mild solvents, and rust for some time.

Because of the interdependence of the bolt, nut, and washer upon the torque used for installation, the twist-off bolt unit is preassembled by the manufacturer. Substitutions of other nuts or washers may adversely affect performance and cause bolt tensions to be too high or too low.

The joint is first snugged using a systematic method, as with all installation procedures previously discussed. Care must be used to make sure that the spline is not twisted off during the snugging operation. Any bolts that twist off during snugging must be replaced. In some cases, deep sockets are used on conventional impact wrenches to snug the joints, therefore protecting the splines. Upon completion of snugging, the snug condition is verified to make sure no improper gaps exist. Upon acceptance of the snugged joint, the installation crew proceeds to tighten each twist-off bolt with the installation wrench until the spline shears off. A systematic pattern should be used for this step.

8.3.5 Alternate design fastener (lock pin and collar) installation method

Another form of alternate design fastener is the lock pin and collar fastener. With this type of fastener, a pin with a set of concentric locking grooves and a break-neck is tensioned with a hydraulic tool. As the tool pulls on the pin, the unit also swages a locking collar onto the locking grooves to retain the pin's pretension. At a point above the required tension for the fastener, the break-neck portion of the pin fractures in tension, stopping the tensioning process. The pin maintains the required residual pretension because of the locking collar.

8.3.6 Bolt inspection procedures

Bolt inspection is a multistep operation that begins before the bolt installation starts. See Sec. 8.2 for steps 1 through 7. The following additional steps should be included in a bolt inspection program:

1. Visually check after snugging to verify that the snug condition has been achieved. For dti's, also verify that the installers check for an adequate number of remaining gaps in their dti's. For twist-off bolts, check that the splines are intact.

2. Observe the installation crew for proper technique. Observation of the installation crew does not mean that the installation of each individual bolt is observed, but that the crew is observed to understand and follow the proper techniques on a uniform basis.

Torque testing is applicable only when there is a dispute, per Section 9(b) of the RCSC bolt specification. The methodology given is not intended as an inspection method, but as a way to resolve claims that the crew may not have followed the proper techniques for a particular joint. The specification states that this section is to be used only when inspection per 9(a) has been performed—the visual observation of the preinstallation testing, checking for snug, and observation of the installation technique of the crew.

For AASHTO bridge work, torque testing is still required for 10% of the bolts in each connection, minimum two per connection, for bolts installed using turn-of-the-nut or calibrated wrench methods. The torque-testing procedures of AASHTO are similar to the RCSC bolt specification Section 9(b), except that only three bolts are used to determine the inspection torque, not five. For dti, twist-off bolt, and lock pin and collar installation, no torque testing is required.

8.4   Bolt Inspection Issues

8.4.1   Hole punching and drilling

The AISC specification addresses the use of punching and drilling for making bolt holes. The size of hole used for a particular size bolt may vary with the type of joint and hole selected by the engineer. Table 8.2 states the given hole sizes for each diameter of bolt, as required by AISC Table J3.3. Metric hole sizes have not yet been established by adopted specifications, but anticipated values are given.

Oversized holes may be used only in slip-critical joints. Slotted holes may be used in shear-bearing joints only when the load is transverse to the direction of the slot. Otherwise, they may be used only in slip-critical joints.

Under AISC Specification Section M2.5, holes may be punched in the material provided the thickness of the material does not exceed the diameter of the hole plus $\frac{1}{8}$ in (3 mm), except for holes to be

Table J3.3 Nominal Hole Dimensions

Bolt Diameter	Hole dimensions, in			
	Standard	Oversized	Short slot	Long slot
$\frac{1}{2}$	$\frac{9}{16}$	$\frac{5}{8}$	$\frac{9}{16} \times \frac{11}{16}$	$\frac{9}{16} \times 1\frac{1}{4}$
$\frac{5}{8}$	$\frac{11}{16}$	$\frac{13}{16}$	$\frac{11}{16} \times \frac{7}{8}$	$\frac{11}{16} \times 1\frac{9}{16}$
$\frac{3}{4}$	$\frac{13}{16}$	$\frac{15}{16}$	$\frac{13}{16} \times 1$	$\frac{13}{16} \times 1\frac{7}{8}$
$\frac{7}{8}$	$\frac{15}{16}$	$1\frac{1}{16}$	$\frac{15}{16} \times 1\frac{1}{8}$	$\frac{15}{16} \times 2\frac{3}{16}$
1	$1\frac{1}{16}$	$1\frac{1}{4}$	$1\frac{1}{16} \times 1\frac{5}{16}$	$1\frac{1}{16} \times 2\frac{1}{2}$
$\geq 1\frac{1}{8}$	$d + \frac{1}{16}$	$d + \frac{5}{16}$	$(d + \frac{1}{16}) \times (d + \frac{3}{8})$	$(d + \frac{1}{16}) \times (2.5\,d)$

Metric Table J3.3 Nominal Hole Dimensions

Bolt Diameter	Hole dimensions, mm			
	Standard	Oversized	Short slot	Long slot
M16	18	20	18 × 22	18 × 40
M20	22	24	22 × 26	22 × 50
M22	24	28	24 × 30	24 × 55
M24	27	30	27 × 32	27 × 60
M27	30	35	30 × 37	30 × 67
M30	33	38	33 × 40	33 × 75
$\geq$ M36	$d + 3$	$d + 8$	$(d + 3) \times (d + 10)$	$(d + 3) \times (2.5\,d)$

punched in quenched and tempered steels such as A514. For these steels, the limit for punching is $\frac{1}{2}$-in (13-mm) thickness.

For bridges, AASHTO Section 11.4.8.1 limits the punching thickness to $\frac{3}{4}$ in (19 mm) for structural steel, $\frac{5}{8}$ in (16 mm) for high-strength steel (50 ksi and over), and $\frac{1}{2}$ in (13 mm) for quenched and tempered steel. If the thickness of the steel exceeds these limits, the holes must be drilled or subpunched or subdrilled and reamed to the proper diameter. AASHTO also requires that holes penetrating through five or more layers of steel be either subdrilled and reamed to the proper diameter or drilled full size while preassembled.

The size of the completed hole may exceed the nominal diameter of the hole by a maximum of $\frac{1}{32}$ in (1 mm). If the hole size is larger, then it must be considered oversized, which may change the design assumption, the design strength for the bolt, and the bearing strength of the steel.

Holes may also be flame-cut, although this method is not addressed by specification except for the slotting of holes. Research* indicates that flame-cut holes in A36 steel to $\frac{1}{2}$-in (13-mm) thickness are suitable provided they meet the hole size requirements, a difficult task. Piercing with the torch and reaming to the proper diameter is suggested.

8.4.2 Bolt storage and control

Bolts, nuts, and washers are typically purchased as commodity items and are placed into inventory. Because the shop bolt list is not completed until the shop detail drawings are done, and the field bolt list is not done until the erection plans and shop details are done, bolts are ordered in advance using estimates of quantities and lengths.

Bolts, nuts, and washers should be maintained in protected storage with the manufacturer's certification available. Because the RCSC bolting specifications require preinstallation testing for most fastener assemblies by production lot, bolts and nuts should not be mixed with others of the same grade and length. If torque control methods are used for installation, inventory control by lot is especially important.

Certificates of compliance with ASTM specifications are required for bolts, nuts, and washers only when requested by the customer. However, ASTM specifications require that such certificates be prepared by the manufacturer, whether or not they are requested by the customer.

Only a few fastener manufacturers place their lot number on the fastener itself. All others place their lot identification on the keg or

---

*Ivankiw, Nestor, and Schlafly, Thomas, "Effect of Hole-Making on the Strength of Double Lap Joints," *AISC Engineering Journal*, vol. 19, no. 3, pp. 170–178, 3d quarter, 1982.

box only. Once removed from the container, lot identification can be maintained only through established shop or field control procedures.

8.4.3 Lubrication

All black (plain) bolts must be lubricated when installed. Most manufacturers apply a water-soluble oily lubricant to black bolts, nuts, and washers as a part of their production operations. If the fasteners are exposed to rain, snow, dew, condensation, or other moisture conditions, this lubricant may be washed off. This lubrication will also evaporate after a period of time when left in open containers.

It is a specification requirement that black fasteners be oily to the touch prior to being installed. When compared to oily fasteners, bolts that have lost their lubrication may require as much as twice the torque to install them, requiring more time and more powerful tools. In addition, the bolt's ductility (ability to stretch) is reduced because of the higher torque used to tighten poorly lubricated fasteners.

Should any of the bolts, nuts, or washers show rust, the rust must be cleaned from the surface of the fastener component, then the component must be relubricated. Dirt, sand, grit, and other foreign material must be cleaned off the bolts prior to installation, with relubrication when necessary.

If a bolt, nut, or washer has lost its lubrication, it is required that the component be relubricated prior to installation. The type of lubrication to be used is not specified, but typically a similar oil-based product, stick wax, bee's wax, liquid wax, or spray lubricant can be used.

The most effective lubrication is placed on the threads of the bolt, the threads of the nut, and the inside face of the nut. As much as 60 to 70% of the torque used to tighten a bolt is used to overcome the friction between nut and washer or steel.

In some cases for black bolts, relubrication mandates the retesting of fasteners in a Skidmore-Wilhelm (or similar device) prior to installation in the structure. This verifies the effectiveness of the relubrication. Highly efficient lubricants can actually increase the risk of thread stripping, so this condition is also checked.

If the calibrated wrench method is used for installation, any relubrication mandates the recalibration of the installation wrenches. Relubrication rarely negatively affects the performance of bolts using the turn-of-the-nut or the dti methods of installation.

Twist-off bolts must also be rechecked for performance. Many twist-off bolts use a special lubricant that is not as oily as common structural bolts. Contacting the manufacturer of the twist-off bolt system is encouraged prior to relubrication. These fasteners are particularly sensitive to inadequate lubrication and overlubrication, and loose bolts or broken bolts may result.

Galvanized fasteners, either hot-dipped galvanized or mechanically galvanized, are lubricated in a manner different than black bolts. They are not oily. The nut is the only lubricated component of the assembly. The nut must receive from the manufacturer a coating that is clean and dry to the touch. Usually a wax-based product is used, but the wax's presence may not always be determined by touch. Often, a dye is added to the lubricant to verify that the nuts have indeed been lubricated. Sometimes, a UV solution is used in the lubricant to make the nut glow under a black light.

If the presence of a lubricant is uncertain, torque testing in a Skidmore-Wilhelm (or similar) device will provide indication of the lubrication's presence. For bridge work, if the torque required to tighten the assembly is less than the maximum torque permitted in the AASHTO rotational-capacity test, then the nut has been adequately lubricated.

If relubrication is required, a wax-based or similar lubricant works well. Apply the lubricant to the threads of the nut and to the inside face of the nut. It is not necessary to lubricate the bolt or washer when this is done. After relubrication, test the assembly in a Skidmore-Wilhelm (or similar) device for torque performance and resistance to stripping.

8.4.4   Bolt stickout

Stickout is the amount of thread sticking out beyond the face of the nut after tightening. The specification requirement is that the end of the bolt be at least flush with the face of the nut. The bolt end *cannot* be below the face of the nut after tightening is completed.

There is no maximum stickout by specification, but excessive stickout indicates a risk that the nut has actually met the thread runout. If this has occurred, pretensioning is questionable for the calibrated wrench and twist-off bolting methods because the nut would cease rotation and the torque would become very high, although the bolt would remain loose. For the turn-of-the-nut method, the required turns could not be applied. For the dti method, the dti gap requirements would not be achieved.

For pretensioned bolts, a second danger of maximum stickout is that the risk of thread stripping is increased. The bolt threads will neck down in a very short region when the bolt is pretensioned, reducing the thread contact between bolt and nut.

Excessive stickout measurement is determined by the actual bolt and nut combination, and can be checked visually using an untightened bolt with the nut run up to the bolt thread runout. Generally, six threads of stickout can be permitted for $\frac{1}{2}$-, $\frac{5}{8}$-, $\frac{3}{4}$-, and $1\frac{1}{8}$-in bolts. For $\frac{7}{8}$-, 1-, $1\frac{1}{4}$-, and $1\frac{3}{8}$-in bolts, five threads of stickout can be permitted; and for $1\frac{1}{2}$-in bolts, four threads can be permitted. Stickout beyond these values should be checked with the comparison set, and may be found acceptable.

Bolt ductility is highest when the nut is flush with the end of the bolt because of the maximum number of threads available for stretching. With maximum stickout, the bolt's ductility is reduced because the stretch is limited to the very short length of thread in the grip.

A traditional "rule of thumb" had been to require two threads of stickout for high-strength bolts. This was a guideline developed for applications when the threads-excluded condition was specified. It is neither a valid indicator that the threads-excluded condition has been achieved, nor is it required by specification; therefore, it should not be part of an inspection requirement.

8.4.5 Washers

The RCSC Specification, Section 7(c), provides the following situations where F436 hardened steel washers and other special washers are required. Washers are suggested, even for cases when not required, to ease installation and provide better consistency for installation and inspection.

1. For shear-bearing joints, if either snug-tight only or pretensioned using the turn-of-the-nut method or the direct tension indicator method, and if only standard holes are present in the outer plies, washers are not required.

2. For shear-bearing joints with slotted holes present in an outer steel ply, either snug-tight only or pretensioned using the turn-of-the-nut method or the direct tension indicator method, an F436 washer or common plate washer is required over the slot.

3. If the slope of the face of the connected part exceeds 1:20, or 3°, relative to the bolt or nut face, a hardened beveled washer must be used between the fastener and the steel to compensate for the slope.

The following provisions apply only to pretensioned bolts:

1. If the calibrated wrench method is used, F436 washers must be used under the turned element.

2. If twist-off bolts are used, the supplier's washer must be used under the nut.

3. If A490 bolts are used in A36 steel (or other steels below 40-ksi yield strength), an F436 washer must be provided below both the bolt head and the nut.

4. If oversized or short-slotted holes are used in an outer steel ply, and the bolts are A325 of any diameter or A490 of 1-in diameter or less, an F436 washer must be placed over the hole or slot.

5. If oversized or short-slotted holes are used in an outer steel ply, and the bolts are A490 over 1-in diameter, an F436 washer of minimum $\frac{5}{16}$-in thickness must be placed under both bolt head and nut. Multiple standard thickness F436 washers cannot be substituted for the thicker single washer.

6. If a long-slotted hole is used in an outer steel ply, and the bolts are A325 of any diameter or A490 of 1-in diameter or less, a plate washer or continuous bar of minimum $\frac{5}{16}$-in thickness with standard holes must be used to cover the slot. The bar or plate material must be of structural grade but need not be hardened.

7. If a long-slotted hole is used in an outer steel ply, and the bolts are A490 of over 1-in diameter, a $\frac{5}{16}$-in F436 washer must be used.

8. If a twist-off bolt having a round head with a diameter at least equal to that of an F436 washer is used, and the preceding provisions (items 3 and 4) call for a standard thickness F436 washer, no washer is required under the bolt head.

8.4.6 Systematic tightening

The specifications require that joints be snugged and tightened in a systematic manner. A pattern should be chosen for tightening the bolts so that the joint is drawn together without undue bending of any of the connection parts, achieving the condition of no gaps between the steel at the bolt holes. The pattern should also be used so that bolts are not inadvertently missed during snugging or tightening.

The joint should be snugged first, starting at the most rigid part of the joint. In a joint with a single or double row of bolts, this would be where the steel is already in contact, working toward the end where the steel may not be in contact. If there is solid contact between the steel at all locations, the direction of tightening does not matter. In a bolt pattern with several rows, such as a large web splice plate in a girder, the bolts in the center of the joint should be snugged first; then proceed to work toward the free edges of the plate.

After the joint has been completely snugged, pretensioning of the bolts should follow the same systematic pattern used for snugging.

8.4.7 Reuse of bolts previously tightened

Occasionally, it may be necessary to remove a previously tightened bolt and later reinstall it. The specification permits reuse of black

A325 bolts only with the engineer's permission. Galvanized bolts and A490 bolts cannot be reused in any case.

Bolts that have been installed to the snug condition, then subsequently loosen when adjacent bolts are snugged, are not considered as reused bolts. Similarly, bolts that are touched up in the pretensioning process are not considered reused. To be considered as reuse, the bolt must be loosened and removed from the hole.

A325 bolts that have been installed only to the snug condition, then removed, can generally be reused. Snugged-only A490 bolts and galvanized A325 bolts should be considered for reuse only if snugged by hand or if very lightly snugged with an impact wrench.

To check previously snugged and previously tightened black A325 bolts to see whether they can be reused, run the nut up the entire length of the bolt threads by hand. If this is possible, the bolt may be reused. Bolts that have yielded from tightening will stretch in the first few threads (nearest the bolt head), preventing the nut from progressing further up the threads. These bolts should not be reused.

Because of the overtapping of the nut threads for galvanized fasteners, this check is not valid for galvanized bolts.

A490 bolts do not have the same ductility as A325 bolts. A490 bolts may not be reused.

8.5 Inspection Prior to Welding

The responsibilities and levels of welding inspection must be established in the contract documents. Neither the AWS or AISC, nor the model building codes, provide a complete listing of welding inspection duties. Inspection duties may be assigned to the contractor (fabrication/erection inspection) or to an inspector who reports to the owner or engineer (verification inspection). Under the special inspection requirements of the *Uniform Building Code* and *National Building Code,* certain welding inspection must take place by an inspector responsible to the owner or engineer. The level of inspection varies according to the type of project, the structural system employed, and the seismic category.

The welding inspection provisions of AWS do not assign specific inspection tasks to either the contractor or verification inspector. Verification inspection is not mandated. Any forms of nondestructive testing, beyond visual inspection, must be specified in the contract documents, as AWS requires only visual inspection by the welder.

Welding inspection is a start-to-finish task. Inspection can be broken into three timing categories: before welding, during welding, and after welding. Both general and specific welding inspection requirements are provided in AWS D1.1. Section 6 on "Inspection" covers procedural matters and acceptance criteria. Some workmanship requirements are found in Section 5. Various inspector checklists have been compiled and published in numerous sources.

8.5.1 Welding processes

Four welding processes predominate in structural steel fabrication. These are:

Shielded-metal arc welding (SMAW)

Flux-cored arc welding (FCAW)

Gas-metal arc welding (GMAW)

Submerged arc welding (SAW)

In addition, three other processes may be used from time to time. Electroslag welding (ESW) and electrogas welding (EGW) may be used for large, thick weldments in some applications. Gas tungsten arc welding (GTAW), also commonly called *tungsten inert gas welding* (TIG), may be used for small root passes and for joining specialty steels.

The first four processes mentioned (SMAW, FCAW, GMAW, and SAW) are considered prequalified welding processes under AWS D1.1 Section 3.2.1. However, the short-circuiting transfer mode of GMAW, abbreviated GMAW-S, is not prequalified. The benefit of prequalified welding processes is that no testing of the process is required for a particular joint provided the joint design and welding procedures fall within the limits of the D1.1 specification. Nonprequalified procedures require additional documentation and testing of the welding procedure specification (WPS).

Shielded-metal arc welding (SMAW). The shielded-metal arc welding process is a common welding process used for a variety of applications. The electrode is a fixed-length wire or rod of a given steel and diameter, covered by a coating that serves as a shielding and fluxing agent (see Fig. 8.1). The electrical current passes through the electrode to the steel, or from steel to electrode, depending upon polarity, forming an arc between the electrode and steel. The heat of the arc melts a portion of the steel and the end of the electrode. The electrode steel and alloying material is transferred to the weld puddle by the electrical forces generated. The electrode material and steel are mixed together by the arc action, then solidify to form the weld metal. The fluxing agent supplied by the electrode coating, as well as impurities generated by the welding itself, solidify on the top surface of the weld in the form of slag.

SMAW electrodes are categorized as low hydrogen and non–low hydrogen. Low hydrogen electrodes have coatings designed for min-

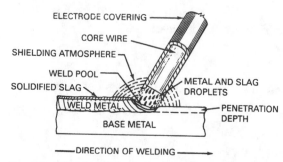

Figure 8.1   Shielded metal arc welding. (*Adapted from American Welding Society* Welding Handbook, *vol. 2.*)

mum moisture absorption from the atmosphere and provide a weld and heat-affected zone with a minimum amount of hydrogen. Hydrogen is considered a cause of low ductility and underbead cracking. Low hydrogen electrodes are identified as EXXX5, EXXX6, and EXXX8. Low hydrogen electrodes also require special care in storage and handling, including baking and drying, to retain their low hydrogen characteristics.

Electrodes can be from AWS/ANSI A5.1 "Specification for Mild Steel Covered Arc Welding Electrodes" or AWS/ANSI A5.5 "Specification for Low Alloy Steel Covered Arc Welding Electrodes." A5.5 electrodes are identified as EXXXX-X. A5.5 electrodes are optional for steels of groups I and II as listed in AWS Table 3.1. They are required for AWS group III steels and the steels listed in AWS annex M, unless WPS qualification testing is performed. The nomenclature used to identify SMAW electrodes is shown in Table 8.3.

TABLE 8.3   Electrode Classification System for Shielded-Metal Arc Welding

EXXXX-X	
E	Electrode
XX	Minimum tensile strength of undiluted weld metal, ksi, in the as-welded condition
X	Permitted position 1. All 2. Horizontal (fillets only) and flat 4. Flat, horizontal, overhead, and vertical down
X	Coating and operating characteristics 5, 6, and 8—low hydrogen
-X	Major alloying elements (A5.5 electrodes)

Flux-cored arc welding (FCAW). Flux-cored arc welding (FCAW) is another popular welding process. The process uses a tubular wire containing fluxing and alloying agents inside the tube. The wire is fed into the weld area using a welding "gun." The heat of the arc vaporizes the fluxing agents in the core, releases the alloying elements, melts the electrode wire to enable transfer by the arc, and melts the steel being welded in the area of the arc (see Fig. 8.2).

One option used with FCAW, typically used as self-shielded, is gas shielding. Such a welding process is designated FCAW-G. FCAW-G is also considered a prequalified welding process. Because of the gas shielding, welding must not be done in wind speeds above 5 mi/h (8 km/h) under the provisions of AWS D1.1 Section 5.12.1.

Flux-cored electrodes for structural steel may be of two types, AWS/ANSI A5.20 "Specification for Carbon Steel Electrodes for Flux Cored Arc Welding" and AWS/ANSI A5.29 "Specification for Low Alloy Steel Electrodes for Flux Cored Arc Welding." Flux-cored electrodes are designated as shown in Table 8.4.

Flux-cored arc welding is known for higher productivity rates because the wire is continuously fed from a coil, compared to the fixed-length electrodes of SMAW, and because of higher deposition rates.

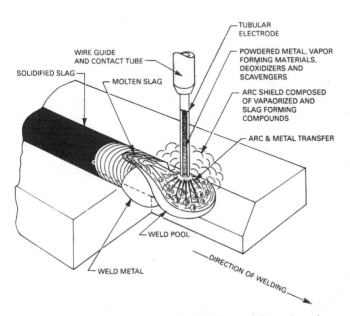

Figure 8.2 Flux-cored arc welding. (*Adapted from American Welding Society* Welding Handbook, *vol. 2.*)

TABLE 8.4 Electrode Classification System for Flux-Cored Arc Welding

EXXTX-X	
E	Electrode
X	Minimum tensile strength of deposited weld metal, 10 ksi, in the as-welded condition
X	Permitted position 0. Flat and horizontal 1. All positions
T	Tubular wire
X	Electrode classification
-X	Alloy information (A5.29 electrodes)

Gas-metal arc welding (GMAW). Gas-metal arc welding (GMAW) is similar to FCAW in that the electrode wire is continuous and welding is performed using a welding "gun." The GMAW wire is a solid wire, rather than tubular like FCAW. In some speciality steels other than carbon and low-alloy structural steels, a metal-cored wire may be used. Metal-cored wires may also be considered FCAW wires. The shielding of the weld region is supplied by an external shielding gas fed into the weld through a hose (see Fig 8.3). Gas-metal arc welding is also commonly referred to as MIG (*metal inert gas*) *welding, or* as MAG (*metal active gas*).

GMAW electrodes can be of two types, AWS/ANSI A5.18 "Specification for Carbon Steel Filler Metals for Gas Shielded Arc Welding" or ANSI/AWS A5.28 "Specification for Low Alloy Steel Filler Metals for Gas Shielded Arc Welding." A5.28 electrode wires must be used for AWS Table 3.1, group III steels and AWS annex M steels, unless otherwise qualified by test. Electrodes for gas-metal arc welding are identified as shown in Table 8.5.

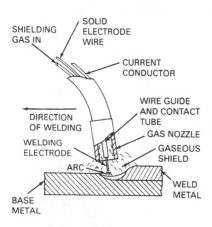

Figure 8.3 Gas-metal arc welding. (*Adapted from American Welding Society Welding Handbook, vol. 2.*)

TABLE 8.5   Electrode Classification System for Gas-Metal Arc Welding

ERXXS-X

ER	Electrode rod (if ER, may also be used as GTAW filler rod; if E, may not be used for GTAW)
XX	Minimum tensile strength of deposited weld metal, ksi, in the as-welded condition
S	Solid wire (C designates metal-cored wire)
-X	Alloy information (A5.28)

Common shielding gases used for structural steel include mixtures of argon and oxygen, argon and carbon dioxide, and straight carbon dioxide. Similar to FCAW-G, GMAW welding cannot be performed in winds greater than 5 mi/h (8 km/h); therefore, it is rarely used in field applications.

GMAW has two advantages for welding structural steel. The first is that there is minimal slag to be removed. The second is that there are minimal welding fumes to be exhausted from the work space.

Submerged arc welding (SAW).   The *submerged arc welding* (SAW) *process* is a unique process that employs both a wire-fed welding gun and a deposit of flux powder. The arc is buried beneath a blanket of granular flux. The flux shields the weld region from atmospheric impurities, as well as provides fluxing and alloying elements (see Fig. 8.4). Of

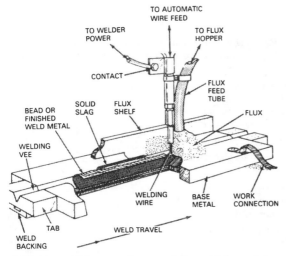

Figure 8.4   Submerged arc welding. (*Adapted from American Welding Society Welding Handbook, Vol. 2.*)

the four principal welding prequalified welding processes, SAW provides the most penetration. As an automatic or semiautomatic process, it is also the fastest once setup is complete. Because the weld puddle is buried beneath the flux, not visible to the welder, establishing proper welding parameters and accurately tracking the root of the weld is critical to the satisfactory completion of the weld.

Submerged arc welding may be performed with multiple welding heads, set either in parallel, with both electrodes controlled by the same feeder and power supply, or in tandem, with separate feeders and controls.

The proper combination of electrode wire and flux is important for quality welding. For this reason, the electrode and flux combination is specified within one document. ANSI/AWS A5.17 covers "Specification for Bare Mild Steel Electrodes and Fluxes for Submerged Arc Welding," and ANSI/AWS A5.23 covers "Specification for Low Alloy Electrodes and Fluxes for Submerged Arc Welding." The nomenclature system used for SAW electrodes and fluxes under ANSI/AWS A5.17 is shown in Table 8.6.

ANSI/AWS A5.23 specifications use a more complex nomenclature system with an additional -XX at the end of the electrode section to provide additional information on the system.

TABLE 8.6 Flux and Electrode Classification System for Submerged Arc Welding

FXXX-EXXX	
F	Flux
X	Strength level intended for flux, 10 ksi units
X	A or P, A for weld deposit strength determined in the as-welded condition, P for strength determined in postweld heat-treated condition
X	Impact properties temperature Z. No impact properties 0. 20 ft-lb at 0°F (27 J at –18°C) 2. 20 ft-lb at –20°F (27 J at –29°C) 4. 20 ft-lb at –40°F (27 J at –40°C) 6. 20 ft-lb at –60°F (27 J at –51°C)
-E	Electrode (EC for composite electrode)
X	Manganese content L. Low manganese (0.6% max.) M. Medium manganese (1.40% max.) H. High manganese (2.25% max.)
X	Nominal carbon content, in 1/1000% Ex: 12=0.12%
K	Killed steel (if applicable)

8.5.2   Welding procedures

The use of written established welding procedure specifications (WPS) is mandated by AWS D1.1 in Section 3.1. The WPS may be either prequalified or qualified by test. A prequalified WPS must fall within the limits prescribed in Section 3 of D1.1 (notably Table 3.7), must use a prequalified welding process, and must be on a joint deemed prequalified in Section 3. All other WPSs must be qualified by test using the procedures set forth in Section 4 of D1.1. Annex E of D1.1 provides example WPS forms.

Welding procedure specifications may be furnished by electrode suppliers, welding equipment suppliers, technical organizations, or consultants, or may be developed by the contractor. WPSs are specific to the following parameters:

Welding process

Base metal (steel classification, strength, type)

Base metal thickness (range)

Electrode classification

Flux classification

Shielding gas

Joint type (butt, tee, corner)

Weld type (groove, fillet, plug)

Joint design details (root opening, groove angle, use of backing)

Use of backgouging

Position (flat, horizontal, vertical, overhead, tubular application)

Using these parameters, the following items are established:

Number and position of passes

Electrode diameter

Polarity

Current or wire-feed speed

Travel speed

Voltage

Technique

Shielding gas (if used) flow rate

Preheat, interpass, and postheat requirements

Cleaning requirements

Inspection requirements

The inspector should check for and review the welding procedure specifications (WPS) to be used for the project. The WPS is specific to the material, welding process, position, type of joint, and configuration of joint. Verify that a proper WPS is available for all welds to be completed, and that the welding personnel follow the WPS as prescribed.

8.5.3 Welding personnel

Welders, tack welders, and welding operators must be qualified by the contractor responsible for the welding prior to the welding being performed, as required by AWS D1.1 Section 4.1.2.

Welders are individuals who manipulate the welding electrode, wire and/or filler metal by hand to make the weld. A tack welder is a fitter who makes small welds as necessary to hold parts together until final welding by a welder or welding operator. A welding operator sets up and adjusts equipment to perform automatic welding.

The fabricator or erector responsible for welding must have each welder, tack welder, and welding operator tested using the methods of D1.1 Section 4, Part C, to prove their capability to make adequate quality welds. These individuals are tested and categorized by:

Welding process

Welding position

Electrode classification

Base metal thickness range

CJP groove welds if made without backing

The testing may be performed by the contractor (fabricator or erector) or by an independent testing laboratory.

Welder performance qualification (WPQ) documentation must be made available for the inspector's review prior to the start of welding. If a previous employer's testing results are to be used, then the engineer must approve the current employer's reliance upon these previous tests.

Welders who perform and pass such testing at an independent testing laboratory accredited by the American Welding Society can have their test records placed on file with the AWS and receive the designation of AWS Certified Welder.

A welder's or welding operator's qualification for a given employer remains in effect indefinitely, as long as that individual continues welding in that given process, although not especially with the tested electrode classification or in the tested position. If the welder fails to use that process for a period exceeding 6 months, the welder must complete and pass a welding test. If the welder's quality becomes subject to question, the welder's qualification may be revoked by the employer, forcing a retest. Tack welders' qualifications remain in effect perpetually, unless there is specific reason to question the tack welder's abilities. Although welder qualification is the responsibility of the contractor, under the provisions of AWS D1.1 Section 6.4.2, the inspector may also force requalification testing if the welder's quality is poor.

8.5.4 Base metal quality

Verify that the quality of the base metal is suitable for welding. The steel to be welded must be clean, smooth, and without surface discontinuities such as tears, cracks, fins, and seams. Such surface discontinuities could propagate into the weld after welding. The surface should also be free of excessive rust, mill scale, slag, moisture, grease, oil, and any other material that could cause welding problems. Some materials may be permitted, such as thin mill scale (mill scale that withstands a vigorous wire brushing), thin rust-inhibitive coatings, and antispatter compounds made specifically for weld-through applications. AWS Section 5.15 provides additional information and exceptions to these provisions.

8.5.5 Joint preparation and fit-up

Fillet-weld fit-up tolerances are given in AWS Section 5.22.1. Gaps between parts of $\frac{1}{16}$ in (1.6 mm) or less are permitted without correction. If the gap exceeds $\frac{1}{16}$ in (1.6 mm) but does not exceed $\frac{3}{16}$ in (5 mm), then the leg size of the fillet weld should be increased to compensate for the gap between the parts. Gaps over $\frac{3}{16}$ in (5 mm) are permitted only in thick materials over 3 in (76 mm). In these cases, the use of a backing material is required as well as compensation in the weld leg size. Such provisions cannot be used for gaps over $\frac{5}{16}$ in (8 mm). Similar provisions are used for partial-penetration groove welds when the welds are parallel to the length of the member.

When groove welds are used, tolerances to the root opening, groove angle, and root face apply. The specific tolerances depend upon the type of groove weld, the presence of backing, and the use of backgouging. AWS Section 5.22.4 and AWS Fig. 5.3 provide these values. Groove tolerances are also provided in the prequalified groove weld details in AWS Figs. 3.3 and 3.4. For tubular joints, Section 5.22.4.2 governs.

Part alignment on butt joints can be critical, depending upon application. AWS Section 5.22.3 requires alignment within 10% of the part thickness, not to exceed $\frac{1}{8}$ in (3 mm), when the parts are restrained from bending from such misalignment. No provisions are given for cases where such restraint does not exist. For girth welds in tubular joints, the alignment tolerances are provided in Section 5.22.4.

8.5.6 Welding equipment

In order to properly follow the parameters of a welding procedure specification (WPS), the equipment used for welding must be in good repair and properly calibrated. The inspector should check the maintenance and testing records of the equipment to be employed, and if necessary, use testing equipment to verify that the equipment settings and the welding machine output are accurate.

8.5.7 Welding consumables

Welding electrodes, fluxes, and shielding gases should be checked to be in conformance with AWS Section 5.3. Low hydrogen SMAW electrodes require special controls, including requirements for baking and storage temperatures and exposure time limits. Fluxes for SAW require dry, contamination-free storage, with the removal of the top 1 in (25 mm) of material from previously opened bags prior to use. Drying of flux from damaged bags may be required.

Shielding gases must be of welding grade and have a dew point of $-40°F$ ($-40°C$) or lower. The gas manufacturer's certification of dew point may be required by the engineer.

Welding materials, such as electrodes, fluxes, and shielding gases, also have manufacturers' certificates of compliance that they meet applicable American Welding Society and ANSI standards. These certificates of compliance may be requested by the owner. Lot numbers may be placed on electrodes; otherwise, identification is limited to the containers. Once welding has been completed, traceability back to a particular lot is impossible. Only strict shop controls can ensure that the lots for which certification is provided have been used for that customer's project.

8.5.8 Welding conditions

For good welding, the welder and the operating equipment must have conditions suitable for welding. The environmental conditions for welding must also be adequate, and limits are given in AWS D1.1, Section 5.12. The temperature of the area immediately surrounding the welding must be above $0°F$ ($-18°C$). The temperature in the general vicinity can be lower, but heating must be provided to raise the temperature immediately around the weld to at least this temperature. The surfaces to be welded must not be wet or exposed to moisture. High winds must be avoided. For GMAW, GTAW, EGW, and gas-shielded FCAW, the wind speed must not exceed 5 mi/h (8 km/h), requiring protective enclosures in most field applications. No maximum wind speed is specified for welding processes requiring no shielding gases, but a practical level is generally around 25 mi/h (40 km/h).

8.5.9 Preheat

Preheating of the steel is necessary for thick steels, certain high-strength steels, and steels when their temperature is below 32°F (0°C). The preheating requirements should appear in the welding procedure specification (WPS). AWS preheat requirements may also be found in Table 3.2, Minimum Prequalified Preheat and Interpass Temperature. In this table, the required preheat is given for a specified steel specification group, welding process, and thickness range of part. When the temperature of the steel is below 32°F (0°C), the steel must be heated to at least 70°F (21°C). The thicker the steel, the higher the required preheat temperature. Higher-strength steels require higher preheats. Certain high-strength steels listed in AWS D1.1 Annex M have preheats limited to a maximum of 400°F (205°C). Preheat requirements may also be modified using the provisions of AWS D1.1 Annex XI, which evaluates the welding process, restraint, and the weldability (carbon equivalency) of the steel.

8.6 Inspection During and After Welding

After checking the welder qualifications, WPS, welding consumables, steel materials, welding conditions, equipment, joint fit-up, and preheat, the welding inspection performed during welding is limited to verifying that the welding procedures are properly followed. This includes the maintenance of interpass temperature during welding, usually the same temperature required for preheat. Each pass should be thoroughly cleaned and visually inspected. Control of electrodes, especially low hydrogen SMAW electrodes, must be maintained. In some cases, nondestructive testing may be performed at various stages during welding.

After the weld has been completed, the final size and location is checked to verify that it meets the plans and specifications of the contract. Visual inspection is performed, and nondestructive testing of the completed weld may be performed if required by the contract documents. If repairs are required, the inspection should include the repair work and reinspection of the repaired weld. The inspector responsible for the completed weld should place an identifying mark near the weld. Written documentation of the weld's quality, including any noted significant discontinuities, should be prepared and submitted.

8.7   Nondestructive Testing

Several methods of nondestructive testing (NDT), also called nondestructive examination (NDE), may be used on a structural steel project. The frequency and location of NDT must be stated in the contract documents. Little NDT is mandatory under the various codes except for certain types of joints in seismic applications.

The first common form of NDT is *visual testing* (VT). Most visual inspection is performed without the use of magnifiers. Magnifying glasses can be used to more closely examine areas that are suspected of cracks and other small, but potentially significant discontinuities. Adequate light and good visual acuity is necessary. Various weld gages are used to determine weld size and various other measurements.

An expanded form of visual inspection is *penetrant testing* (PT). The weld surface and surrounding steel is thoroughly cleaned. A penetrating liquid dye is applied to the weld surface and allowed time to penetrate cracks, pores, and other surface discontinuities. After an allotted time (dwell time), the penetrant is removed and a developer is applied to the surface. The developer draws the penetrant back to the surface of the weld. The developer is of a color (usually white) that contrasts with the color of the dye in the penetrant. The inspector observes the dye in the developer, then removes the developer and dye to more closely inspect the weld surface visually. Some penetrant testing uses an ultraviolet solution, rather than a dye, to aid in visibility when a UV lamp is available. Penetrant testing can detect surface discontinuities only.

Magnetic particle testing (MT) can be used to detect surface and slightly subsurface discontinuities. The general limit to the depth of inspection is around $5/16$ in (8 mm). Electricity is induced into the region of the weld through the use of prods or a yoke. The electricity generates a magnetic field on and near the surface of the steel. Magnetic particles such as fine iron particles are then applied to the surface of the steel. These particles may be in the form of a dry powder or may be in a liquid emulsion.

When cracks or other discontinuities are on, or near, the surface, the flux lines generated by the current are interrupted. In essence, the interruption has created two new magnetic poles in the steel, attracting the particles. The inspector then observes and interprets the position and nature of the accumulated particles, judging them to indicate a crack or other discontinuity on the surface or subsurface.

For best performance, the flux lines must flow perpendicular to the discontinuity. Therefore, the MT technician must rotate the yoke, prods, or other source at 90° angles along the length of the weld to inspect it for both longitudinal and transverse discontinuities. Although MT is generally a visual technique, permanent records of discovered defects can be made with the use of adhesive tape placed over attracted magnetic particles during testing.

Ultrasonic testing (UT) is a very popular method of nondestructive testing. It is capable of testing weldments from approximately $\frac{5}{16}$ to 8 in (8 to 200 mm) in thickness. The most common method of testing uses a pulse-echo mode similar to radar or sonar. The control unit sends high-frequency electronic signals into a transducer made of piezoelectric material. The electrical energy is transformed by the transducer into vibration energy. The vibration is transmitted into the weldment through a coupling liquid. The vibration carries through the steel until a discontinuity or other interruption, such as an edge or end of the material, disrupts the vibration. The disruption causes the vibration to return the ultrasound wave back toward the transducer. The return vibration is then converted back into electrical energy by the transducer, sending a signal to the display unit. The return signal's configuration, strength and time delay are then interpreted by the testing technician.

The interpretation by the technician uses the height of the response signal to indicate size and severity. Using a calibration setup, the distance on the display unit from the initial pulse to the reflection is used to determine the distance from the transducer to the discontinuity. The operator can also manipulate the transducer in various patterns to determine a better understanding of the location, length, depth, orientation, and nature of the discontinuity.

AWS D1.1 Table 6.6 prescribes the testing procedures for butt, tee, and corner joints of various thicknesses. The search angle and faces to be used are given. Annex K of AWS D1.1 gives alternative techniques for ultrasonic testing and the evaluation of weld discontinuities.

Radiographic testing (RT) is another common method of NDT for welds in structural steel. RT is performed, using either x-rays or gamma rays, sending energy into the steel weldment. Film is placed on the side of the weldment opposite the energy source. The steel and weld metal absorb energy, preventing it from exposing the film, but weld discontinuities allow more energy to get to the film. This exposes the film more, producing a darkened area on the film to be interpreted by the radiographer.

Radiographic testing is effective in steels up to about 9 in (230 mm) in thickness. X-ray machine capabilities depend upon the voltage setting of the machine, with 2000 kV required for an 8-in (200-mm) thickness. Gamma ray machine capabilities depend upon the isotope used, usually cobalt-60 or iridium-192, but sometimes cesium-137. Exposure time and film selection are varied according to conditions and thicknesses. Image quality indicators (IQIs), either wire-type or hole penetrameter, are used to verify the sharpness and sensitivity of the film image, as well as to provide a measurement scale on the exposed film. Because of the radiation exposure hazards and time and equipment involved, radiographic testing is typically the most expensive of the methods previously mentioned.

8.8   Weld Acceptance Criteria

The acceptance criteria to be used for the required weld quality is to be established by the engineer. Commonly, the quality and inspection criteria found in AWS D1.1 Sections 5 and 6 are adopted. However, the use of alternate criteria is both accepted and encouraged. Section 6.8 of AWS states that "The fundamental premise of the Code is to provide general stipulations applicable to most situations. Acceptance criteria for production welds different from those specified in the Code may be used for a particular application, provided they are suitably documented by the proposer and approved by the Engineer." The Commentary to Section 6.8 provides additional insights into the development and use of alternate acceptance criteria.

The visual acceptance criteria for welds is summarized in Table 6.1 of AWS D1.1. This table is broken down into three categories of connections: statically loaded nontubular, cyclically loaded nontubular, and tubular. Generally, cyclically loaded and tubular connections require higher standards of quality. These values also apply when penetrant testing (PT) and magnetic particle testing (MT) are used.

When ultrasonic testing is used, AWS Table 6.2 is used for statically loaded nontubular connections and Table 6.3 for cyclically loaded nontubular connections. For tubular connections, use AWS Section 6.13.3. Alternately, the techniques of AWS Annex K may be employed when approved by the engineer.

When radiographic testing is used, AWS Fig. 6.1 is used for statically loaded nontubular connections, Fig. 6.4 for cyclically loaded nontubular tension connections, and Fig. 6.5 for cyclically loaded nontubular compression connections. For tubular connections, AWS Sections 6.12.3 and 6.18 apply.

Because many of the acceptance criteria found in AWS D1.1 are based upon what a qualified welder can provide, rather than the quality necessary for structural integrity, alternate acceptance criteria can be used to save both time and money. In addition, repairs to some welds with innocuous discontinuities may result in more damage to the material in the form of additional discontinuities, lower toughness, larger heat-affected zones, more distortion, and higher residual stresses.

Alternative acceptance criteria have been published by several organizations in various forms. In the United States, the Electric Power Research Institute has published "Visual Weld Acceptance Criteria," Document NP-5380, for use in reinspections of welds in existing nuclear power plant facilities. This weld acceptance criteria was accepted for use by the Nuclear Regulatory Commission. The Welding Research Council has published several WRC bulletins providing suggested criteria. The ASME *Boiler and Pressure Vessel Code,* Section IX, provides acceptance criteria for welds that can be also used for structural welds. The International Institute of Welding has published several documents providing suggested acceptance criteria, with considerable research documentation justifying the criteria. British Standards Institution document PD 6493:1991, "Guidance on Methods for Assessing the Acceptability of Flaws in Fusion Welded Structures," is one of the most thorough documents currently available.

8.9 Inspector Certification Programs

Welding inspectors must be qualified to perform the work on the basis of the contract documents and the building code. The engineer has the responsibility to require any inspector credentials exceeding those of AWS D1.1 Section 6.1.3.1. Under this provision, a welding inspector may be qualified by virtue of being an AWS Certified Welding Inspector, a welding inspector certified under the Canadian Welding Bureau, or "an engineer or technician who, by training or experience, or both, in metals fabrication, inspection and testing, is competent to perform inspection of the work." Section 6.1.3.4 requires an eye examination, with or without corrective lenses.

The inspector's qualification to perform the inspection remains effective indefinitely, unless there is specific reason to question the inspector's abilities. In this case, the engineer or similar individual should reevaluate the inspector's abilities and credentials.

For individuals performing only nondestructive testing work, the inspector need not be generally qualified for welding inspection. However, the individual must be qualified using the provisions of the American Society for Nondestructive Testing "Recommended Practice No. SNT-TC-1A." This document provides recommendations for the training, experience level, and testing of NDT technicians. A suitable alternative to the "Recommended Practice," although not referenced in D1.1, is the ASNT "Standard for Qualification and Certification of Nondestructive Testing Personnel."

These documents provide specific listings applicable to several areas of NDT:

Radiographic testing

Magnetic particle testing

Ultrasonic testing

Liquid penetrant testing

Electromagnetic testing

Neutron radiographic testing

Leak testing

Acoustic emission testing

Visual testing (SNT-TC-1A only)

NDT technicians are placed into four categories. Formal definitions vary between the *recommended practice* and the *standard.* Using the definitions of the *standard,* the Level III technician has the "skills and knowledge to establish techniques; to interpret codes, standards and specifications; to designate the particular technique to be used; and to verify the adequacy of procedures." This individual is responsible for the training and testing of other NDT personnel in the individual's area of certification. The Level II technician has "the skills and knowledge to set up and calibrate equipment, to conduct tests, and to interpret, evaluate, and document results in accordance with procedures approved by an NDT Level III." The Level I technician has "the skills and knowledge to properly perform specific calibrations, specific tests, and with prior written approval of the Level III, perform specific interpretations and evaluations for acceptance or rejection and document the results." The trainee is a technician who works under the supervision of a Level II or III, and cannot independently conduct any tests or report any test results.

Many Level III technicians have taken and passed a nationally administered ASNT examination in the particular field of NDT. However, it is possible for an individual to be named by the employer to a Level III designation, based upon his or her experience and knowledge in the field.

Steel Deck Connections

Richard B. Heagler
Nicholas J. Bouras, Inc.
Summit, NJ

Fastening deck is an important design function which requires the attention of the design professional. Unlike structural steel, the fastening of steel deck has little or nothing to do with its fabrication so the deck supplier has no responsibility for choosing the type of fastening or the spacing. However, the deck supplier, or the Steel Deck Institute, can aid the designer by providing information that can be helpful in the selection process.

For construction purposes the deck is almost always used as a working platform. It is, therefore, quite important that the deck be quickly and adequately attached as it is placed. Additionally, the fastened deck acts to stabilize joists and brace beams. Although the construction process is usually not part of the design, safety of the working platform is obviously important.

The factors that most often affect the fastening are the anticipated wind and earthquake loads. These cause both horizontal (diaphragm) and vertical (uplift) forces to be applied to the fasteners and, as a result, are of most interest to designers. Some Underwriters Laboratories' fire-rated constructions also specify fastening types and spacings, and must be consulted.

Shear and uplift strengths are the parameters most needed. Table 9.1 shows the ultimate tensile strength of arc puddle welds through

432   Chapter Nine

(Courtesy of The Steel Institute of New York.)

TABLE 9.1   Tensile Strength of Arc Puddle Welds, lb

The weld tensile strengths shown in the table are based on the <u>lowest</u> weld strengths obtained using the range of steel properties of roof deck (and floor deck). The AISI Specifications are the basis of the table. The strengths are the nominal (ultimate) values. For LRFD apply a ϕ factor of 0.60; and, for ASD, use a safety factor of 2.5. For ASD it may be appropriate to take advantage of the 1/3 increased allowed for temporary wind loading.

Case 1. Single deck thickness.
Case 2. Two layers of deck such as at an end lap.
Case 3. At a side lap (on structural steel or bar joist).

Case	Gage	Visible Weld Diameter				Profile
		.5	.625	.75	1.0	
1	22	550	690	840	1130	
	20	660	830	1010	1360	
	18	850	1080	1310	1780	
	16	1040	1330	1630	2220	
2	22	870	1150	1440	2000	
	20	980	1330	1670	2360	
	18	850	1080	1310	1780	
	16	1200	1780	2350	3500	
3	22	380	490	590	790	
	20	460	580	710	950	
	18	590	760	920	1250	
	16	730	930	1140	1550	

steel deck. Three types of welds are illustrated: type 1 is through a single deck thickness, type 2 would be at a deck end lap or through cellular deck and is through two thicknesses of metal, and type 3 is at a deck edge lap and its lower values are the result of the eccentric loading at the edge. The formulas in the American Iron and Steel Institute (AISI) specification were used to develop the table values. If an allowable stress design method is used to check wind uplift, a safety factor of 2.5 is suggested; however, the temporary nature of the uplift load may allow a one-third increase to develop the design load.

Table 9.2 shows the ultimate shear strength of welds. These are to be used for *diaphragm* loads and are based on the formulas given in the Steel Deck Institute (SDI) *Diaphragm Design Manual*. For allowable stress design a safety factor of 2.75 is suggested for welds; no one-third increase for temporary loading is allowed by the SDI. A quality arc spot weld should have at least 75% of the perimeter attaching the deck to the structural steel. A quality control procedure is shown in Fig. 9.1.

Weld washers are only recommended for attaching deck to the structural frame or bar joists when the deck steel is less than 0.028 in (0.71 mm) thick. The purpose of the weld washers is to provide a heat sink and keep the weld burn from consuming too much of the thin steel. The weld washer then forms a "head" on the weld button and provides the uplift and shear strengths as shown in Table 9.3. Common weld washers furnished by deck manufacturers are made of 16 gage material [0.057 in (1.44 mm)] and have a $\frac{3}{8}$-in (10-mm)-diameter hole. The weld should slightly overfill the hole to produce a visible weld diameter of about $\frac{1}{2}$ in (13 mm).

Self-drilling screws are frequently used as deck-to-frame attachments. These are installed with an electric screw gun that has a clutch and a depth-limiting nose piece to prevent overtorquing.

TABLE 9.2   Weld Shear Strengths, lb (for Diaphragm Calculations)*

These values are based on the formulas from the Steel Deck Institute Diaphragm Design Manual, Second Edition. Pn Values shown: Suggested ϕ = 0.6: Suggested S.F. = 2.75, F_y = 33 ksi, F_u = 45 ksi, F_{xx} = 70 ksi

Metal Thickness	Visible Weld Diameter			
	$\frac{5}{8}"$	$\frac{3}{4}"$	$\frac{7}{8}"$	1"
0.0295	1739	2104	2469	2834
0.0358	2088	2531	2974	3417
0.0418	2413	2931	3448	3965
0.0474	2710	3297	3884	4470
0.0598	3346	4086	4826	5566

Courtesy of the Steel Deck Institute.

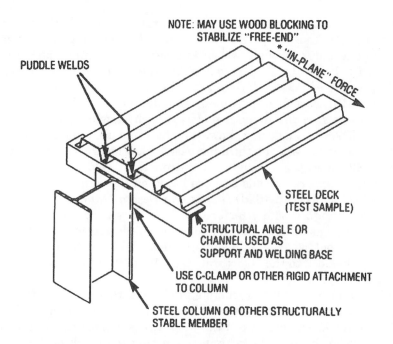

NOTE: MAY USE WOOD BLOCKING TO
STABILIZE "FREE-END"

*"IN-PLANE" FORCE

PUDDLE WELDS

STEEL DECK
(TEST SAMPLE)

STRUCTURAL ANGLE OR
CHANNEL USED AS
SUPPORT AND WELDING BASE

USE C-CLAMP OR OTHER RIGID ATTACHMENT
TO COLUMN

STEEL COLUMN OR OTHER STRUCTURALLY
STABLE MEMBER

A preliminary check for welding machine settings and operator qualifications can be made through a simple field test by placing a pair of welds in adjacent valleys at one end of a panel. The opposite end of the panel can then be rotated, which places the welds in shear. Separation leaving no apparent external weld perimeter distresses, but occurring at the sheet-to-structure plane, may indicate insufficient welding time and poor fusion with the substrate. Failure around the external weld perimeter, showing distress within the panel but with the weld still attached to the substrate, would indicate a higher quality weld.

Figure 9.1   Weld quality control check. (*Courtesy of the Steel Deck Institute.*)

Screws are #12s or $\frac{1}{4}$ in with the drill point selected to drill through the total metal thickness of deck and beam (or joist) flange. Uplift (pullover and pullout) values are shown in Table 9.4. Self-drilling screws are available for the special application of attaching steel deck to wood framing.

In recent years, excellent deck-fastening methods have been developed which compete with traditional welds and screws. These fastening methods use powder or air pressure to drive pins through the deck into structural steel. Strength values and diaphragm tables are published by the manufacturers of these products and they also provide technical assistance for designers. The Steel Deck Institute advises that, "No substitution of fastener type or pattern should be made without the approval of the designer." Fastener manufacturers can provide the data needed to make substitutions with their products.

Shear studs can be welded through the deck into the steel framing

TABLE 9.3   Weld Washer Strengths

Case	Gage	Pn, Uplift (Tensile) Values, Lbs. [1]		Shear, Lbs.[2]	Profile
1	28	0.0149	1390	1200	
	26	0.0179	1430	1550	
	24	0.0239	1520	2350	
	23	0.0269	1560	2830	
2	28	0.0149	1590	1200	
	26	0.0179	1670	1550	
	24	0.0239	1840	2350	
	23	0.0269	1780	2830	
3	28	0.0149	960	1200	
	26	0.0179	990	1550	
	24	0.0239	1050	2350	
	23	0.0269	1090	2830	

(1) A suggested safety factor (ASD) is 2.5; the recommended ϕ factor (LRFD) is 0.60.
(2) A recommended safety factor (ASD) is 2.75; the recommended ϕ factor (LRFD) is 0.55.
The table is based on typical form deck material (F_y = 80 ksi); a 70 ksi electrode strength was used. Washers are 16 gage.

with an automatic stud "gun." The primary function of the studs is to make the beam act compositely with the concrete but they also act as the fastening of the deck to the frame. Shear studs can be welded through two well-mated thicknesses of steel such as cellular deck. But, for deck heavier than 16 gage, consultation with the stud manufacturer is advised. The American Welding Society (AWS) provides a quality control check for welded studs. Figure 9.2 shows the patterns and pattern nomenclature of deck-to-frame connections.

Deck-to-deck connections at side laps are sometimes called "stitch connections." Screws, welds, and button punches are the usual ways to accomplish the connection. The primary purpose of side-lap attachments is to let adjacent sheets help in sharing vertical and horizontal loads. Stitch screws are usually of the self-drilling type; #8s though $\frac{1}{4}$-in diameter can be used but screws smaller than #10 diameter are not recommended. The installer must be sure that the underlying sheet is drawn tightly against the top sheet. Again, as when screws are used as the frame attachment, special screw-driving guns are used to prevent overtorquing.

Manual button punching of side laps requires a special crimping tool. Button punching requires the worker to adjust his or her weight so the top of the deck stays level across the joint. Since the quality of the button punch attachment depends on the strength and care of the tool operator, it is important that a consistent method be developed. Automatic power-driven crimping devices are rarely seen on deck jobs but should not be ruled out as a fastening method.

TABLE 9.4   Uplift Values for Screwed Deck

SCREW DATA				
Screw Size	d dia.	d_w nom. head dia.	Avg. tested tensile strength, kips	
10	0.190	0.415 or 0.400	2.56	
12	0.210	0.430 or 0.400	3.62	
¼	0.250	0.480 or 0.520	4.81	

Pull Over Values, kips

$P_{not} = 1.5\, t_1 d_w F_u;\; d_w < 0.50"$.

d_w	Gage						
	16	18	20	22	24	26	28
0.400	1.61	1.28	0.97	0.80	0.86	0.64	0.54
0.415	1.68	1.33	1.00	0.83	0.89	0.67	0.56
0.430	1.74	1.38	1.04	0.86	0.92	0.69	0.58
0.480	1.94	1.54	1.16	0.96	1.03	0.77	0.64
0.500	2.02	1.60	1.21	1.00	1.08	0.81	0.67

The table pull over strengths kips, are based on $F_u = 45$ ksi for 16 thru 22 gage, and 60 ksi for 24 thru 28 gage.
The safety factor for pull over (ASD) is 3, but for wind loading the 1/3 load increase may be proper. The ϕ factor (LRFD) is 0.5.

Pull Out Values, kips

$P_{not} = 0.85 t_2 d F_u;$ Metal thickness = t_2

Screw	Gage									
	¼"	3⁄16"	10	⅛"	12	14	16	18	20	22
# 10	2.34	1.76	0.98	1.17	0.76	0.55	0.44	0.35	0.26	0.22
# 12	2.66	2.00	1.12	1.33	0.87	0.62	0.50	0.38	0.29	0.24
# ¼	3.08	2.31	1.29	1.54	1.00	0.72	0.57	0.45	0.34	0.28

The table pullout strengths kips, are based on $F_u = 45$ for 10, 12 thru 22 gage, and 58 ksi for ¼", 3⁄16", and ⅛".

Deck can be screwed to structural steel, bar joists, or light gage steel framing. The lowest steel strength was used to produce the tabulated values. For bar joists and structural steel, a tensile strength (F_u) of 58 ksi was used which is the lowest value for A36 steel. For gage supports, $F_u = 45$ ksi was used which is the lowest provided in ASTM A653 Structural Quality grade 33. Deck materials furnished in gages 24, 26, 28 are usually grade 80 steels which use a tensile strength (F_u) of 60 ksi as limited by the AISI specifications. Either pull out of the screw or pullover of the deck will normally control. The values are based on the equations provided by the AISI Specifications. These specifications call for a safety factor of 3 to be applied to the table values for ASD and a ϕ of 0.5 for LRFD. If it is known that the tensile strength of the support steel or the sheet steel is greater that the values used for the tables, the tabulated ultimate strengths may be increased by a straight line ratio.

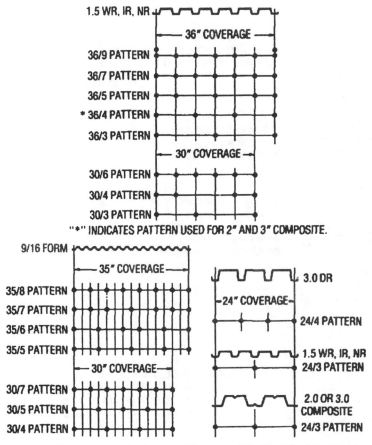

Figure 9.2   Frame connection layouts. (*Courtesy of the Steel Deck Institute.*)

Good metal-to-metal contact is necessary for side-lap welds. Burn holes are the rule rather than the exception and an inspector should not be surprised to see them in the deck. The weld develops its strength by holding around the perimeter and a good weld will have 75% or more of its perimeter working. On occasion, side-lap welds will be specified for deck that has the button punchable side-lap arrangement (see Fig. 9.3 for comments on this subject; see Fig. 9.4 for welding these deck units to the frame). Welding side laps is not recommended for 22 gage deck (0.028 in minimum) or lighter. Weld washers should never be used at side laps between supports. The SDI recommends that side laps be connected at a maximum spacing of 36 in (1 m) for deck spans greater than 5 ft (1.5 m). This minimum spacing could be increased to enhance diaphragm values.

Accessories attached to the deck are welded, screwed, pop riveted, or (rarely) glued. Usually the choice is left to the erector and many

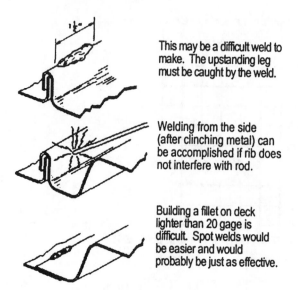

This may be a difficult weld to make. The upstanding leg must be caught by the weld.

Welding from the side (after clinching metal) can be accomplished if rib does not interfere with rod.

Building a fillet on deck lighter than 20 gage is difficult. Spot welds would be easier and would probably be just as effective.

Figure 9.3   Sheet-to-sheet welds between supports. (*Courtesy of the Steel Deck Institute.*)

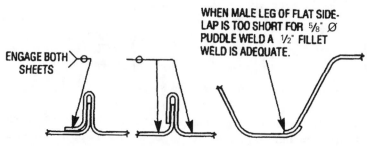

WHEN MALE LEG OF FLAT SIDE-LAP IS TOO SHORT FOR ⅝" ∅ PUDDLE WELD A ½" FILLET WELD IS ADEQUATE.

ENGAGE BOTH SHEETS

Figure 9.4   Side-lap welds at supports. (*Courtesy of the Steel Deck Institute.*)

times is simply the result of the tools available at the time. The importance of fastening accessories can be either structural or architectural, and the designer may need to become involved. For instance, the attachment of reinforcement around penetrations, and the fastening of pour stops, may have a great deal to do with the expected performance of the accessory and care must be taken to see that sufficient attachment is done. If the deck is to be exposed to view, then architectural considerations might be of concern and the fasteners may be selected accordingly.

Frequently the expression "tack welding" is used to describe attaching accessories to deck or to structural steel. A *tack weld* is defined by the AWS as "a weld made to hold parts of a weldment in proper alignment until the final welds are made." The term, when applied to accessories, means a weld of unspecified strength or size simply used to hold the accessory securely in its proper position. When floor deck accessories are tack-welded, the concrete is usually the medium that will hold the parts in their final place. The accessories shown in Fig. 9.5 can be tack-welded or screwed as is appropriate. The one exception is the case of pour stops; the SDI calls for 1-in fillet welds at 12-in oc to the structural steel.

Some additional details on steel joist bearing and connections are shown in Figs. 9.6 and 9.7.

Composite beam details of how metal deck is connected to steel beams is shown in Fig. 9.8. Additional details of commonly used metal deck construction are shown in Figs. 9.9 through 9.12.

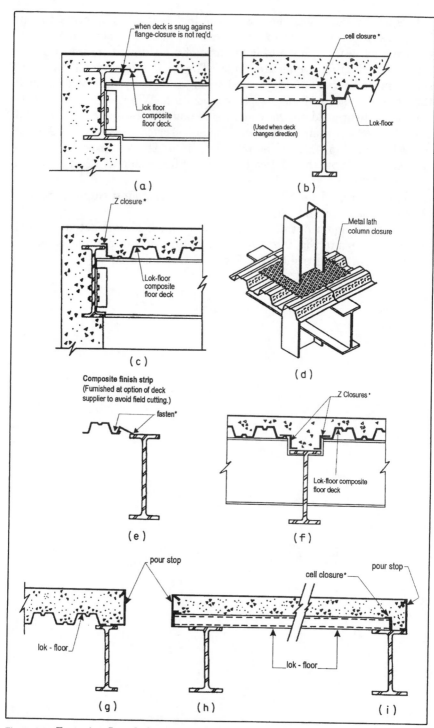

Figure 9.5 Fastening floor deck accessories.

TYPE	K SERIES	LH/DLH SERIES
BOLTED CONNECTIONS 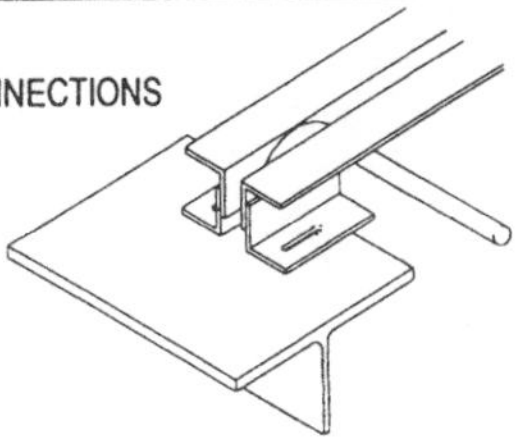	Slotted holes in bearing plates are furnished whenever bolted connections are required. Bolts ($\frac{1}{2}$-in diameter) are not furnished by the joist manufacturer. Minimum bearing on structural steel supports is $2\frac{1}{2}$-in.	Slotted holes in bearing plates are furnished whenever bolted connections are required. Bolts ($\frac{3}{4}$-in diameter) are not furnished by the joist manufacturer. Minimum Bearing on structural steel supports is 4-in.
WELDED CONNECTIONS 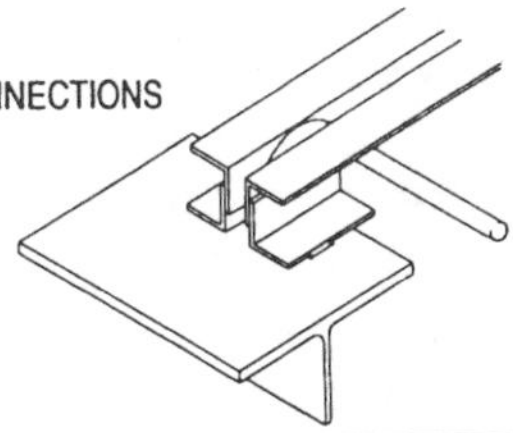	Ends of K Series joists are normally anchored by two 1/8-in fillet welds 1-in long. Minimum bearing on structural steel supports is $2\frac{1}{2}$-in.	Ends of LH/DLH Series joists are normally anchored by two 1/4-in fillet welds 2-in long. Minimum bearing on structurtal steel supports is 4-in.
TYPICAL MASONRY BEARING The setting plates should always be anchored to the masonry wall. The setting plate must be located not more than ½" from the face of the wall. The design professional must design the bearing plate and must take into account the forces acting on the concrete or masonry. 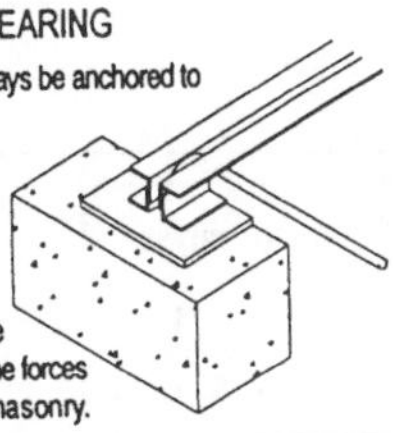	Minimum bearing is 4-in.	Minimum bearing is 6-in.

Figure 9.6 Joist bearing details.

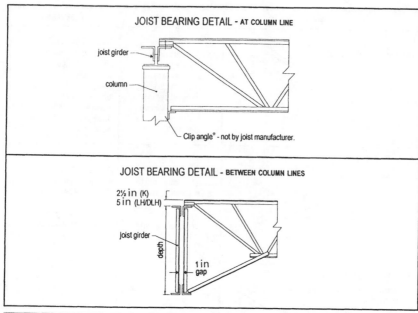

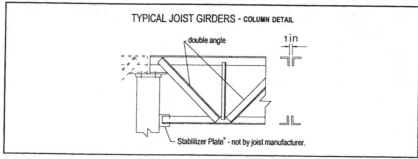

* Weld only after dead load has been applied.

Figure 9.7   Joist bearing on joist girders.

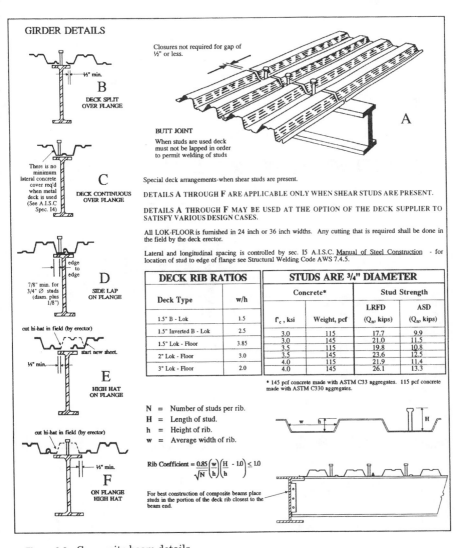

Figure 9.8   Composite beam details.

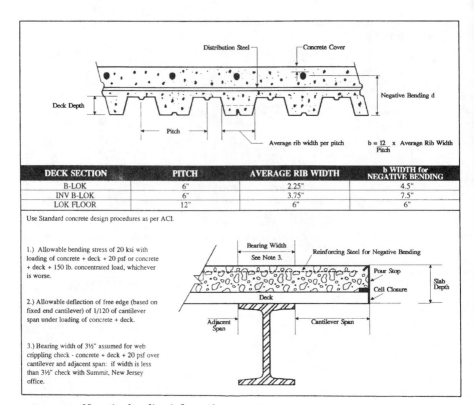

Figure 9.9   Negative bending information.

The following is the content of Figure 9.9:

DECK SECTION	PITCH	AVERAGE RIB WIDTH	b WIDTH for NEGATIVE BENDING
B-LOK	6"	2.25"	4.5"
INV B-LOK	6"	3.75"	7.5"
LOK FLOOR	12"	6"	6"

$b = \dfrac{12}{\text{Pitch}} \times \text{Average Rib Width}$

Use Standard concrete design procedures as per ACI.

1.) Allowable bending stress of 20 ksi with loading of concrete + deck + 20 psf or concrete + deck + 150 lb. concentrated load, whichever is worse.

2.) Allowable deflection of free edge (based on fixed end cantilever) of 1/120 of cantilever span under loading of concrete + deck.

3.) Bearing width of 3½" assumed for web crippling check - concrete + deck + 20 psf over cantilever and adjacent span: if width is less than 3½" check with Summit, New Jersey office.

FLOOR DECK CANTILEVERS

NORMAL WEIGHT CONCRETE (150 PCF)

Slab Depth	UNITED STEEL DECK, INC. DECK PROFILE															
	B-LOK				1.5 LOK-FLOOR				2.0 LOK-FLOOR				3.0 LOK-FLOOR			
	22	20	18	16	22	20	18	16	22	20	18	16	22	20	18	16
4.00"	1'11"	2'3"	2'10"	3'4"	1'11"	2'4"	3'0"	3'6"								
4.50"	1'10"	2'2"	2'9"	3'3"	1'10"	2'3"	2'10"	3'4"	2'6"	2'11"	3'8"	4'3"				
5.00"	1'10"	2'2"	2'8"	3'2"	1'10"	2'3"	2'9"	3'3"	2'5"	2'10"	3'6"	4'1"	3'8"	4'3"	5'3"	6'0"
5.50"	1'9"	2'1"	2'7"	3'0"	1'9"	2'2"	2'9"	3'2"	2'4"	2'9"	3'5"	4'0"	3'7"	4'1"	5'0"	5'9"
6.00"	1'9"	2'0"	2'6"	2'11"	1'9"	2'1"	2'8"	3'1"	2'3"	2'8"	3'4"	3'10"	3'5"	3'11"	4'10"	5'7"
6.50"	1'8"	2'0"	2'6"	2'11"	1'9"	2'1"	2'7"	3'0"	2'3"	2'8"	3'3"	3'9"	3'4"	3'10"	4'8"	5'5"
7.00"	1'8"	1'11"	2'5"	2'10"	1'8"	2'0"	2'6"	2'11"	2'2"	2'7"	3'2"	3'8"	3'3"	3'9"	4'6"	5'3"
7.50"	1'8"	1'11"	2'4"	2'9"	1'8"	2'0"	2'6"	2'10"	2'2"	2'6"	3'1"	3'7"	3'2"	3'8"	4'5"	5'1"
8.00"	1'7"	1'11"	2'4"	2'8"	1'7"	1'11"	2'5"	2'10"	2'1"	2'5"	3'0"	3'6"	3'1"	3'6"	4'3"	4'11"

Figure 9.10   Floor deck cantilevers.

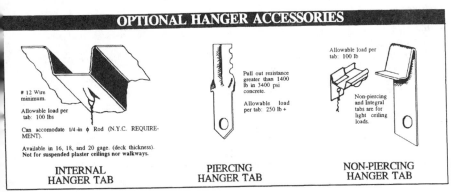

OPTIONAL HANGER ACCESSORIES

INTERNAL HANGER TAB

12 Wire minimum.
Allowable load per tab: 100 lbs

Can accomodate 1/4 -in ⌀ Rod (N.Y.C. REQUIRE-MENT).

Available in 16, 18, and 20 gage. (deck thickness).
Not for suspended plaster ceilings nor walkways.

PIERCING HANGER TAB

Pull out resistance greater than 1400 lb in 3400 psi concrete.

Allowable load per tab: 250 lb +

NON-PIERCING HANGER TAB

Allowable load per tab: 100 lb

Non-piercing and Integral tabs are for light ceiling loads.

Figure 9.11 Optional hanger accessories.

SLAB DEPTH (inches)	OVERHANG (inches) POUR STOP TYPES												
	0	1	2	3	4	5	6	7	8	9	10	11	12
4.00	20	20	20	20	18	18	16	14	12	12	12	10	10
4.25	20	20	20	18	18	16	16	14	12	12	12	10	10
4.50	20	20	20	18	18	16	16	14	12	12	12	10	10
4.75	20	20	18	18	16	16	14	14	12	12	10	10	10
5.00	20	20	18	18	16	16	14	14	12	12	10	10	
5.25	20	18	18	16	16	14	14	12	12	12	10	10	
5.50	20	18	18	16	16	14	14	12	12	12	10	10	
5.75	20	18	16	16	14	14	12	12	12	12	10	10	
6.00	18	18	16	16	14	14	12	12	12	10	10	10	
6.25	18	18	16	14	14	12	12	12	12	10	10		
6.50	18	16	16	14	14	12	12	12	12	10	10		
6.75	18	16	14	14	14	12	12	12	10	10	10		
7.00	16	16	14	14	12	12	12	12	10	10	10		
7.25	16	16	14	14	12	12	12	10	10	10			
7.50	16	14	14	12	12	12	12	10	10	10			
7.75	16	14	14	12	12	12	10	10	10				
8.00	14	14	12	12	12	12	10	10	10				
8.25	14	14	12	12	12	10	10	10					
8.50	14	12	12	12	12	10	10	10					
8.75	14	12	12	12	12	10	10	10					
9.00	14	12	12	12	10	10	10						
9.25	12	12	12	12	10	10	10						
9.50	12	12	12	10	10	10							
9.75	12	12	12	10	10	10							
10.00	12	12	10	10	10								
10.25	12	12	10	10	10								
10.50	12	12	10	10	10								
10.75	12	10	10	10									
11.00	12	10	10	10									
11.25	12	10	10										
11.50	10	10	10										
11.75	10	10											
12.00	10	10											

TYPES	DESIGN THICKNESS
20	0.0358
18	0.0474
16	0.0598
14	0.0747
12	0.1046
10	0.1345

This Selection Chart is based on following criteria:

1. Normal weight concrete (150 pcf).

2. Horizontal and vertical deflection is limited to 1/4 -in maximum for concrete dead load.

3. Design stress is limited to 20 ksi for concrete dead load temporarily increased by one-third for the construction live load of 20 pcf.

4. Pour Stop Selection Chart does not consider the effect of the performance, deflection, or rotation of the pour stop support which may include both the supporting composite deck and/or the frame.

5. Vertical leg return lip is recommended.

6. This selection is not meant to replace the judgement of experienced Structural Engineers and shall be considered as a reference only.

Use return lip for composite construction.

1/2 -in min.

* See note 5.

2 -in weld @12 in

Reinforcing Steel (not by USD or NUB)
* Pour Stop

Slab Thickness

Composite floor deck by United Steel Deck, Inc.

2 -in Overhang

Figure 9.12 Pour stop selection chart.

References

AISC, *Manual of Steel Construction*, American Institute for Steel Construction, Chicago, IL, 1994.

AISI, *Specification for the Design of Cold Formed Steel Structural Members*, American Iron and Steel Institute, Washington, DC, 1996.

AWS, *Structural Welding Code*, American Welding Society, Miami, FL, 1988.

FMRC, *Approval Guide*, Factory Mutual System, Norwood MA, 1996.

Luttrell, L.D., *Diaphragm Design Manual*, 2d ed., Steel Deck Institute, Fox River Grove, IL, 1995.

SDI, *Manual of Construction with Steel Deck*, Steel Deck Institute, Fox River Grove, IL, 1992.

SDI, *Design Manual (Publication 29)*, Steel Deck Institute, Fox River Grove, IL, 1995.

UL, *Fire Resistance Directory*, Underwriters' Laboratory, Northbrook, IL, 1996.

USD, *Steel Decks for Floors and Roofs*, United Steel Deck Inc., Summit, NJ, 1997.

Chapter

10

Connections to Composite Members

Atorod Azizinamini
University of Nebraska
Lincoln, NE
and
National Bridge Research Organization (NaBRO)
University of Nebraska-Lincoln

Bahram Shahrooz
University of Cincinnati
Cincinnati, OH

Ahmed El-Remaily
University of Nebraska
Lincoln, NE

Hassan Astaneh
University of California
Berkeley, CA

(Courtesy of The Steel Institute of New York.)

10.1   Introduction

The combined use of steel and concrete to form composite structures has been used widely. The introduction of new composite building systems has allowed the design and construction of more efficient mid-

and high-rise composite buildings. In most composite building systems, the main problem facing designers has been the selection of an appropriate and economical connection.

This chapter provides suggestions for connection details for three types of composite structure elements: (1) connection details for connecting coupling steel beams to reinforced concrete shear walls (Sec. 10.3), (2) joint details for connecting steel beams to reinforced concrete columns (Sec. 10.4), and (3) connection details for attaching steel beams to rectangular and circular concrete-filled steel-tube columns (Sec. 10.5).

10.2   General Design Considerations

The design of connections, in general, requires consideration of stiffness, strength, stability, serviceability, and cyclic behavior. Following is a brief discussion of each of these items.

10.2.1   Strength and stiffness

When connections are subjected to applied moment, they will cause rotation at the member end. For instance, if a beam is attached to a column using top and seat angles, the applied moment to the beam end generated by vertical or lateral loads will cause the beam end to rotate with respect to the column face. The amount of this rotation is dependent on the stiffness of the connected elements. Experimental results indicate that all connections exhibit some level of rotation and, therefore, one could argue that all connections are semirigid. For design purposes, however, design manuals divide connections into three categories: (1) connections that exhibit relatively large end rotations (simple connections), (2) connections that result in very small rotation (rigid connections), and (3) those which exhibit end rotations between simple and rigid connections, referred to as semirigid connections.

To date, the majority of efforts in the development of connection details for composite members has been focused on rigid-type connections.

Design of connections for strength requires knowledge about the capacity of each connection element at the ultimate strength limit state. To ensure satisfactory performance of connections at ultimate strength limit strength, failure of connection elements must be prevented or controlled. The objective in design is to prevent damage to the connection at its ultimate strength limit state and shift the failure locations to other parts of the structure. Connections could finally fail if the applied load level exceeds a certain limit. As a result, it is desirable to proportion the connection so that it will fail in a "controlled" and "desirable" manner. For instance, design of connection elements could be "controlled" through proportioning such that at the strength limit state the connection elements fail by yielding and not weld fracture. Yielding and, finally, fracture of steel elements of connections are usually "desirable" modes of failure in comparison to weld fracture, which could take place without warning.

For most composite connections, another major consideration is the ability to inspect the connection after a major event. For instance, after an earthquake, one should be able to inspect the connection and make judgments as to the safety of the connection. Unfortunately, most elements of composite connections are not easily accessible and their full inspection is not feasible. Therefore, the designer needs to proportion the connection elements such that the failure of "hidden" elements is prevented.

10.2.2 Stability

Connection elements could fail as a result of buckling (elastic or inelastic) of connection elements. This mode of failure, however, is not usually a major concern in connection design.

10.2.3 Serviceability

Connections, as with any other member of the structure, should perform satisfactorily at different limit states. At service load levels, performance of connections should not adversely affect the behavior of the structure. For instance, at service load levels connections could be subjected to a large number of load cycles. These loads could be generated by wind loads or machinery in the case of industrial buildings. Although these loads could be substantially lower than the ultimate load-carrying capacity of each connection element, the connection could develop fatigue cracking, which could result in failure.

Large flexibility at the connection level could result in large interstory drift and member deflections. Therefore, the selection of the connection types at various floor levels could be dictated by the service limit state.

10.2.4 Cyclic behavior

Connections could fail under a large (high-cycle fatigue) or small (low-cycle fatigue) number of cyclic loadings. In the case of high-cycle fatigue, the magnitude of the applied stress is relatively low. Cracking in bridge elements is caused by high-cycle fatigue. On the other hand, the level of applied stress in the case of low-cycle fatigue is relatively high and could approach the yield strength of the connected elements. During major earthquakes, connections in buildings could experience a few cycles of loading with relatively high stress levels at each cycle. Failure of connections by low-cycle fatigue is confined to earthquake-type loading. The amount of available information on low-cycle fatigue characteristics of connections is limited. This is especially true for composite connections. Principles of fracture mechanics and fatigue could be used to establish life of connections under variable cyclic loading. Two approaches could be undertaken. Full-scale testing of connections under constant and variable amplitude loading provides the most reliable information. In the absence of such information, designers could identify the high stress points within the connection and possible load histories that that particular point within the connection could experience during an earthquake. Information on cyclic behavior of different materials, obtained from simple tension-type specimens, is available. Knowledge of the cyclic load history for the portion of the connection with the highest stress and available damage models for particular materials could then be used to estimate the life of connections under cyclic loading.

However, it should be noted that predicting the life of connections under cyclic loading is a very complex process and its accuracy, in many cases, depends on the experience and judgment of the designer. One of the major questions is estimating the load history that the connection could experience during an earthquake. In addition, it is necessary to conduct nonlinear dynamic time-history analyses, incorporating connection behavior (through inclusion of moment-rotation characteristics of the connection). Fortunately, in general, connections in major earthquake events are subjected to a very few cycles of loading with high stress levels. In general, bolted connections demonstrate better cyclic behavior than welded connections. Behavior of welded connections depends, to a large extent, on quality control and workmanship.

10.3 Beam-to-Wall Connections

10.3.1 Introductory remarks

Structural walls/cores are commonly used for lateral load strength and stiffness. For low- to moderate-rise buildings, up to 25 to 30 stories, the walls/cores can be used to provide a majority of the lateral force resistance. For taller buildings, the use of dual systems is more common, where the perimeter frames are engaged with the walls/cores. Outrigger beams are framed between the core walls and columns (which may be all steel or composite) in the perimeter frame. Core walls can effectively be formed by coupling individual wall piers,

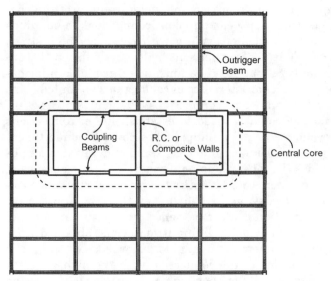

Figure 10.1   Structural components of core wall frame systems.

which may be slip-formed to accelerate construction, with the use of reinforced concrete or steel/steel-concrete composite coupling beams. The floor plan of a representative hybrid building is shown in Fig. 10.1. The walls may be reinforced conventionally, that is, consisting of longitudinal and transverse reinforcement, or may include embedded structural steel members in addition to conventional reinforcing bars. The successful performance of such hybrid structural systems depends on the adequacy of the primary individual components which are the walls/cores, steel frames, and frame-core connections. The focus of this section is on the connections between outrigger beams and walls and the connections between steel/steel-concrete composite coupling beams and walls. Issues related to design of steel/steel-concrete composite coupling beams and connections between floor diaphragms and walls are also discussed.

10.3.2   Qualitative discussion about outrigger beam-wall connection and coupling beam-wall connection

Connections between walls and steel/composite coupling beams or outriggers depend on whether the wall boundary element is reinforced conventionally or contains embedded structural steel columns, the level of forces to be developed, and whether the walls are slip-formed or cast conventionally. A summary of possible connections is provided in the following.

10.3.2.1 Coupling beam-wall connection. Well-proportioned coupling beams above the second floor are expected to dissipate a majority of the input energy during severe earthquakes. Coupling beams will, therefore, undergo large inelastic end rotations and reversals, and adequate connection between coupling beams and wall piers becomes a critical component of the overall system behavior. The connection varies depending on whether reinforced concrete or steel/steel-concrete composite coupling beams are used. A comprehensive discussion for reinforced concrete coupling beams and their connections to walls is provided elsewhere (for example, Aktan and Bertero, 1981; Barney et al., 1978; Paulay, 1980 and 1986; Paulay and Santhakumar, 1976; and Paulay and Binney, 1975).

10.3.2.2 Steel/steel-concrete coupling beams. Structural steel coupling beams provide a viable alternative, particularly in cases where height restrictions do not permit the use of deep reinforced concrete beams, or where the required capacities and stiffness cannot be developed economically by a concrete beam. The member may be encased with a varying level of longitudinal and transverse reinforcement.

If the wall boundary elements include embedded structural steel columns, the wall-coupling beam connection is essentially identical to steel beams and columns but with some modifications. For steel boundary columns located farther away than approximately 1.5 to 2 times the beam depth from the edge, the beam forces can be transferred to the core by the bearing mechanism mobilized by the beam flanges, as illustrated in Fig. 10.2. In such cases, the beam-column connection becomes less critical, and the necessary embedment length can be computed based on a number of available methods, as discussed in Sec. 10.3.3.3. If the embedded steel boundary column is located within approximately 1.5 times the beam depth from the wall edge, the forces can be transferred by mobilizing the internal couple involving the column axial load and bearing stresses near the face, as shown in Fig. 10.3. Clearly, the beam-column connection becomes crit-

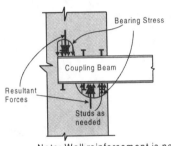

Note: Wall reinforcement is not shown for clarity.

Figure 10.2 Transfer of coupling beam forces through bearing.

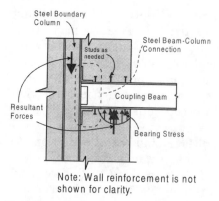

Note: Wall reinforcement is not
shown for clarity.

Figure 10.3 Transfer of coupling beam
forces through bearing and beam-column
connections.

ical in mobilizing this mechanism. The connection between the coupling beam and steel boundary column is expected to be enhanced by
the presence of concrete encasement as indicated by a recent study
(Leon et al., 1994) which shows improved performance of encased riveted beam column. Due to insufficient data, however, it is recommended to ignore the beneficial effects of the surrounding concrete,
and to follow standard design methods for steel beam-column connections. Outrigger beams may also be directly attached to columns
which are closer to the core face and protruded beyond the column.
This detail is illustrated in Fig. 10.4. Considering the magnitude of
typical coupling beam forces, the steel boundary column may deform
excessively, particularly if the column is intended to serve as an erec-

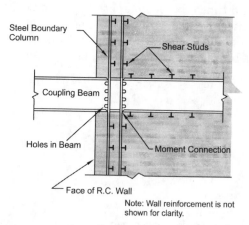

Note: Wall reinforcement is not
shown for clarity.

Figure 10.4 Transfer of coupling beam forces
through a direct beam-column connection.

tion column, leading to splitting of the surrounding concrete and loss of stiffness. Adequate confinement around the column and headed studs improves the behavior by preventing separation between the steel column and surrounding concrete.

If the wall boundary element is reinforced with longitudinal and transverse reinforcing bars, a typical connection involves embedding the coupling beam into the wall and interfacing it with the boundary element, as illustrated in Fig. 10.5. The coupling beam has to be embedded adequately inside the wall such that its capacity can be developed. A number of methods may be used to calculate the necessary embedment length (Mattock and Gaafar, 1982, and Marcakis and Mitchell, 1980). These methods are variations of Precast Concrete Institute guidelines (PCI, 1992) for design of structural steel brackets embedded in precast reinforced concrete columns. Additional details regarding the design methodology are provided in Sec. 10.3.3.3. A second alternative is possible, particularly when core walls are slip-formed. Pockets are left open in the core to later receive coupling beams. After the forms move beyond the pockets at a floor, steel beams are placed inside the pockets and grouted. This detail is illustrated schematically in Fig. 10.6. Calculation of the embedment length is similar to that used for the detail shown in Fig. 10.5.

10.3.2.3 Outrigger beam-wall connection. In low-rise buildings, up to 30 stories, the core is the primary lateral load–resisting system, the

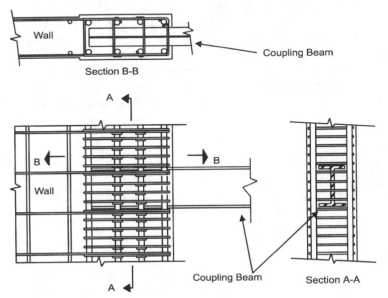

Figure 10.5 Coupling beam-wall connection for conventionally reinforced walls.

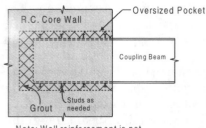

Figure 10.6 A possible coupling beam-wall connection for slip-formed walls.

perimeter frame is designed for gravity loads, and the connection between outrigger beams and cores is generally a *shear connection*. A typical shear connection is shown in Fig. 10.7. Here, a steel plate with shear studs is embedded in the wall/core during casting, which may involve slip-forming. After casting beyond the plate, the web of the steel beam is welded to the stem of a steel plate (*shear tab*) which is already welded to the plate. Variations of this detail are common.

In taller buildings, moment connections are needed to engage the perimeter columns as a means of reducing lateral deformation of the structural system. For short-span outrigger beams, a sufficient level of stiffness can be achieved by a single structural member (either a built-up or a rolled section). In such cases, a number of different moment-resisting connection details are possible. The detail shown in Fig. 10.8 is suitable for developing small moments (clearly not the full moment capacity of the beam) as found by Roeder and Hawkins (1981) and Hawkins et al. (1980). A larger moment can be resisted by embedding the outrigger beam in the wall during construction, similar to that shown in Fig. 10.5 or 10.6, or by using the detail shown in Fig. 10.9. In the latter option, the outrigger beam is welded to a plate which is anchored in the wall by an embedded structural element similar to the

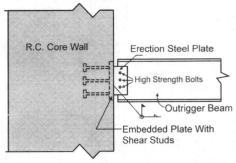

Figure 10.7 Shear connection between outrigger beams and walls.

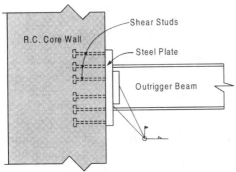

Figure 10.8 Moment connection between outrigger beams and walls (small moments).

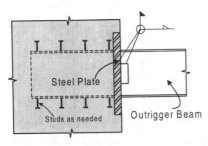

Figure 10.9 Moment connection between outrigger beams and walls (large moments).

outrigger beam. The latter detail is suitable for slip-formed core walls, as well as for conventional construction methods. These details rely on developing an internal couple due to bearing of the beam flanges against the surrounding concrete or grout. If the wall boundary is reinforced with a structural column, the outrigger beam can be directly attached to the wall, as shown in Fig. 10.3 or 10.4.

The span of most outrigger beams is such that a single girder does not provide adequate stiffness, and other systems are needed. Story-deep trusses are a viable choice. As shown schematically in Fig. 10.10, the connection between the top and bottom chords is essentially similar to that used for shear connections between outrigger beams and wall piers.

10.3.2.4 Floor-wall connection. A common component for either of the connections discussed previously is the connection between the floor and walls. In hybrid structures, the floor system consists of a composite metal deck. When the metal deck corrugations are parallel to the core, continuous bent closure plates are placed to prevent slippage of concrete during pouring. These plates may consist of continuous angles, as shown in Fig. 10.11, which are either attached to weld

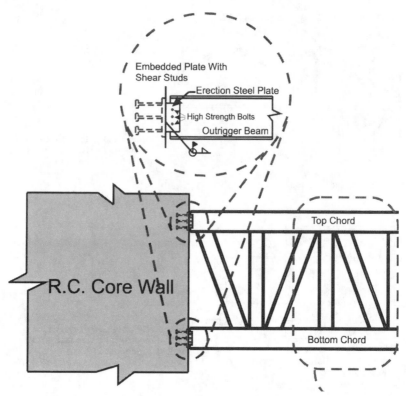

Figure 10.10   Connection between story-deep trusses and walls.

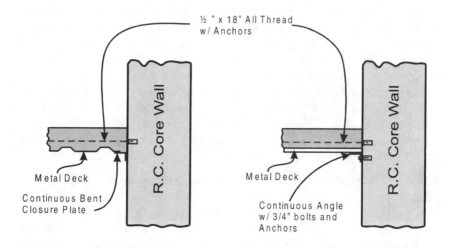

(a) Deck is parallel to core Wall   (b) Deck is perpendicular to core Wall

Figure 10.11   Representative connection between composite floors and core walls.

plates already cast in the wall, or anchored directly to the wall. When the metal deck corrugations are perpendicular to the core, the deck is supported by steel angles which are attached to the core typically at 12 to 24 in on center (Fig. 10.11b). In addition, dowels at regular intervals (18 in on center is common) are used to transfer lateral loads into the core. Note that for encased coupling beams (that is, steel-concrete composite members), the floor system is a reinforced concrete slab or posttensioned system. For these cases, the floor-wall connection is similar to reinforced concrete slab or posttensioned floor-wall connections.

10.3.3 Design of steel or steel-concrete composite coupling beam-wall connections

10.3.3.1 Analysis. Accurate modeling of coupled wall systems is a critical step, particularly when steel or steel-concrete composite coupling beams are used. Previous studies (Shahrooz et al., 1992, 1993; Gong and Shahrooz, 1998) suggest that steel or steel-concrete composite coupling beams are not fixed at the face of the wall. As part of design calculations, the additional flexibility needs to be taken into account to ensure that wall forces and lateral deflections are computed reasonably well. Based on experimental data (Shahrooz et al., 1992, 1993, and Gong and Shahrooz, 1998), the *effective fixed point* of steel or steel-concrete composite coupling beams may be taken as one-third of the embedment length from the face of the wall. The corresponding design model is illustrated in Fig. 10.12.

Stiffness of coupling beams needs to be estimated properly as the design forces and hence detailing of coupling beam-wall connection are impacted. For steel coupling beams, standard methods are used to calculate the stiffness. The stiffness of steel-concrete composite coupling beams needs to account for the increased stiffness due to encasement. Stiffness based on gross transformed section should be used to calculate the upper-bound values of demands in the walls, most notably wall axial force. Cracked transformed section moment of inertia may be used when deflection limits are checked and to compute the maximum wall overturning moment. Note that a previous study suggests that the additional stiffness due to floor slab is lost shortly after composite coupling beams undergo small deformations (Gong and Shahrooz, 1998). Until additional experimental data become available, it is recommended to include the participation of the floor slab for calculating wall axial force. Effective flange width for T beams, as specified in American Concrete Institute 8.10.2 (ACI 318-95), may be used for this purpose. The participation of floor slab toward the stiffness of steel-concrete composite coupling beams may

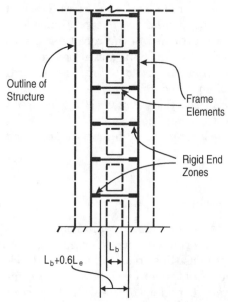

Figure 10.12 Design model for coupled wall systems using steel or composite coupling beams.

be ignored when drift limits are checked or when the maximum wall overturning moments are computed.

10.3.3.2 Design of coupling beam

Steel coupling beams. Well-established guidelines for shear links in eccentrically braced frames (AISC, 1992 and 1994) may be used to design and detail steel coupling beams. The level of coupling beam rotation angle plays an important role in the number and spacing of stiffener plates which may have to be used. This angle is computed with reference to the collapse mechanism shown in Fig. 10.13 which corresponds to the expected behavior of coupled wall systems, that is, plastic hinges form at the base of walls and at the ends of coupling beams. The value of θ_p is taken as $0.4R\theta_e$ in which elastic interstory drift angle θ_e is computed under code level lateral loads (for example, UBC, 1994, and NEHRP, 1994). The minimum value of the term $0.4R$ is 1.0. Knowing the value of θ_p, shear angle γ_p is calculated from Eq. (10.1).

$$\gamma_p = \frac{\gamma_p L}{L_b + 0.6\,L_e} \tag{10.1}$$

Note that in this equation, the additional flexibility of steel/composite coupling beams is taken into account by increasing the length of the

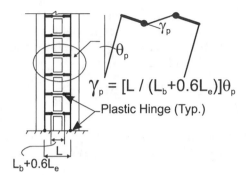

$$\gamma_p = [L / (L_b + 0.6L_e)]\theta_p$$

Plastic Hinge (Typ.)

Figure 10.13   Model for calculating shear angle of steel or composite coupling beams.

coupling beam to $L_b + 0.6L_e$. This method is identical to that used for calculating the expected shear angle in shear links of eccentrically braced frames with the exception of the selected collapse mechanism.

Steel-concrete composite coupling beams.   Previous research on steel-concrete composite coupling beams (Shahrooz et al., 1992 and 1993, and Gong and Shahrooz, 1998) indicates that nominal encasement around steel beams provides adequate resistance against flange and web buckling. Therefore, composite coupling beams may be detailed without web stiffener plates. Due to inadequate data regarding the influence of encasement on local buckling, minimum flange and web thicknesses similar to steel coupling beams need to be used.

10.3.3.3 Connection design.   The connection becomes more critical when steel or steel-concrete composite coupling beams are used. For the details shown in Fig. 10.3 or 10.4, standard design methods for steel beam-column connections can be followed. If the connection involves embedding the coupling beam inside the wall (see Figs. 10.5 or 10.6), the required embedment length is calculated based on mobilizing the moment arm between bearing forces C_f and C_b, as shown in Fig. 10.14. This model was originally proposed by Mattock and Gaafar (1982) for steel brackets embedded in precast concrete

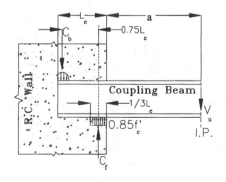

Figure 10.14   Model for computing embedment length.

columns. Previous studies (Shahrooz et al., 1992 and 1993, and Gong and Shahrooz, 1998) have shown the adequacy of this model for steel or steel-concrete composite coupling beams. This model calculates the required embedment length, L_e, from Eq. (10.2):

$$V_u = 48.6 \sqrt{f'_c} \left(\frac{t_{\text{wall}}}{b_f}\right)^{0.66} \beta_1 b_f L_e \left(\frac{0.58 - 0.22\beta_1}{0.88 + (a/L_e)}\right) \quad (10.2)$$

For the detail shown in Fig. 10.6, the value of f'_c in Eq. (10.2) is to be taken as the minimum of the compressive strength of the wall and grout.

The value of V_u in Eq. (10.2) should be selected to ensure that the connection does not fail prior to fully developing the capacity of the coupling beam. For steel coupling beams, V_u is taken as the plastic shear capacity of the steel member as computed from Eq. (10.3):

$$V_p = 0.6 F_y (h - 2t_f) t_w \quad (10.3)$$

To account for strain hardening, it is recommended that F_y be taken as 1.25 times the nominal yield strength.

The contribution of encasement toward shear capacity of composite coupling beams needs to be taken into account when the embedment length is calculated. Embedment length should be adequate such that most of the input energy is dissipated through formation of plastic hinges in the beam and not in the connection region (Shahrooz et al., 1992 and 1993, and Gong and Shahrooz, 1998). In lieu of fiber analysis, shear capacity of composite coupling beams V_u can be computed from Eqs. (10.4), which has been calibrated based on a relatively large number of case studies (Gong and Shahrooz, 1998):

$$V_u = 1.56 (V_{\text{steel}} + V_{\text{RC}})$$

$$V_{\text{steel}} = 0.6 F_y t_w (h - 2t_f) \quad (10.4)$$

$$V_{\text{RC}} = 2\sqrt{f'_c} b_w d + \frac{A_v f_y d}{s}$$

In this equation, the nominal values of F_y and f'_c (in psi) are to be used because the equation has been calibrated to account for strain hardening and material overstrength.

Additional bars attached to the beam flanges (*transfer bars*) can contribute toward load resistance. These bars can be attached through mechanical half couplers which have been welded onto the flanges. The embedment length as computed by Eq. (10.2) can be modified to account for the additional strength (Gong and Shahrooz, 1998). However, to ensure that the calculated embedment length is

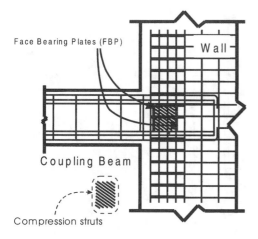

Face Bearing Plates (FBP)

Wall

Coupling Beam

Compression struts

Figure 10.15   Face-bearing plates.

sufficiently large to avoid excessive inelastic damage in the connection region, it is recommended that the contribution of transfer bars be neglected.

A pair of stiffener plates (on both sides of the web) placed along the embedment length will mobilize compression struts in the connection region as depicted schematically in Fig. 10.15. These stiffener plates are commonly referred to as *face-bearing plates*. The first face-bearing plate should be inside the confined core of the wall boundary element. The distance between the face-bearing plates should be such that the angle of compression struts is approximately 45° (hence, the distance between the plates should be about equal to the clear distance between the flanges). To ensure adequate contribution of the face-bearing plates, the width of each face-bearing plate should be equal to the flange width on either side of the web. The thickness of the face-bearing plates can be established based on available guidelines for the detailing of shear links in eccentrically braced frames (AISC, 1992 and 1994).

10.3.3.4 Design example.   An example is used to illustrate the procedure for computing the required embedment length of steel or steel-concrete composite coupling beams. A representative connection at floor 7 of the structure shown in Fig. 10.16 is designed in this example. The building in this example has 20 floors. The coupling beams are encased, that is, composite, and the walls are assumed to be reinforced only with longitudinal and transverse reinforcement. The clear span of the coupling beam is 8 ft. The thickness of the wall boundary element, t_{wall}, is 22 in. The material properties are: f'_c (for the encasement as well as the wall) = 4 ksi, F_y (yield strength of the web of the steel coupling beam) = 40 ksi, and f_y (yield strength of reinforcing

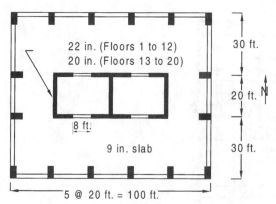

Figure 10.16   Floor plan of example structure.

bars in the encasement) = 60 ksi. The cross section of the coupling beam is shown in Fig. 10.17. The effective depth for the concrete element is taken as 21.5 in. The goal is to compute the required embedment length of the steel coupling beam inside the reinforced concrete wall.

The embedment length needs to develop $V_u = 1.56 \, (V_{steel} + V_{RC})$. The values of V_{RC} and V_{steel} are computed in Eq. (10.5):

$$V_{RC} = 2 \sqrt{f'_c} \, b_w \, d + \frac{A_s f_y d}{s}$$

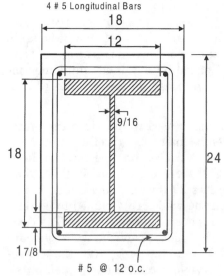

Figure 10.17   Cross section of composite coupling beam.

$$V_{RC} = 2\,\frac{\sqrt{4000}}{1000}\,(18)(21.5) + \frac{2(0.31)(60)(21.5)}{12}$$

$$V_{RC} = 116\text{ kips}$$

$$V_{steel} = 0.6F_y t_w (d - 2t_f)$$

$$V_{steel} = 0.6(40)(9/16)[18 - (2)(1.875)]$$

$$V_{steel} = 193\text{ kips} \tag{10.5}$$

The embedment length is designed to develop $V_u = 1.56(116 + 193) = 480$ kips.

$$V_u = 48.6\,\sqrt{f_c'}\left(\frac{t_{wall}}{b_f}\right)^{0.66} \beta_1\, b_f L_e \left(\frac{0.58 - 0.22\beta_1}{0.88 + (a/L_e)}\right) \tag{10.6}$$

Therefore,

$$480 = 48.6\,\frac{\sqrt{4000}}{1000}\left(\frac{22}{12}\right)^{0.66}(0.85)(12)\,L_e \left(\frac{0.58 - 0.22(0.85)}{0.88 + (48/L_e)}\right) \tag{10.7}$$

By solving Eq. (10.7), the required embedment length is 48.6 in, say 49 in. The final detail is shown in Fig. 10.18. Note that 1.5-in transfer bars have been added to the top and bottom flanges as shown.

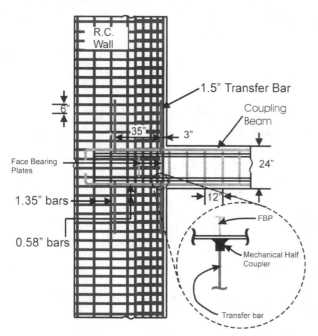

Figure 10.18 Connection detail.

10.3.4   Design of outrigger beam-wall connections

10.3.4.1   Shear connections. As explained in Sec. 10.3.2.3, outrigger beams are typically connected to core walls through *shear connections* similar to that shown in Fig. 10.7. Although this connection provides some moment resistance, it is generally accepted that the connection is flexible and does not develop large moments. The main design issues are: (1) the connection between the steel outrigger beam and shear tab which is welded onto the embedded plate and (2) the transfer of forces, which are gravity shear force and diaphragm force, as shown in Fig. 10.19, to the wall. Note that the diaphragm force may be tensile or compressive, and the line of action of gravity shear is assumed to be along the bolts according to standard practice. The outrigger beam-shear tab connection is a typical shear connection, and common design methods for steel structures (AISC, 1994) are followed for this purpose. The most critical part is the transfer of forces, particularly tensile diaphragm forces, from the shear tab to the core wall, which is achieved by headed studs. To ensure adequate safety against stud failure, the following design methodology is recommended. This method is based on a research conducted by Wang (1979):

1. Based on an assumed layout of studs, establish the tensile capacity as the lesser of the strength of the stud or concrete cone. Available guidelines (for example, PCI, 1992) can be used for this purpose.

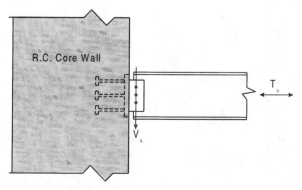

Note: Floor deck and wall reinforcement are not shown for clarity.

Figure 10.19 Forces on shear connection between outrigger/collector beams and core walls.

2. Assuming that all the applied shear is resisted by the studs in the compression region, calculate the required number of studs. The shear capacity is taken as the smaller of: (a) shear capacity of a single stud, which can be calculated based on available guidelines (for example, PCI, 1992) and (b) tensile capacity calculated in step 1. Use the same number of studs in the tension zone. Once the required number of studs is known, compute shear strength governed by concrete failure.

3. Using the stud arrangement obtained in step 2, compute tensile capacity of the stud group.

4. Increase the value of T_u by 50% to ensure adequate ductility.

5. Based on the model shown in Fig. 10.20 and the formulation shown in Eq. (10.8), calculate the depth of compression region, k_d:

$$k_d = \frac{T_{\text{capacity}} - 1.5T_u}{(0.85f'_c)b} \qquad (10.8)$$

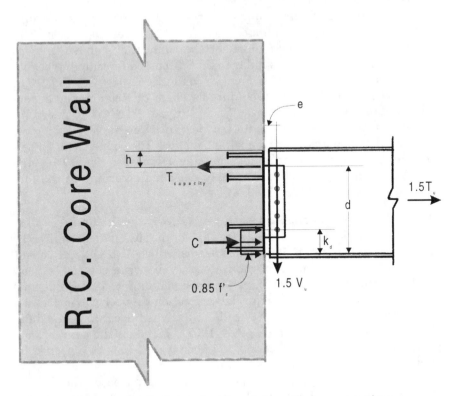

Figure 10.20 Design model for design of outrigger beam-wall shear connections.

6. Calculate the required depth of the embedded plate from Eq. (10.9):

$$d = \frac{1.5eV_u + 0.425bf'_c k_d^2 + 0.75T_u h}{0.85\ bf'_c k_d + 0.75T_u} \tag{10.9}$$

$$\text{depth} = d + h$$

Note that in this equation, the value of gravity shear V_u is amplified by 1.5 to ensure a ductile mode of failure.

7. Check the capacity of studs under combined actions of tension and shear. For this purpose, the shear may be assumed to be resisted equally by the tension and compression studs, but the tensile force is resisted by tension studs. Available interaction equations in PCI guidelines (PCI, 1992) can be used for this purpose.

10.3.4.2 Moment connections. As mentioned previously in Sec. 10.3.2.3, outrigger beams may be attached to core walls through moment connections to enhance the overall structural stiffness. The basic force-transfer mechanism for the connections shown in Fig. 10.2, 10.5, 10.6, or 10.9 is similar to that discussed for coupling beams embedded inside core walls. For the connection shown in Fig. 10.8, the aforementioned design procedure for shear connections can generally be followed, but the term, $1.5eV_u$, in Eq. (10.9) is replaced by $1.5M_u$. Once again, the calculated design moment, M_u, has been increased by 50% to ensure a ductile behavior. The connections for top and bottom chords of story-deep outrigger trusses (Fig. 10.10) are similar to shear connections, and are designed according to the formulation described in Sec. 10.3.4.1.

10.3.4.3 Floor-wall connections. In a structure with the floor plan shown in Fig. 10.1, it is possible to transfer diaphragm forces directly to core walls through the outrigger beams, which also serve as collector elements. In such cases, the connection between composite floor systems and core walls, which were discussed in Sec. 10.3.2.4, has to simply resist the gravity shear. The connection between the necessary supporting elements and core walls is designed according to established guidelines (for example, PCI, 1992). To reduce the demands on outrigger beam-wall connections, the floor system may be designed to participate in the transfer of diaphragm forces to the core walls. Dowels at regular spacing can be used for this purpose. The dowels have to be embedded adequately in the floor slab, and be anchored to the wall so that their capacity can be developed. These dowels have to

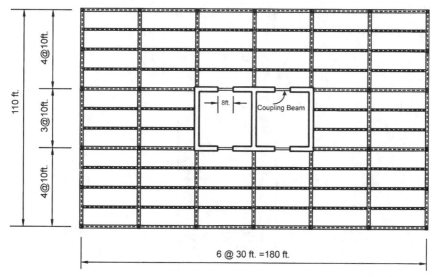

Figure 10.21 Plan view of design example.

resist the portion of tensile diaphragm force not resisted by the collector element.

10.3.4.4 Design example for shear connections. An example of shear connections between outrigger beams and core walls is illustrated in this section. The example is with reference to a 15-story building with the floor plan shown in Fig. 10.21. The calculated forces for the outrigger beam in floor 5 are $T_u = 40$ kips and $V_u = 93$ kips. The outrigger beam is W 24 × 55, and the core walls are 18 in thick. The concrete compressive strength of the wall is 6000 psi.

- *Design of shear tab:* The shear tab is designed and detailed by following standard design practice for steel structures (AISC, 1994). The shear tab dimensions are 15.5 in deep × 4.5 in wide × ½ in thick. The shear tab is welded to the embedded steel plate through ¼-in fillet weld. Five 1-in A490 bolts are used to connect the outrigger beam to the shear tab.

- *Design of embedded steel plate:* Try ¾-in-diam studs with 7 in of embedment:

  1. *Tensile capacity of studs:*

$$\phi P_s = \phi 0.9 A_b f_y = (1)(0.9)(0.4418)(60) = 23.85 \text{ kips}$$

Assuming that the stud is located as shown below, the tensile strength governed by concrete failure is

$$\phi P_c = 10.7 l_e \, (l_e + d_h) \sqrt{f'_c} \, \frac{d_e}{l_e}$$

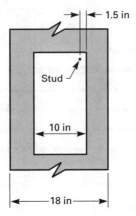

in which $l_e = 7\frac{3}{8} = 6.625$ in

$ d_e = 5.5$ in

$ \phi P_c = 10.7(6.625)(6.625 + 1.25) \sqrt{6000} \, (5.5/6.625)$

$ \phi P_c = 35,900$ lb $= 35.9$ kips

Therefore, use ϕP_s.

2. *Shear strength:*

$$\phi V_s = \phi 0.75 A_b f_y = (1)(0.75)(0.4418)(60) = 19.9 \text{ kips}$$

The shear strength is the smaller of ϕV_s and tensile strength. Hence, shear strength = 19.9 kips. The number of required studs = 93/19.9 = 4.7, say 5 studs. Compute shear strength governed by concrete failure. Since the edge distance > $15d_b$ (= 11.25 in),

$$\phi V_c = \phi 800 \, A_b \sqrt{f'_c} \, n$$

Therefore, $\phi V_c = 0.85(800)(0.4418)(\sqrt{6000})5 = 116354$ lb $= 116.4$ kips, which is larger than V_u, ok. To have an even number, use six studs in both tension and compression zones.

3. *Tensile strength of stud groups:* Assuming the stud pattern shown below, the capacity is computed from the following equation.

$$\phi P_c = \phi(4)\sqrt{f'_c}(x + d_{e1} + d_{e2})(y + 2 l_e)$$

$$\phi P_c = 0.85(4)(\sqrt{6000})(6 + 6 + 6)[3 + 2(6.625)] = 77,033 \text{ lb}$$

$$\phi P_c = 77.0 \text{ kips} > 1.5 T_u, \text{ ok}$$

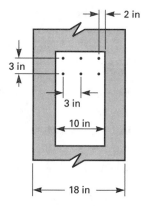

4. *Size the embedded plate:*   From Eq. (10.8),

$$k_d = \frac{77.0 - 1.5(40)}{0.85(6)(10)} = 0.334 \text{ in}$$

Assuming that the plate extends 1 in above the top stud, the value of h in Eq. (10.9) is 2.5 in. As seen from Fig. 10.22, the value of $e = 2.75$ in. Use Eq. (10.9) to solve for d.

$$d = \frac{1.5(93)(2.75) + 0.425(6)(0.334)^2\,(10) + 0.75(40)(2.5)}{0.75(40) + 0.85(6)(0.334)(10)} = 9.76 \text{ in}$$

Therefore, the depth of the embedded plate is $d + h = 2.5 + 9.76 = 12.3$ in, say 12.5 in. Note that this depth is less than that required for the shear tab. Assuming that the embedded plate

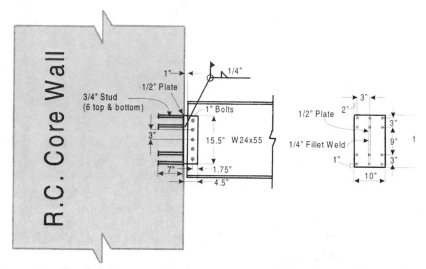

Figure 10.22   Detail of shear connection between outrigger/collector beam and core wall.

extends ¾ in beyond the shear tab, the required depth of the embedded plate is $0.75 + 15.5 + 0.75 = 17$ in.

According to PCI guidelines, the plate thickness is taken as two-thirds of the diameter of the stud. Hence, the plate thickness is 0.5 in.

The final design is sketched in Fig. 10.22.

5. *Check the studs for combined effects of shear and tension:* Use the following interaction equations recommended by PCI (1992):

$$\frac{1}{\phi}\left[\left(\frac{T}{P_c}\right)^2 + \left(\frac{V_u}{V_c}\right)^2\right] \le 1.0$$

$$\left[\left(\frac{T}{P_s}\right)^2 + \left(\frac{V_u}{V_s}\right)^2\right] \ge 1.0$$

Using the free-body diagram shown in Fig. 10.23, the value of T can be computed as follows:

$$\Sigma F_x = 0;$$

$$0.85f'_c k_d b + 1.5T_u - T = 0.85(6)(10) k_d + 1.5(40) - T = 0;$$

$$T = 51 k_d + 60$$

$$\Sigma M_{\text{about } T} = 0;$$

$$1.5T_u (14.5 - 8.5) + 0.85f'_c k_d b \left(14.5 - \frac{k_d}{2}\right) - 1.5V_u e = 0;$$

$$1.5(40)(6) + 0.85(6)(k_d)10(14.5 - 0.5k_d) - 1.5(93)(2.75) = 0;$$

$$25.5k_d^2 - 739.5k_d + 23.63 = 0;$$

$$k_d = 0.032 \text{ in}$$

Therefore, $T = 51(0.0322) + 60 = 61.6$ kips

$$P_c = 4 \sqrt{f'_c} (x + d_{e1} + d_{e2})(y + 21_e)$$

$$= 4\sqrt{6000}(6 + 6 + 6)[3 + 2(6.625)] = 906,280 \text{ lb}$$

$$P_c = 90.6 \text{ kips}$$

$$P_s = 0.9(A_s f_y)n = 0.9(0.4418)(60)6$$

$$= 143 \text{ kips (six studs are in tension; } n = 6)$$

$$V_c = 800 A_b \sqrt{f'_c}\, n = 800(0.4418)\sqrt{6000}(12) = 328,530 \text{ lb}$$

$$= 329 \text{ kips}$$

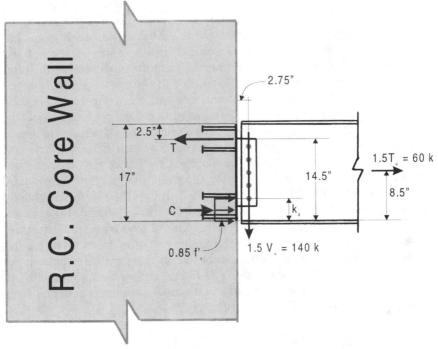

Figure 10.23 Free-body diagram to check final design.

$$V_s = 0.75\,A_b f_y\, n = 0.75(0.4418)(60)(12)$$

$$= 239 \text{ kips (shear is resisted by all studs; } n = 12)$$

Therefore,

$$\frac{1}{0.85}\left[\left(\frac{61.6}{90.6}\right)^2 + \left(\frac{140}{329}\right)^2\right] = 0.76 < 1.0, \text{ ok}$$

$$\left[\left(\frac{61.6}{143}\right)^2 + \left(\frac{140}{239}\right)^2\right] = 0.53 < 1.0, \text{ ok}$$

The final design shown in Fig. 10.22 is adequate.

10.4 Joints between Steel Beams and Reinforced Concrete Columns

10.4.1 Introduction

Composite frames consisting of steel beams and reinforced concrete columns constitute a very cost-effective structural system, especially in tall buildings where the columns have to sustain high axial loads. Concrete columns are known to be more cost-effective than structural steel columns under axial loads. On the other hand, steel beams have the advantages of faster construction and no formwork or shoring required. The combination of concrete columns and steel beams in one system results in the most efficient use of the materials. However, to achieve the full advantage of such system, the beam-column connection must be properly detailed and designed. Due to the current separation of the concrete and steel specifications, the need arises for guidelines to design such connections. The ASCE Task Committee (1994) on Design Criteria for Composite Structures in Steel and Concrete presented guidelines for the design moment resisting joints where the steel beams are continuous through the reinforced concrete column. These guidelines are based on the experimental study by Sheikh et al. (1989) and Deierlein et al. (1989) where 15 two-thirds scale joint specimens were tested under monotonic and cyclic loading. The recommendations were also based on relevant information from existing codes and standards. The following sections summarize the ASCE guidelines. For more information, the reader is referred to the paper by the ASCE Task Committee (1994).

10.4.2 Joint behavior

The joint behavior depends on joint details that activate different internal force transfer mechanisms. Failure of the joint can happen in either one of the two primary failure modes shown in Fig. 10.24. The first mode is the panel shear failure, which results from the transmission of the horizontal flange forces through the joint. Both the steel web and concrete panel contribute to the horizontal shear resistance

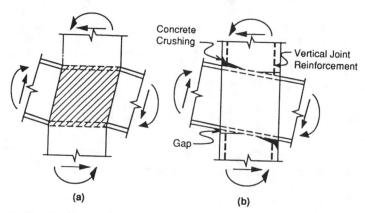

(a)   (b)

Figure 10.24 Joint failure modes: (a) panel shear and (b) vertical bearing (ASCE, 1994).

in the joint. Attachments that mobilize the concrete panel are discussed in the next section. The second mode is the vertical bearing failure, which results from the high bearing stresses of the compression flange against the column. The joint should be detailed and designed to eliminate the possibility of joint failure and force the failure to occur in the connected members.

10.4.3 Joint detailing

Several configurations of attachments can be used to improve the joint strength (see Fig. 10.25). Details shown in Figs. 10.25a and b enhance the joint shear capacity through mobilizing a greater portion of the concrete panel. The concrete panel is divided into inner and outer panels. The inner panel is mobilized by the formation of a compression strut through bearing against the FBPs between the beam flanges. Figure 10.26 shows the mobilization of the outer panel by the formation of compression field through bearing against the extended FBPs or steel columns above and below the joint. The FBP may vary in width and may be split for fabrication ease. The ASCE recommendations require that when significant moment is transferred through the beam-column connection, at least FBPs should be provided within the beam depth with the width no less than the flange width. The vertical joint reinforcement shown in Fig. 10.25c enhances the joint bearing capacity.

10.4.4 Joint forces

Various forces are transferred to the joint by adjacent members, including bending, shear, and axial loads as shown in Fig. 10.27. Existing data indicate that axial compressive forces in the column can improve the joint strength by delaying the formation of cracks. To simplify the design, and since it is conservative, the axial forces in the column are ignored. Since the axial forces in the beam are generally small, they are also neglected. Accordingly, the design forces are reduced to those shown in Figs. 10.28a and b. Considering moment equilibrium, the following equation is obtained:

$$\Sigma M_c = \Sigma M_b + V_b h - V_c d \qquad (10.10)$$

where

$$\Sigma M_b = (M_{b1} + M_{b2}) \qquad (10.11)$$

$$V_b = \frac{V_{b1} + V_{b2}}{2} \qquad (10.12)$$

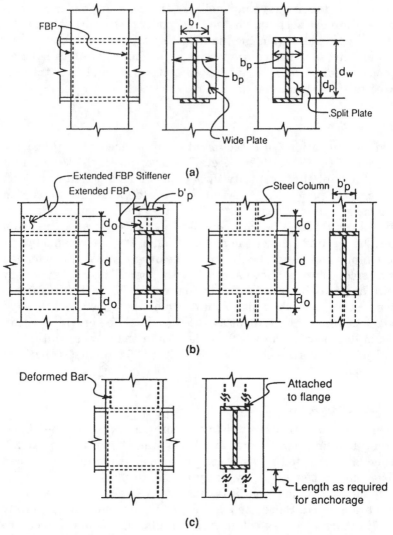

Figure 10.25 Joint details: (a) FBP; (b) extended FBP and steel column; (c) vertical joint reinforcement (*ASCE, 1994*).

$$V_c = \frac{V_{c1} + V_{c2}}{2} \qquad (10.13)$$

$$\Sigma M_c = (M_{c1} + M_{c2}) \qquad (10.14)$$

and

$$\Delta V_b = V_{b2} - V_{b1} \qquad (10.15)$$

$$\Delta V_c = V_{c2} - V_{c1} \qquad (10.16)$$

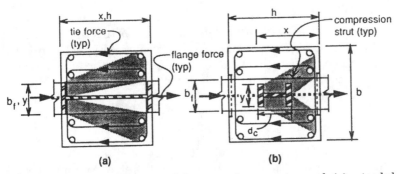

Figure 10.26 Transfer of horizontal force to outer concrete panel: (*a*) extended FBP and (*b*) steel column (*ASCE, 1994*).

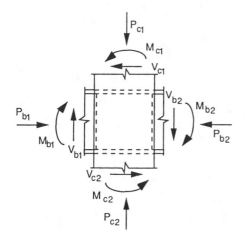

Figure 10.27 Forces acting on joint (*ASCE, 1994*).

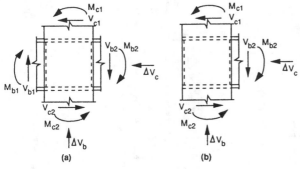

Figure 10.28 Joint design forces: (*a*) interior and (*b*) exterior (*ASCE, 1994*).

10.4.5 Effective joint width

The *effective width of the joint* is defined as the portion of the concrete panel effective in *resisting joint* shear. The concrete panel is divided into inner and outer panels. As shown in Fig. 10.29, the effective joint width, b_j, is equal to the sum of the inner and outer panel widths, b_i and b_o, and can be expressed as

$$b_j = b_i + b_o \qquad (10.17)$$

The inner width, b_i, is taken equal to the greater of the FBP width, b_p, or the beam flange width, b_f. Where neither the steel columns nor the extended FBPs are present, the outer panel width, b_o, is taken as zero. Where extended FBPs or steel columns are used, b_o is calculated according to the following:

$$b_o = C(b_m - b_i) < 2d_o \qquad (10.18)$$

$$b_m = \frac{b_f + b}{2} < b_f + h < 1.75 b_f \qquad (10.19)$$

$$C = \frac{x}{h}\frac{y}{b_f} \qquad (10.20)$$

where b = the concrete column width measured perpendicular to the beam
h = the concrete column depth
y = the greater of the steel column or extended FBP width

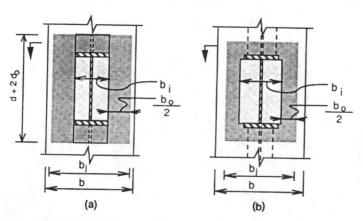

Figure 10.29 Effective joint width (*a*) extended FBP and (*b*) wide FBP and column (*ASCE, 1994*).

$x = h$ where extended FBPs are present or $x = h/2 + d_c/2$ when only the steel column is present (see Fig. 10.26)

d_c = steel column width

d_o = additional effective joint depth provided by attachment to the beam and is determined as follows: When a steel column is present, $d_o = 0.25d$ where d = beam depth; when extended FBPs are used, d_o should be taken as the lesser of $0.25d$ or the height of the extended FBPs

10.4.6   Strength requirements

The joint strength is based on the two possible modes of failure mentioned earlier. Joint design strength is obtained by multiplying the nominal strength by a resistance factor, ϕ. Unless otherwise noted, ϕ should be taken equal to 0.70.

10.4.6.1   Vertical bearing. Vertical forces in the joint are resisted by concrete bearing and by joint reinforcement. The equilibrium of the vertical bearing forces is shown in Fig. 10.30, where the moments in the upper and lower columns, M_{c1} and M_{c2}, are replaced with the corresponding forces in the joint reinforcement and the vertical bearing force. To obtain the joint bearing strength, the forces C_c, T_{vr}, and C_{vr} are replaced by their nominal values. The bearing strength of the joint is checked according to the following:

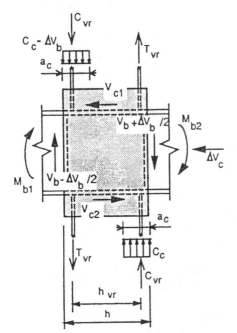

Figure 10.30   Vertical bearing forces (*ASCE, 1994*).

$$\Sigma M_c + 0.35h\Delta V_b \leq \phi[0.7hC_{cn} + h_{vr}(T_{vrn} + C_{vrn})] \qquad (10.21)$$

where ΣM_c = net column moments transferred through the joint
ΔV_b = net vertical beam shear transferred into the column
C_{cn} = the nominal concrete bearing strength
T_{vrn} = the nominal tension strength of the vertical joint reinforcement
C_{vrn} = the nominal compression strength of the vertical joint reinforcement
h_{vr} = the distance between the bars

C_{cn} is calculated using a bearing stress of $2f'_c$ over a bearing area with length $a_c = 0.3h$ and width b_j. The values of $2f'_c$ and $0.3h$ are based on test data. T_{vrn} and C_{vrn} are based on the connection between the reinforcement and steel beam, development of the reinforcement through bond or anchorage to concrete, and the material strength of reinforcement. To avoid overstressing the concrete within the joint, the contribution of the vertical reinforcement is limited by Eq. (10.22):

$$T_{vrn} + C_{vrn} \leq 0.3f'_c b_j h \qquad (10.22)$$

To ensure adequate concrete confinement in bearing regions, three layers of ties should be provided within a distance of $0.4d$ above and below the beam (see Fig. 10.31). The minimum requirement for each layer is given by the following:

$b \leq 500$ mm	four 10-mm bars
500 mm $< b \leq 750$ mm	four 12-mm bars
$b > 750$ mm	four 16-mm bars

These ties should be closed rectangular ties to resist tension parallel and perpendicular to the beam.

10.4.6.2 Joint shear. As described in Secs. 10.4.2 and 10.4.3, shear forces in the joint are resisted by the steel web and the inner and outer concrete panels. The three different mechanisms are shown in Fig. 10.32. The horizontal shear strength is considered adequate if the following equation is satisfied.

$$\Sigma M_c - V_b jh \leq \phi[V_{sn}d_f + 0.75V_{csn}d_w + V_{cfn}(d + d_o)] \qquad (10.23)$$

where V_{sn} = steel panel nominal strength
V_{csn} = the inner concrete compression strut nominal strength
V_{cfn} = the outer concrete compression field nominal strength
V_b = antisymmetric portion of beam shears

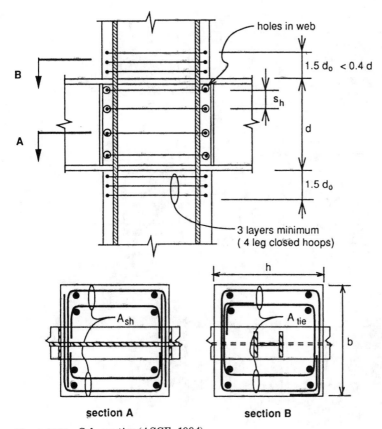

Figure 10.31 Column ties (*ASCE, 1994*).

d_f = the center-to-center distance between the beam flanges

d_w' = the depth of the steel web

d_o = additional effective joint depth provided by attachment to the beam

jh = horizontal distance between bearing force resultant and is given by the following:

$$jh = \frac{\sum M_c}{\phi(T_{vrn} + C_{vrn} + C_c) - \Delta V_b/2} \geq 0.7h \qquad (10.24)$$

in which

$$C_c = 2f_c'b_j a_c \qquad (10.25)$$

$$a_c = \frac{h}{2} - \sqrt{\frac{h^2}{4} - K} \leq 0.3h \qquad (10.26)$$

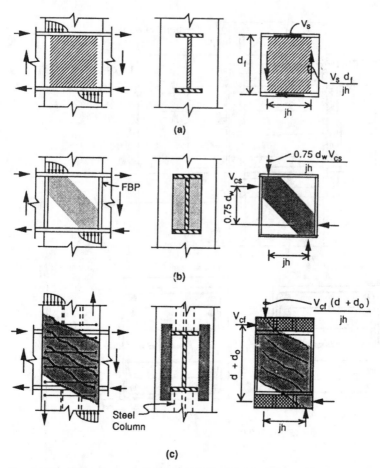

Figure 10.32 Joint shear mechanism (*ASCE, 1994*).

$$K = \frac{1}{\phi 2 f_c' b_j} \left[\sum M_c + \frac{\Delta V_b h}{2} - \phi(T_{vrn} + C_{vrn})h_{vr} \right] \quad (10.27)$$

In Eq. (10.23), it is assumed that the contributions of the mechanisms are additive. The following sections describe the individual contribution of each of the three different mechanisms.

Steel panel. The steel contribution is given as the capacity of the beam web in pure shear. Assuming the effective panel length to be equal to jh and the average shear yield stress is $0.6F_{ysp}$, the nominal strength of the steel panel, V_{sn}, is expressed as follows:

$$V_{sn} = 0.6F_{ysp}t_{sp}jh \quad (10.28)$$

where F_{ysp} = the yield strength of the steel panel and t_{sp} = the thickness of the steel panel.

The vertical shear forces in the steel web cause the beam flanges to bend in the transverse direction. To prevent beam flanges failure, the thickness should satisfy the following:

$$t_f \geq 0.30 \sqrt{\frac{b_f t_{sp} F_{ysp}}{h F_{yf}}} \qquad (10.29)$$

where F_{yf} is the yield strength of the beam flanges.

Concrete strut. The nominal strength of the concrete compression strut mechanism, V_{csn}, is calculated as follows:

$$V_{csn} = 1.7 \sqrt{f_c'} b_p h \leq 0.5 f_c' b_p d_w \qquad (10.30)$$

$$b_p \leq b_f + 5t_p \leq 1.5 b_f \qquad (10.31)$$

where f_c' = the concrete compressive strength, in MPa
b_p = the effective width of FBP, and is limited by Eq. (10.31)
t_p = the thickness of the FBP and should meet the following conditions:

$$t_p \geq \frac{\sqrt{3}}{b_f F_{up}} (V_{cs} - b_f t_w F_{yw}) \qquad (10.32)$$

$$t_p \geq \frac{\sqrt{3} V_{cs}}{2 b_f F_{up}} \qquad (10.33)$$

$$t_p \geq 0.2 \sqrt{\frac{V_{cs} b_p}{F_{yp} d_w}} \qquad (10.34)$$

$$t_p \geq \frac{b_p}{22} \qquad (10.35)$$

$$t_p \geq \frac{b_p - b_f}{5} \qquad (10.36)$$

where F_{up} = the specified tensile strength of the bearing plate and V_{cs} = the horizontal shear force carried by the concrete strut.

Where split FBPs are used, the plate height, d_p, should not be less than $0.45 d_w$.

Compression field. The nominal strength of the concrete compression field mechanism, V_{cfn}, is calculated as follows:

$$V_{cfn} = V_c' + V_s' \leq 1.7\,\sqrt{f_c'}\,b_o h \tag{10.37}$$

$$V_c' = 0.4\,\sqrt{f_c'}\,b_o h \tag{10.38}$$

$$V_s' = \frac{A_{sh}F_{ysh}\,0.9h}{s_h} \tag{10.39}$$

where f_c' = the concrete compressive strength, in MPa
$\quad V_c'$ = the concrete contribution to nominal compression field, $V_c' = 0$ where the column is in tension
$\quad V_s'$ = the contribution provided by the horizontal ties to nominal compression field strength
$\quad A_{sh}$ = the cross-sectional area of reinforcing bars in each layer of ties spaced at s_h through the beam depth, $A_{sh} \geq 0.004bs_h$

Where extended FBP and/or steel columns are used, they should be designed to resist a force equal to the joint shear carried by the outer compression field, V_{cf}. The thickness of column flanges or the extended FBP is considered adequate if the following equation is satisfied:

$$t_f \geq 0.12\,\sqrt{\frac{V_{cf}b_p'}{d_o F_y}} \tag{10.40}$$

where V_{cf} = is the horizontal shear force carried by the outer compression field
$\quad b_p'$ = the flange width of the steel column or the width of the extended FBP
$\quad F_y$ = the specified yield strength of the plate

In addition to the preceding requirement, the thickness of the extended FBP should not be less than the thickness of the FBP between the beam flanges.

Ties above and below the beam should be able to transfer the force, V_{cf}, from the beam flanges into the outer concrete panel. In addition to the requirements in Sec. 10.4.6.1, the minimum total cross-sectional area should satisfy the following:

$$A_{\text{tie}} \geq \frac{V_{cf}}{F_{ysh}} \tag{10.41}$$

where F_{ysh} = the yield strength of the reinforcement

A_{tie} = the total cross-sectional area of ties located within the vertical distance $0.4d$ of the beam (see Fig. 10.31)

10.4.6.3 Vertical column bars. To limit the slip of column bars within the joint, the size of the bar should satisfy the following requirements:

$$d_b < \frac{d + 2d_o}{20} \qquad (10.42)$$

where, for single bars, d_b = the vertical bar diameter, and, for bundled bars, d_b = the diameter of a bar of equivalent area to the bundle.

Exceptions to Eq. (10.42) can be made where it can be shown that the change in force in vertical bars through the joint region, ΔF_{bar}, satisfies the following:

$$\Delta F_{bar} < 80(d + 2d_o)\sqrt{f_c'} \qquad (10.43)$$

where f_c' is in MPa.

10.4.7 Limitations

The ASCE recommendations are limited to joints where the steel beams are continuous through the reinforced concrete column. Although this type of detail has been successfully used in practice, the guidelines do not intend to imply or recommend the use of this type over other possible details. Both interior and exterior joints can be designed using the recommendations; however, top-interior and top-corner joints are excluded because supporting test data are not available. For earthquake loading, the recommendations are limited to regions of low-to-moderate seismic zones. The ratio of depth of concrete column, h, to the depth of the steel beam, d, should be in the range of 0.75 to 2.0. For the purpose of strength calculation, the nominal concrete strength, f_c', is limited to 40 MPa (6 ksi) and only normal-weight concrete is allowed, the reinforcing bars yield stress is limited to 410 MPa (60 ksi), and the structural steel yield stress is limited to 345 MPa (50 ksi).

10.5 Connections to Concrete-Filled Tube (CFT) Columns

10.5.1 Introduction

Steel tubes of relatively thin wall thickness filled with high-strength concrete have been used in building construction in the United States and far east Asian countries. This structural system allows the designer to maintain manageable column sizes while obtaining increased stiffness and ductility for wind and seismic loads. Column shapes can take the form of tubes or pipes as required by architectural restrictions. Additionally, shop fabrication of steel shapes helps ensure quality control.

In this type of construction, in general, at each floor level a steel beam is framed to these composite columns. Often, these connections are required to develop shear yield and plastic moment capacity of the beam simultaneously.

10.5.2 Current practice

In current practices, there are very limited guidelines for selecting or designing connections for attaching steel beams to CFT columns. In these instances, heavy reliance is made on the judgment and experience of individual designers.

The majority of available information on steel beams to CFT columns has been developed as a result of the U.S.–Japan Cooperative Research Program on Composite/Hybrid Structures (1992). It should be noted that the information developed under this initiative is targeted toward highly seismic regions. Nevertheless, the information could be used to design connection details in nonseismic regions.

One of the distinct categories of connection details suggested is attaching the steel beams using full-penetration welds, as practiced in Japan. Japanese practice usually calls for a massive amount of field and shop welding. Figure 10.33 shows some of the connection details suggested in Japan. In general, the type of details that are used in Japan are not economical for U.S. practice.

10.5.3 Problems associated with welding beams to CFT columns

When beams are welded or attached to steel tubes through connection elements, complicated stiffener assemblies are required in the joint area within the column. However, welding of the steel beam or connecting element directly to the steel tube of composite columns could produce potential problems, some of which are outlined in the following:

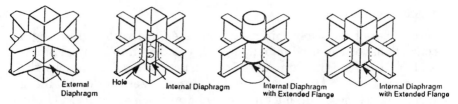

External   Hole   Internal Diaphragm   Internal Diaphragm   Internal Diaphragm
Diaphragm   with Extended Flange   with Extended Flange

Figure 10.33 Typical connection details suggested in Japan.

1. Transfer of tensile forces to the steel tube can result in separation of the tube from the concrete core, thereby overstressing the steel tube. In addition, the deformation of the steel tube will increase connection rotation, decreasing its stiffness. This is especially important if the connection is required to develop full plastic moment capacity of the beam.

2. Welding of the thin steel tube results in large residual stresses because of the restraint provided by other connection elements.

3. The steel tube is designed primarily to provide lateral confinement for the concrete. Further, in building construction where CFT columns are utilized, the steel tube portion of the column also acts as longitudinal reinforcement. Transferring additional forces from the beam to the steel tube could result in overstressing the steel tube portion of the column.

10.5.4 Possible connection detail

With these considerations in mind, Azizinamini and Parakash (1993) and Azizinamini, Yerrapalli, and Saadeghvaziri (1995) suggest two general types of connections. Figure 10.34 shows one alternative in

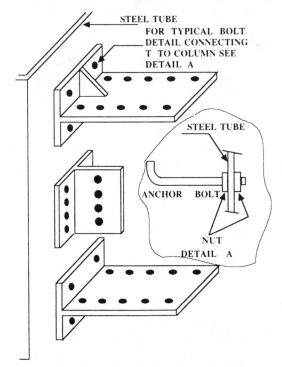

Figure 10.34 Connection detail using anchor bolts.

which forces are transmitted to the core concrete via anchor bolts connecting the steel elements to the steel tube. In this alternative, all elements could be preconnected to the steel tube in the shop. The nut inside the steel tube is designed to accomplish this task. The capacity of this type of connection would be limited to the pull-out capacity of the anchor bolts and local capacity of the tube.

Another variation of the same idea is shown in Fig. 10.35, where connecting elements would be embedded in the core concrete via slots cut in the steel tube. In this variation, slots must be welded to connection elements after beam assembly for concrete confinement. The ultimate capacity of this detail also would be limited to the pull-out capacity of the connection elements and the concrete in the tube. The types of connections shown in Figs. 10.34 and 10.35 could be suitable to nonseismic applications, at the story levels, where the level of forces is relatively small.

Another suggested type of connection (Azizinamini and Parakash, 1993, and Azizinamini, Yerrapalli, and Saadeghvaziri, 1995) is to pass the beam completely through the column, as shown in Fig. 10.36. In this type of detail, a certain height of column tube, together with a short beam stub passing through the column and welded to the tube, could be shop-fabricated to form a "tree column" as shown in Fig. 10.37. The beam portion of the tree column could then be bolted to girders in the field.

Alostaz and Schneider (1996) report tests on six different connection details for connecting steel beams to circular CFT columns. The objectives of these tests were to examine the feasibility of different connection details for use in highly seismic areas and suitable to U.S. practice.

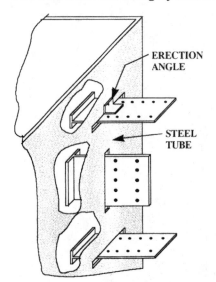

ERECTION
ANGLE

STEEL
TUBE

Figure 10.35 Connection detail using embedded elements.

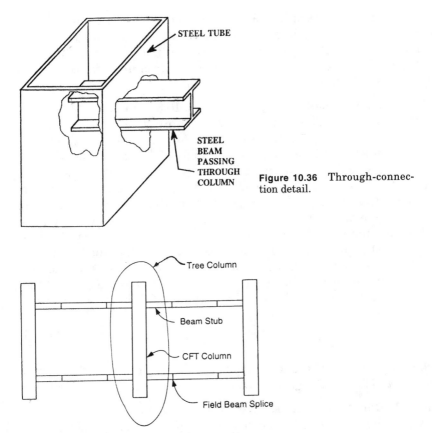

Figure 10.36 Through-connection detail.

Figure 10.37 Tree column construction concept.

These connections ranged from a very simple detail that attached the beam to the tube skin as in connection type I to a more rigid detail in which the girder was passed through the tube core as represented by connection type VII. All connections were designed with a beam stub. The beams were bolted and/or welded to these stubs. The specimens had a T configuration, thus representing an exterior joint in a building. Each specimen consisted of a 14- × ¼-in (356- × 6.4-mm)-diam pipe and W 14 × 38 beam. The concrete compressive strength varied between 7.8 and 8.3 ksi (53.8 and 57.2 MPa). The pipe yield strength was 60 ksi (420 MPa). The stub flanges and web yield strengths were 50 and 40 ksi (350 and 280 MPa), respectively. This resulted in a column-to-beam bending capacity ratio of approximately 2.6. This relatively high column-to-beam capacity ratio is not desirable when one attempts to investigate connection behavior. At the extreme, very high column moment capacity will force the plastic hinge to form at the end of the beam, preventing the investigation of behavior of joints. Despite this shortcoming, Alostaz and Schneider's data provide valuable information that could be used to develop connection details suitable for seismic as well as nonseismic applications. Following is a brief discussion of the behavior of different connection details tested by Alostaz and Schneider (1996).

10.5.4.1 Simple connection, type I. Figure 10.38 illustrates the details of this specimen. The flange and web plates of the connection stub were welded directly to the steel pipe. At the tube face, the flange plates were flared to form a central angle of 120°, and the width of the plates was decreased gradually over a 10-in (254-mm) distance to match that of the girder flanges. Figure 10.39 shows the load-displacement relationship. Failure was due to fracture at the flange tip on the connection stub and pipe wall tearing. The connection survived a limited number of inelastic cycles and it could not develop the plastic flexural strength of the girder. This connection had the lowest flexural strength and was the most flexible of all connections tested. This connection had a ductility ratio of 1.88, which was the lowest of all connections tested. The flexural ductility ratio (FDR) was defined as

$$FDR = \frac{\delta_{max}}{\delta_{yield}} \qquad (10.44)$$

where δ_{max} is the maximum displacement at the girder tip prior to failure and δ_{yield} is the yield displacement obtained experimentally.

10.5.4.2 Continuous web plate connection, type IA. In an attempt to improve the behavior of connection type I, the web plate was extended through the concrete core. To continue the web through the tube, a vertical slot was cut on opposite sides of the tube wall. The web plate was fillet-welded to the tube. Figure 10.40 illustrates the details of this specimen. Figure 10.41 shows the load-displacement relationship. The hysteretic behavior of this modified connection exhibited significant improvement compared to the original simple connection. This connection was able to develop approximately 1.26 times the flexural plastic strength of the girder and the initial stiffness was comparable to the ideal rigid connection. However, the strength deteriorated rapidly and only 50% of the girder bending strength remained at the end of the test. This connection had a ductility ratio of 2.55.

10.5.4.3 Connection with external diaphragms, type II. Behavior of the simple connection was improved by expanding the connection stub flanges to form external diaphragms. The diaphragm was fillet-welded to the pipe wall on both sides of the plate. Figure 10.42 illustrates the details of this specimen. Figure 10.43 shows the load-displacement relationship. The hysteretic performance of this connection

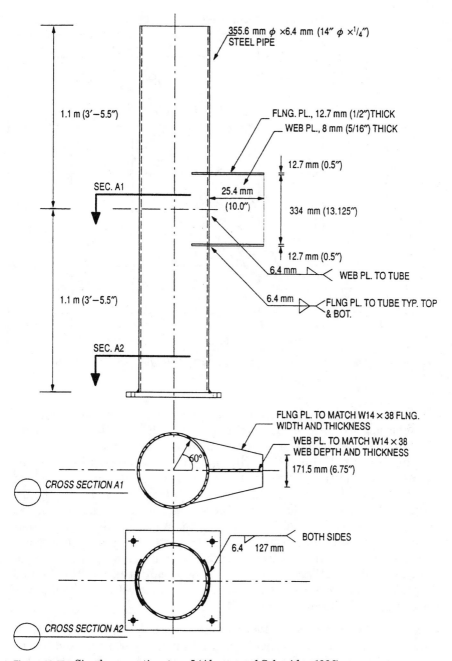

Figure 10.38 Simple connection, type I (*Alostaz and Schneider, 1996*).

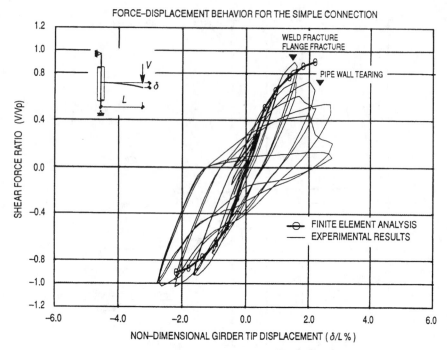

FORCE–DISPLACEMENT BEHAVIOR FOR THE SIMPLE CONNECTION

Figure 10.39 Load-displacement behavior of connection type I (*Alostaz and Schneider, 1996*).

improved relative to the simple connection type I. This resulted in a connection strength of approximately 17% higher than the girder bending strength. The geometry of the diaphragm was a critical issue in the behavior of this detail. The sharp reentrant corner between the diaphragm and the girder created a large stress concentration which initiated fracture in the diaphragm. This fracture caused rapid deterioration in the connection performance. Significant tearing was noted through the welded region of the diaphragm plates. Although connection type IA had higher strength, its strength deteriorated at a faster rate compared to the connection with external diaphragms. This connection had a ductility ratio of 2.88. Analytically, this detail exhibited significant improvement when the girder was shifted further away from the CFT column face.

10.5.4.4   Connection with deformed bars, type III.   This specimen is identical to connection type I, except that holes were drilled in the pipe to insert weldable deformed bars into the core of the tube. Four #6 (19-mm) deformed bars were welded to each flange. Figure 10.44 illustrates the details of this specimen. Figure 10.45 shows the force-displacement relationship. This connection exhibited stable

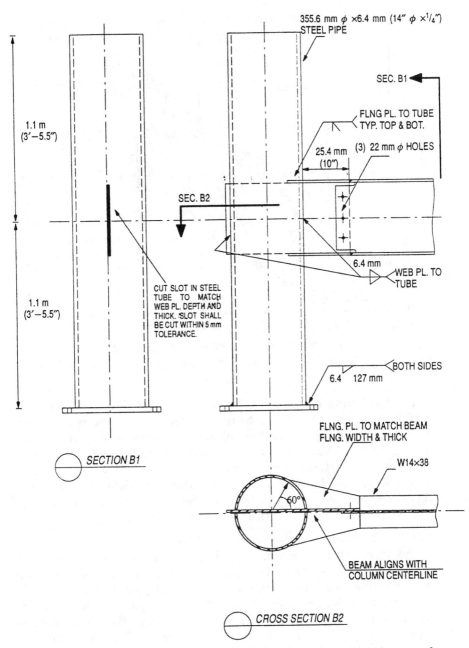

Figure 10.40 Simple connection with continuous web plate, type IA (*Alostaz and Schneider, 1996*).

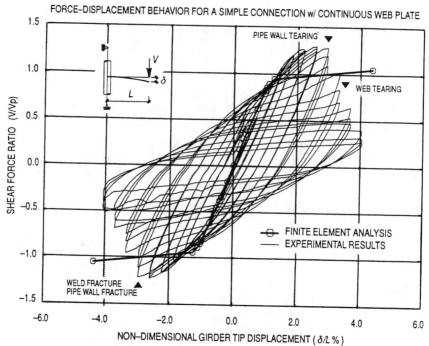

Figure 10.41 Load-displacement behavior of connection type IA (*Alostaz and Schneider, 1996*).

strain-hardening behavior up to failure, and it developed approximately 1.5 times the girder bending strength. Failure was sudden and occurred by rupture of three of the four deformed bars in the connection detail, while the fourth bar failed by pull-out of the concrete core. The connection ductility was approximately 3.46 compared to only 1.88 for an identical connection without the deformed bars. The clearance, weldability of the deformed bars, and the configuration of the weld on the bars are critical issues in this detail.

10.5.4.5 Continuous flanges, type VI. To resolve the problems of connection type III, the connection stub flanges were continued through the pipe and fillet-welded to the pipe wall. A shear tab was fillet-welded to the tube skin. No effort was made to enhance the bond between the embedded flanges and the concrete core. Figure 10.46 illustrates the details of this specimen. Figure 10.47 shows the force-displacement relationship. The fillet weld attaching the flanges to the tube wall fractured at low amplitude cyclic deformations. The embedded flanges slipped through the concrete core without significant resistance. The hysteretic curves were quite pinched and it is likely that this connection may not perform well during a severe seismic event.

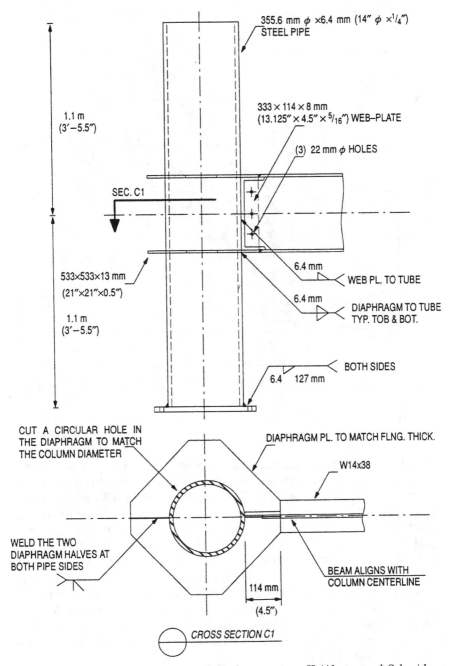

Figure 10.42 Connection with external diaphragms, type II (*Alostaz and Schneider, 1996*).

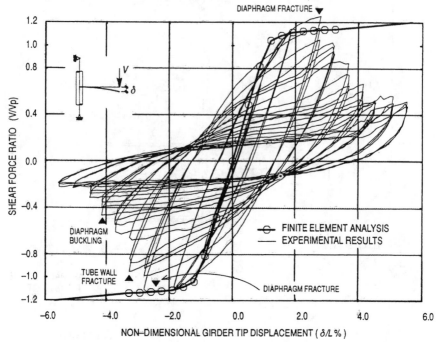

FORCE–DISPLACEMENT BEHAVIOR FOR A CONNECTION WITH EXTERNAL DIAPHRAGMS

Figure 10.43 Load-displacement behavior of connection type II (*Alostaz and Schneider, 1996*).

10.5.4.6 Through-beam connection detail, type VII. Alostaz and Schneider (1996) also tested one specimen with the through-beam connection detail suggested by Azizinamini et al. (1993 and 1995). In this detail, the full cross section of the girder was continued through the tube core. An I-shaped slot was cut in the tube wall and the beam stub was passed through the pipe. The beam stub was fillet-welded to the pipe. Figure 10.48 illustrates the details of this specimen. Figure 10.49 shows the force-displacement relationship. The flexural strength of this connection exceeded 1.3 times the plastic bending strength of the girder. This detail had a ductility ratio of 4.37, the highest of all connections tested. It also had a satisfactory hysteretic performance. Table 10.1 shows a summary of the flexural characteristics of the tested connections.

Results of Alostaz and Schneider tests indicated that the through-beam connection detail had the best performance, especially for seismic regions.

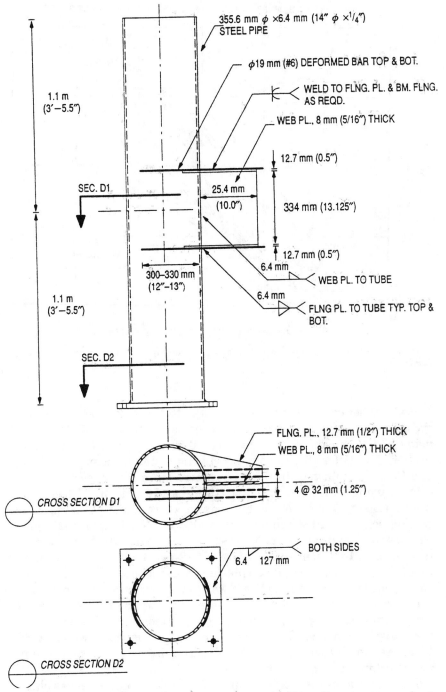

Figure 10.44 Connection with embedded deformed bars, type III (*Alostaz and Schneider, 1996*).

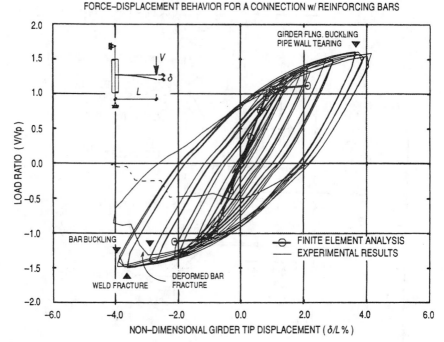

Figure 10.45 Load-displacement behavior of connection type III (*Alostaz and Schneider, 1996*).

10.5.4.7 Other connection details. Ricles et al. (1997) report results of cyclic tests conducted on beams attached to rectangular CFT columns using bolted or welded tees. Figures 10.50 through 10.52 show connection details for three of the specimens tested (specimens C4, C5, and C6). The split tees in these specimens were posttensioned to the column using 14-A490 bolts after curing of the concrete. These bolts were passed through the column using PVC conduits placed prior to casting concrete. In specimens C4 and C5, 22-mm-diam A325 bolts with 2-mm oversized bolt holes were used to attach the beam flanges to split tees. In specimen C6, however, 12-mm fillet welds were used to attach the beam flanges to split tees. In specimens C5 and C6, the shear tabs for attaching the beam web to CFT column were omitted.

Figures 10.53 through 10.55 give plots of applied beam moment versus the resulting plastic rotation at the connection level for the three test specimens. These specimens were designed based on AISC LRFD seismic provisions (1994 version), following the weak beam-strong column configuration.

Test observations indicated that damage to the joint area was eliminated. Some elongation of A490 bolts was observed. This was attrib-

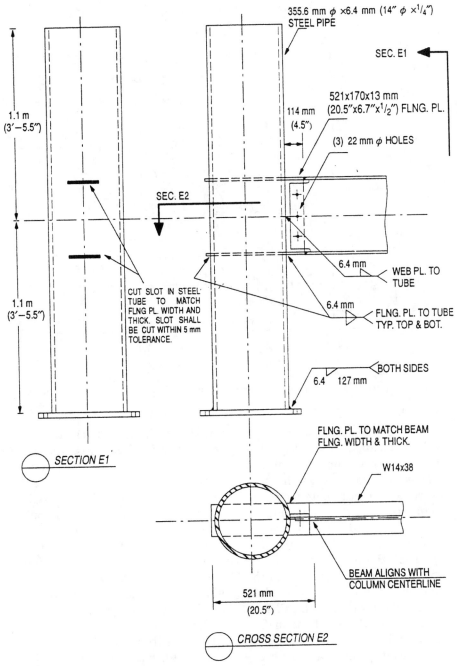

Figure 10.46 Continuous flanges, type VI (*Alostaz and Schneider, 1996*).

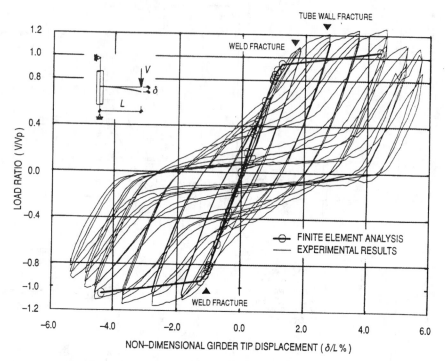

FORCE–DISPLACEMENT BEHAVIOR FOR A CONNECTION w/ EMBEDDED FLANGES

Figure 10.47   Load-displacement behavior of connection type VI (*Alostaz and Schneider, 1996*).

uted to compressive bearing forces transferred from split tees to CFT columns, causing distortion of the joint area in CFT columns. Another major observation was the slippage of the stem of split tees with respect to beam flanges in specimens C4 and C5. Ricles et al. (1997) were able to eliminate this slippage by welding washers to the beam flanges. The washers, acting as reinforcing material around the bolt hole, prevented bolt hole elongation and elimination of the slippage.

In this type of detail, attention should be directed to shear transfer between the beam end and CFT column. The load path for transferring the beam shear force to the CFT column is as follows. The beam end shear is first transferred as axial force from the beam end to the steel tube portion of the CFT column. This axial compressive or tensile force could only be transferred to the concrete portion of the CFT column if composite action between the steel tube and the concrete core exists. There are several ways through which this composite action could be developed. Friction due to bending or use of shear studs are two possible mechanisms. The guidelines for such shear

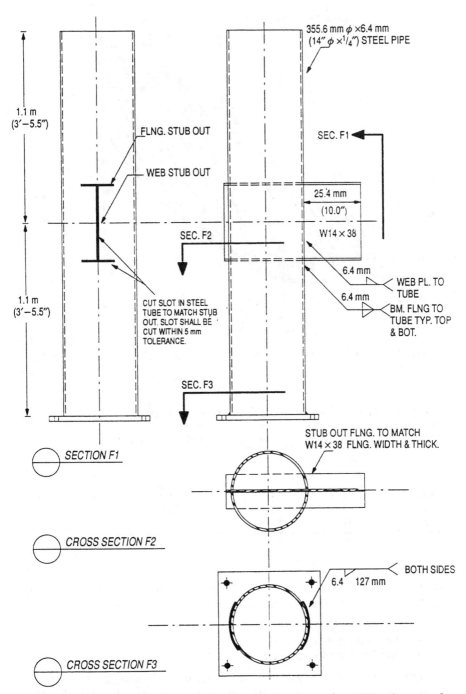

355.6 mm φ ×6.4 mm
(14″ φ ×¹/₄″) STEEL PIPE

1.1 m
(3′−5.5″)

FLNG. STUB OUT

SEC. F1

WEB STUB OUT

25.4 mm
(10.0″)

SEC. F2

W14 × 38

6.4 mm   WEB PL. TO
                         TUBE
6.4 mm
1.1 m   CUT SLOT IN STEEL   BM. FLNG TO
(3′−5.5″)   TUBE TO MATCH STUB   TUBE TYP. TOP
                OUT. SLOT SHALL BE   & BOT.
                CUT WITHIN 5 mm
                TOLERANCE.

SEC. F3

SECTION F1

STUB OUT FLNG. TO MATCH
W14 × 38 FLNG. WIDTH & THICK.

CROSS SECTION F2

                                    BOTH SIDES
                         6.4   127 mm

CROSS SECTION F3

Figure 10.48 Continuation of the girder through the column, type VII (*Alostaz and Schneider, 1996*).

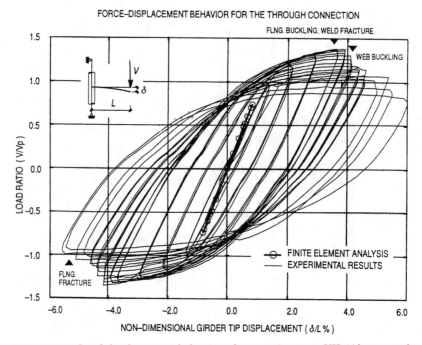

FORCE–DISPLACEMENT BEHAVIOR FOR THE THROUGH CONNECTION

Figure 10.49 Load-displacement behavior of connection type VII (*Alostaz and Schneider, 1996*).

TABLE 10.1   Flexural Characteristics of the Tested Connections

Detail	Ductility	M_{max}/M_p	Initial stiffness ratio
I	1.88	0.97	85
IA	2.55	1.26	100
II	2.83	1.17	100
III	3.46	1.56	106
VI	3.76	1.23	100
VII	4.37	1.37	100

transfer mechanisms are still lacking. Ongoing research by Roeder (1997) attempts to resolve this issue.

10.5.5   Force transfer mechanism for through-beam connection detail

A combination of analytical and experimental investigations were undertaken to approximate the force transfer mechanism for the

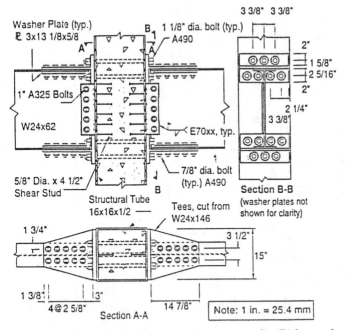

Figure 10.50 Split tee connection detail, specimen C4 (*Ricles et al.*, *1997*).

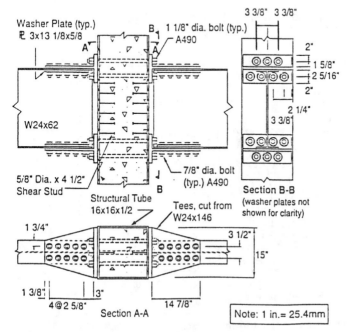

Figure 10.51 Split tee connection detail, specimen C5 (*Ricles et al.*, *1997*).

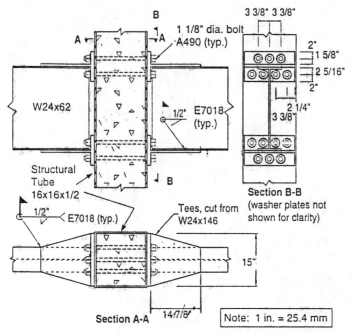

Figure 10.52   Split tee connection detail, specimen C6 (*Ricles et al., 1997*).

West Beam Moment-Plastic Rotation Response at End of Conn., Spec C4

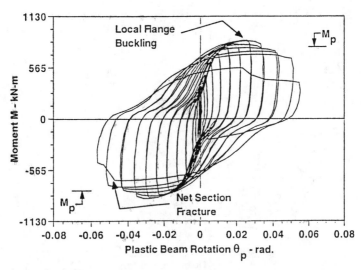

Figure 10.53   Moment-plastic rotation response, specimen C4 (*Ricles et al., 1997*).

East Beam Moment-Plastic Rotation Response at End of Conn., Spec C5

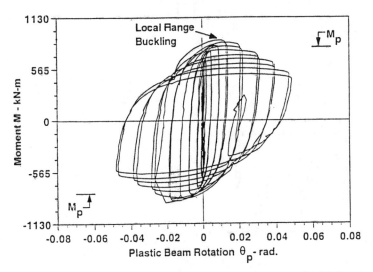

Figure 10.54 Moment-plastic rotation response, specimen C5 (*Ricles et al., 1997*).

East Beam Moment-Plastic Rotation Response at End of Conn., Spec C6

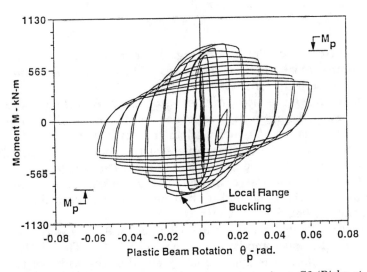

Figure 10.55 Moment-plastic rotation response, specimen C6 (*Ricles et al., 1997*).

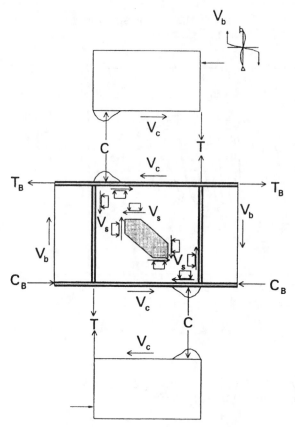

Figure 10.56 Force transfer mechanism for through-beam connection detail.

through-beam connection detail utilizing both circular and rectangular CFT columns (Azizinamini and Parakash, 1993, and Azizinamini, Yerrapalli, and Saadeghvaziri, 1995).

Figure 10.56 shows the force transfer mechanism. A portion of the steel tube between the beam flanges acts as a stiffener, resulting in a concrete compression strut which assists the beam web within the joint in carrying shear. The effectiveness of the compression strut increases to a limit by increasing the thickness of the steel plate.

The width of the concrete compression strut on each side of the beam web in the direction normal to the beam web was approximately equal to half the beam flange width.

A compressive force block was created when beam flanges were compressed against the upper and lower columns (Fig. 10.56). The width of this compression block was approximately equal to the width of the beam flange. In the upper and lower columns, shown in Fig. 10.56, the compressive force C is shown to be balanced by the tensile force in the steel pipe. In Azizinamini and Parakash (1993), rods embedded in the concrete and welded to the beam flanges were provided to assist the steel tube in resisting the tensile forces and to minimize the tensile stresses in the steel tube. For small columns this may be necessary, however, for relatively larger columns there may not be a need for placing such rods. Ongoing research at the University of Nebraska—Lincoln is investigating this and other aspects of the force transfer mechanism. The next section suggests design provisions for the through-beam connection detail. These provisions are tentative and are applicable for both circular and rectangular CFT columns.

10.5.6 Tentative design provisions for through-beam connection detail

This tentative design procedure is in the form of equations relating the applied external forces to the connection details such as the thickness of the steel pipe. The design procedure follows the general guidelines in the AISC LRFD manual (1994). In developing the design equations the following assumptions were made:

1. Externally applied shear forces and moments at the joints are known.

2. At the ultimate condition, the concrete stress distribution is linear and the maximum concrete compressive stress is below its limiting value.

The joint forces implied in assumption 1 could be obtained from analysis and require the knowledge of the applied shear and moment at the joint at failure. These quantities are assumed to be related as follows:

$$V_c = \alpha V_b$$
$$M_b = l_1 V_b$$
$$M_c = l_2 V_c$$

where V_b and M_b are the ultimate beam shear and moment, respectively, and V_c and M_c are the ultimate column shear and moment, respectively. Figure 10.57 shows these forces for an isolated portion of a structure subjected to lateral loads.

Assumption 2 is valid for the cases where the moment capacity of columns is relatively larger than the beam capacity.

Figure 10.58 shows the free-body diagram (FBD) of the beam web within the joint and upper column at ultimate load. With reference to Fig. 10.58 the following additional assumptions are made in deriving the design equations:

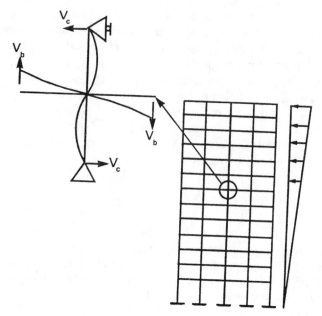

Figure 10.57 Assumed forces on an interior joint in a frame subjected to lateral loads.

1. The concrete stress distribution is assumed to be linear. The width of the concrete stress block is assumed to equal b_f, the beam flange width.

2. As shown in Fig. 10.58, the strain distribution over the upper column is assumed to be linear.

3. The steel tube and concrete act compositely.

4. The portion of the upper column shear, V_c, transferred to the steel beam is assumed to be βC_c, where C_c is the resultant concrete compressive force bearing against the beam flange and ß is the coefficient of friction.

5. Applied beam moments are resolved into couples concentrated at beam flanges.

6. The resultant of the concrete compression strut is along a diagonal as shown in Fig. 10.58.

Considering the preceding assumptions and the strain distribution shown for the upper column in Fig. 10.58, the maximum strain in concrete, ε_c, could be related to ε_l, the steel pipe strain in tension:

$$\varepsilon_c = \frac{a}{d_c - a}\,\varepsilon_l \qquad (10.45)$$

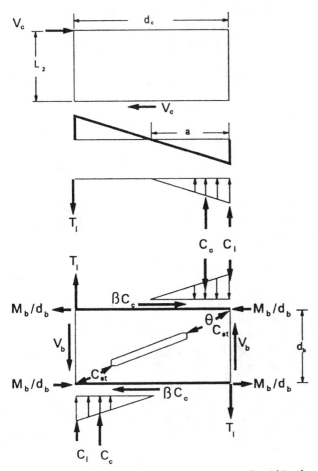

Figure 10.58 FBD of upper column and beam web within the joint.

The maximum stress in the concrete and the stresses in the steel tube can be calculated as follows:

$$f_c = E_c \varepsilon_c \qquad (10.46)$$

$$f_{lc} = E_s \varepsilon_c \qquad (10.47)$$

$$f_{lt} = E_s \varepsilon_l \qquad (10.48)$$

where f_c, f_{lc}, and f_{lt} are the maximum concrete compressive stress, the stress in the steel pipe in compression, and the stress in the steel pipe in tension, respectively.

Substituting Eq. (10.45) into Eqs. (10.46) to (10.48) and multiplying Eqs. (10.46) to (10.48) by the corresponding area, the resultant forces for different connection elements can be calculated as follows:

$$C_c = \left(\frac{1}{2}\right)\left(\frac{1}{\eta}\right)\xi b_f \frac{a^2}{d_c - a} f_{yl} \tag{10.49}$$

$$C_l = \gamma \xi b_f t_l \frac{a}{d_c - a} f_{yl} \tag{10.50}$$

$$T_l = \xi \gamma b_f t_l f_{yl} \tag{10.51}$$

Using the FBD of the upper column, shown in Fig. 10.58, Eqs. (10.49) to (10.51), and satisfying the vertical force equilibrium, the following equation could be obtained:

$$t_l = \frac{a^2}{d_c - 2a} \frac{1}{2\gamma\eta} \tag{10.52}$$

where d_c = diameter of the steel tube
  a = depth of neutral axis
  η = ratio of modulus of elasticity for steel over modulus of elasticity of concrete
  t_l = thickness of steel tube
  γ = factor reflecting portion of steel tube effective in carrying tensile forces. Experimental data for square tubes indicated that it could be assumed that $\gamma = 2$. The same value is assumed for circular columns.

Considering the moment equilibrium of the FBD of the upper column shown in Fig. 10.58, the following expression can be derived:

$$\frac{1}{d_c - a}\left[\frac{a^3 d_c}{d_c - 2a} + a^2\left(d_c - \frac{a}{3}\right)\right] = \frac{2\eta}{\xi} \frac{\alpha l_2}{b_f f_{yl}} V_b \tag{10.53}$$

where f_{yl} is the yield strength of the steel tube.

In Eq. (10.53), ξf_{yl}, is the stress level that the steel tube is allowed to approach at the ultimate condition. Based on the experimental data and until further research is conducted, it is suggested that a value of 0.75 be used for ξ.

Equations (10.52) and (10.53) relate the externally applied force, V_b, directly and the externally applied forces, V_c and M_c, indirectly (through the coefficients α and l_2) to different connection parameters.

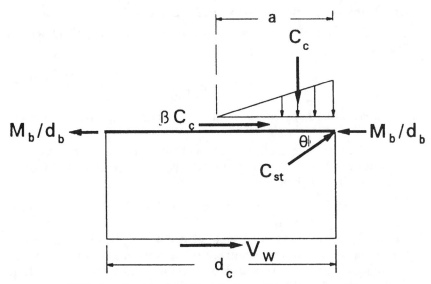

Figure 10.59 FBD of portion of web within joint area.

10.5.6.1 Design approach. Before designing the through-beam connection detail, additional equations will be derived to relate the shear stress in the beam web within the joint to the compressive force in the concrete compression strut and the externally applied forces.

Considering the FBD of a portion of the beam web within the joint area as shown in Fig. 10.59 and satisfying the horizontal force equilibrium, the following equation can be derived:

$$V_w + C_{st}\cos\theta + \text{\ss}C_c - \frac{2M_b}{d_b} = 0 \qquad (10.54)$$

where V_w is the shear force in the beam web at the ultimate condition and $\theta = \arctan(d_b/d_c)$. Equations (10.52) to (10.54) can be used to proportion the through-beam connection detail.

Until further research is conducted, the following steps are suggested for designing the through-beam connection detail following the LRFD format:

1. From analysis, obtain the factored joint forces.

2. Select b_f, d_c, and f_{yl}.

3. Solving Eq. (10.53), obtain a, the depth of the neutral axis.

4. Solving Eq. (10.52), obtain t_p, the required thickness of the pipe steel.

5. Check stress in different connection elements.
6. From the vertical equilibrium requirement of the FBD shown in Fig. 10.59:

$$C_{st} = \frac{C_c}{\sin \theta} \qquad (10.55)$$

Using Eq. (10.49), calculate C_c and then using Eq. (10.55) calculate C_{st}.

7. Using Eq. (10.54) calculate V_w, the shear force in the beam at the ultimate condition and compare it to V_{wy}, the shear yield capacity of the beam web given by

$$V_{wy} = 0.6F_{yw}t_w d_c \qquad (10.56)$$

where F_{yw} is the beam web yield stress and t_w is the thickness of the beam web. If necessary increase the thickness of the web within the joint region. In this design procedure the assumption is that at the factored load level, the web starts to yield.

8. Check the shear stress in the concrete in the joint area. The limiting shear force could be assumed to be as suggested by ACI 352 (1985):

$$V_u = \phi R \sqrt{f_c'} A_e \qquad (10.57)$$

where $\phi = 0.85$
   $R = 20$, 15, and 12 for interior, exterior, and corner joints, respectively
   $f_c' =$ concrete compressive strength

It is suggested that the value of f_c' be limited to 70 MPa, implying that in the case of 100-MPa concrete, for instance, f_c' be taken as 70 MPa rather than 100 MPa.

10.5.6.2 Design example. Design a through-beam connection detail with the following geometry and properties:

Given steps 1 and 2:

$$b_f = 139.7 \text{ mm}$$
$$d_b = 368.3 \text{ mm}$$
$$d_c = 406.4 \text{ mm}$$
$$f_{yl} = 248.22 \text{ MPa}$$
$$F_{yw} = 248.22 \text{ MPa}$$
$$t_w = 6.35 \text{ mm}$$

$$\alpha = 0.85$$
$$l_2 = 812.8 \text{ mm}$$
$$V_b = 351.39 \text{ kN}$$
$$M_b = 187.54 \text{ kN} \cdot \text{m}$$
$$\beta = 0.5$$
$$\xi = 0.75$$
$$\eta = 4.3$$
$$f_c' = 96.53 \text{ MPa}$$
$$E_s = 200 \text{ GPa (modulus of elasticity of steel)}$$
$$E_c = 46 \text{ GPa (modulus of elasticity of concrete)}$$

Step 3: Using Eq. (10.53), calculate a, the depth of the neutral axis. Equation (10.53) will result in a third-degree polynomial which can be shown to have only one positive, real root. For this example Eq. (10.53) results in $a = 149.35$ mm.

Step 4: Using Eq. (10.52), calculate the required thickness of the steel pipe (use $t_l = 12.0$ mm).

$$t_l = \frac{149.35^2}{406.4 - 2(149.35)} \frac{1}{2(2)(4.3)} = 12.04 \text{ mm}$$

Step 5: Check stresses in different connection elements against their limit values. First calculate the tensile strain in the steel tube

$$\varepsilon_l = \frac{\xi f_{yl}}{E_s} = \frac{0.75(248.22)}{200,000} = 0.000931 \text{ mm/mm}$$

Using Eqs. (10.45) and (10.46), calculate f_c

$$f_c = 24.90 \text{ MPa} < f_c' = 96.53 \text{ MPa}$$

Using Eqs. (10.47) and (10.48), calculate the stresses in the other connection elements. This yields

$$f_{lc} = 108.1 \text{ MPa} < \phi_c F_y = 0.85 \times 248.22 = 211 \text{ MPa}$$
$$f_{lt} = 186.2 \text{ MPa} < \phi_t F_y = 0.9 \times 248.22 = 223.4 \text{ MPa}$$

Step 6: Using Eqs. (10.49) and (10.55), calculate the compressive force in the concrete compression strut

$$\theta = \arctan \frac{368.3}{406.4} = 42.2°$$

$$C_c = \left(\frac{1}{2}\right)(1/\eta)\zeta b_f \left(\frac{a^2}{d_c - a}\right) f_{yl}$$

$$C_c = \frac{\frac{1}{2}(0.23)(0.75)(139.7)(149.35^2)/(406.4 - 149.35) \times 248.22}{1000}$$

$$= 262.42 \text{ kN}$$

$$C_{st} = \frac{C_c}{\sin \theta} = \frac{262.42}{\sin 42.2} = 390.67 \text{ kN}$$

Step 7: Using Eq. (10.54), compute V_w

$$V_w + C_{st} \cos \theta + \beta C_c - \frac{2M_b}{d_b} = 0$$

$$V_w + 390.67 \cos (42.2) + 0.5(262.42) - \frac{2 \times 187.54 \times 10^3}{368.3} = 0$$

$$V_w = 597.79 \text{ kN}$$

From Eq. (10.56) the shear yield capacity of the beam is

$$V_{wy} = \frac{0.6 \times 248.22 \times 6.35 \times 406.4}{1000} = 384.3 \text{ kN} < 597.79 \text{ kN}$$

Since the shear yield capacity of the web within the joint is not sufficient, using Eq. (10.56), increase the web thickness to

$$t_w = \frac{597.79}{0.6 \times 248.22 \times 406.4/1000} = 9.88 \text{ mm}$$

$$t_w = 10 \text{ mm}$$

Step 8: The shear force carried by concrete within the joint between the beam flanges is assumed to be the horizontal component, C_{st}

$$V_c = C_{st} \cos \theta$$
$$V_c = 390.67 \cos (42.2) = 289.41 \text{ kN}$$

For the interior joint the shear capacity is

$$V_u = \phi(20)\sqrt{f_c}(2b_f)(d_c)$$

$$V_u = 0.85(20)6.895 \times 10^{-3} \times 100 \times \frac{(2 \times 139.7)(406.4)}{1000}$$

$$= 1330.95 \text{ kN} > 289.41 \text{ kN}$$

10.6 References

ACI-ASCE Committee 352, "Recommendations for Design of Beam-Column Joints in Monolithic Reinforced Concrete Structures," *ACI Journal,* May–June, 1985, pp. 266–283.

AISC, "Seismic Provisions for Structural Steel Buildings," American Institute of Steel Construction, Chicago, IL, 1992.

AISC, *Manual of Steel Construction,* "Load and Resistance Factor Design," American Institute of Steel Construction, Chicago, IL, 1994.

Aktan, A. E., and Bertero, V. V., "The Seismic Resistant Design of R/C Coupled Structural Walls," Report No. UCB/EERC-81/07, Earthquake Engineering Research Center, University of California, Berkeley, June 1981.

Alostaz, Y. M., and Schneider, S. P., "Connections to Concrete-Filled Steel Tubes," A Report on Research Sponsored by the National Science Foundation, University of Illinois at Urbana–Champaign, October 1996.

ASCE Task Committee on Design Criteria for Composite Structures in Steel and Concrete, "Guidelines for Design of Joints between Steel Beams and Reinforced Concrete Columns," *Journal of Structural Engineering,* ASCE, vol. 120, no. 8, August, 1994, pp. 2330–2357.

Azizinamini, A., and Parakash, B., "Tentative Design Guidelines for a New Steel Beam Connection Detail to Composite Tube Columns," *AISC Engineering Journal,* vol. 22, 3d quarter, 1993, pp. 108–115.

Azizinamini, A., Yerrapalli, S., and Saadeghvaziri, M. A., "Design of Through Beam Connection Detail for Circular Composite Columns," *Engineering Structures,* vol. 17, no. 3, 1995, pp. 209–213.

Barney, G. B., et al., "Earthquake Resistant Structural Walls—Test of Coupling Beams," Report to NSF, submitted by Portland Cement Association, Research and Development, Skokie, IL, January, 1978.

Deierlein, G. G., Sheikh, T. M., Yura, J. A., and Jirsa, J. O., "Beam-Column Moment Connections for Composite Frames: Part 2," *Journal of Structural Engineering,* ASCE, vol. 115, no. 11, November, 1989, pp. 2877–2896.

Gong B., and Shahrooz B. M., "Seismic Behavior and Design of Composite Coupled Wall Systems," Report No. UC-CII 98/01, Cincinnati Infrastructure Institute, Cincinnati, OH, 1998.

Hawkins, N. M., Mitchell, D., and Roeder, C. W., "Moment Resisting Connections for Mixed Construction," *Engineering Journal,* American Institute of Steel Construction, 1st quarter, vol. 17, no. 1, 1980, pp. 1–10.

Leon, R., Forcier, G. P., Roeder, C. W., and Preece, F. R., "Cyclic Performance of Riveted Connections," *Proceedings of the ASCE Structures Congress,* 1994, pp. 1490–1495.

Marcakis, K., and Mitchell, D., "Precast Concrete Connections With Embedded Steel Members," *Prestressed Concrete Institute Journal,* vol. 25, no. 4, 1980, pp. 88–116.

Mattock, A. H., and Gaafar, G. H., "Strength of Embedded Steel Sections as Brackets," *ACI Journal,* vol. 79, no. 2, 1982, pp. 83–93.

NEHRP (National Earthquake Hazards Reduction Program), *Recommended Provisions for the Development of Seismic Regulations for New Buildings,* 1994 ed., Building Seismic Safety Council, Washington, DC, 1994.

Paulay, T., "Earthquake Resisting Shear Walls—New Zealand Design Trends," *ACI Journal,* vol. 77, no. 3, May–June 1980, pp. 144–152.

Paulay, T., "The Design of Ductile Reinforced Concrete Structural Walls for Earthquake Resistance," *Earthquake Spectra,* vol. 2, no. 4, 1986, pp. 783–823.

Paulay, T., and Santhakumar, A. R., "Ductile Behavior of Coupled Shear Walls,"

Journal of the Structural Division, ASCE, vol. 102, no. ST1, January, 1976, pp. 93–108.

Paulay, T., and Binney, J. R., "Diagonally Reinforced Concrete Beams for Shear Walls," *Shear in Reinforced Concrete*, Publication SP 42, American Concrete Institute, Detroit, MI, 1975, pp. 579–598.

PCI Design Handbook—Precast and Prestressed Concrete, Precast Concrete Institute, Chicago, IL, 1992.

Ricles, J., Lu, L., and Peng, S., "Split-Tee Seismic Connections for CFT Column-WF Beam MRFs," *Proceedings of Structures Congress XV*, vol. 2, ASCE, 1997, pp. 959–963.

Roeder, C. W., "CFT Research in the U.S.–Japan Program," *Proceedings of Structures Congress XV*, vol. 2, ASCE, 1997, pp. 1101–1105.

Roeder, C. W., and Hawkins, N. M, "Connections between Steel Frames and Concrete Walls," *Engineering Journal*, AISC, vol. 18, no. 1, 1981, pp. 22–29.

Shahrooz, B. M., Remmetter, M. A., and Qin, F., "Seismic Response of Composite Coupled Walls," *Composite Construction in Steel and Concrete II*, ASCE, 1992, pp. 429–441.

Shahrooz, B. M., Remmetter, M. A., and Qin, F., "Seismic Design and Performance of Composite Coupled Walls," *Journal of the Structural Division*, ASCE, vol. 119, no. 11, November 1993, pp. 3291–3309.

Sheikh, T. M., Deierlein, G. G., Yura, J. A., and Jirsa, J. O., "Beam-Column Moment Connections for Composite Frames: Part 1," *Journal of Structural Engineering*, ASCE, vol. 115, no. 11, November 1989, pp. 2858–2876.

UBC, *Uniform Building Code*, vol. 2, *Structural Engineering Design Provisions*, International Conference of Buildings Officials, Whitier, CA, May 1994.

U.S.–Japan Cooperative Research Program—Phase 5 Composite and Hybrid Structures, Report No. UMCEE 92-29, Department of Civil and Environmental Engineering, University of Michigan, Ann Arbor, MI, November 1992.

Wang, M. T., "The Behavior of Steel Structure to Shear Wall Connections," Master of Science submitted to the University of Washington, Seattle, 1979, 120 pp.

10.7   Notations (for Sec. 10.3)

a	shear span taken as one-half of coupling beam, in
A_b	cross-sectional area of stud
A_v	total area of web reinforcement in concrete encasement around steel coupling beam
A_{vd}	total area of reinforcement in each group of diagonal bars
b	width of embedded steel plate
b_f	steel coupling beam flange width
b_w	web width of encasing element around steel coupling beam
d	distance from the extreme compression fiber to centroid of longitudinal tension reinforcement in the encasing element around steel coupling beam
d'	distance from the extreme compression fiber to centroid of longitudinal compression reinforcement
d_e	distance from the stud axis to the edge of wall
d_{e1} and d_{e2}	distance from the axis of extreme studs to the edge of wall
d_h	diameter of stud head
e	eccentricity of gravity shear measured from centerline of bolts to face of wall

f_c'	concrete compressive strength, psi [for Eq. (10.2) this is for the concrete used in wall piers]
f_y	yield stress of reinforcing bars or studs
F_y	yield strength of web
h	steel coupling beam depth/distance from the center of resistance of tension studs to edge of the embedded steel plate
H	overall depth of coupling beam
I_g	gross concrete section moment of inertia
k_d	depth of concrete compression block
l_e	embedment length of studs (stud length—thickness of head)
l_n	clear span of coupling beam measured from face of wall piers
L	distance between centerlines of wall piers
L_b	clear distance between wall piers
L_e	embedment length of steel coupling beams inside wall piers
M_u	ultimate coupling beam moment
n	number of studs
P_c	tensile strength of studs based on concrete
P_s	tensile strength of studs based on steel
R	code-specified response modification factor
s	spacing of web reinforcement in encasing element around steel coupling beam
t_f	flange thickness
t_w	web thickness
t_{wall}	wall thickness, in
$T_{capacity}$	tensile capacity of studs
T_u	calculated tensile force in collector/outrigger beam
V_P	plastic shear capacity
V_s	shear strength of stud governed by steel
V_u	ultimate coupling beam shear force/calculated gravity shear in outrigger or collector beam
x	horizontal distance between outermost studs
y	vertical distance between studs
α	angle between diagonal reinforcement and longitudinal steel
β_1	ratio of the average concrete compressive strength to the maximum stress as defined by ACI Building Code
γ_p	coupling beam plastic shear angle
θ_e	elastic interstory drift angle
θ_p	plastic interstory drift angle
ϕ	strength reduction factor taken as 0.85

Structural Shapes— Dimensions and General Information*

Structural Shape Size Groupings

Structural shape sizes per tensile group classifications

Group 1	Group 2	Group 3	Group 4	Group 5
		W shapes		
W 24 × 55	W 44 × 198	W 44 × 248	W 40 × 362 to W 40 × 655	W 36 × 848
W 24 × 62	W 44 × 224	W 44 × 285	W 36 × 328 to W 36 × 798	W 14 × 605 to W 14 × 730
W 21 × 44 to W 21 × 57	W 40 × 149 to W 40 × 268	W 40 × 277 to W 40 × 328	W 33 × 318 to W 33 × 619	
W 18 × 35 to W 18 × 71	W 36 × 135 to W 36 × 210	W 36 × 230 to W 36 × 300	W 30 × 292 to W 30 × 581	
W 16 × 26 to W 16 × 57	W 33 × 118 to W 33 × 152	W 33 × 201 to W 33 × 291	W 27 × 281 to W 27 × 539	
W 14 × 22 to W 14 × 53	W 30 × 90 to W 30 × 211	W 30 × 235 to W 30 × 261	W 24 × 250 to W 24 × 492	
W 12 × 14 to W 12 × 58	W 27 × 84 to W 27 × 178	W 27 × 194 to W 27 × 258	W 21 × 248 to W 21 × 402	
W 10 × 12 to W 10 × 45	W 24 × 68 to W 24 × 162	W 24 × 176 to W 24 × 229	W 18 × 211 to W 18 × 311	
W 8 × 10 to W 8 × 48	W 21 × 62 to W 21 × 147	W 21 × 166 to W 21 × 223	W 14 × 233 to W 14 × 550	
W 6 × 9 to W 6 × 25	W 18 × 76 to W 18 × 143	W 18 × 158 to W 18 × 192	W 12 × 21 to W 12 × 336	

*From Mouser, *Welding Codes, Standards, and Specifications*, pp. 309–374.

Structural Shape Size Groupings (*Continued*)
Structural shape sizes per tensile group classifications

Group 1	Group 2	Group 3	Group 4	Group 5
		W shapes		
W5 × 16	W16 × 67 to	W14 × 145 to		
W5 × 19	W16 × 100	W14 × 211		
W4 × 13	W14 × 61 to	W12 × 120 to		
	W14 × 132	W12 × 190		
	W12 × 65 to			
	W12 × 106			
	W10 × 49 to			
	W10 × 112			
	W8 × 58			
	W8 × 67			
		M shapes		
To 37.7 lb/ft				
		S shapes		
To 35 lb/ft				
		HP shapes		
	To 102 lb/ft	>102 lb/ft		
		Standard channel		
To 20.7 lb/ft	>20.7 lb/ft			
		Miscellaneous channel		
To 28.5 lb/ft	>28.5 lb/ft			
		Angle iron		
To ½ in	>½ to ¾ in	>¾ in		

Wide-Flange Dimensions, in

Designation	Depth	Web thickness	Flange Width	Flange Thickness
		W-shape dimensions		
W 44 × 285	44	1	$11\frac{3}{4}$	$1\frac{3}{4}$
× 248	$43\frac{5}{8}$	$\frac{7}{8}$	$11\frac{3}{4}$	$1\frac{9}{16}$
× 224	$43\frac{1}{4}$	$\frac{13}{16}$	$11\frac{3}{4}$	$1\frac{7}{16}$
× 198	$42\frac{7}{8}$	$\frac{11}{16}$	$11\frac{3}{4}$	$1\frac{1}{4}$
W 40 × 328	40	$\frac{15}{16}$	$17\frac{7}{8}$	$1\frac{3}{4}$
× 298	$39\frac{3}{4}$	$\frac{13}{16}$	$17\frac{7}{8}$	$1\frac{9}{16}$
× 268	$39\frac{3}{8}$	$\frac{3}{4}$	$17\frac{3}{4}$	$1\frac{7}{16}$
× 244	39	$\frac{11}{16}$	$17\frac{3}{4}$	$1\frac{1}{4}$
× 221	$38\frac{5}{8}$	$\frac{11}{16}$	$17\frac{3}{4}$	$1\frac{1}{16}$
× 192	$38\frac{1}{4}$	$\frac{11}{16}$	$17\frac{3}{4}$	$\frac{13}{16}$
W 40 × 655	$43\frac{5}{8}$	2	$16\frac{7}{8}$	$3\frac{9}{16}$
× 593	43	$1\frac{13}{16}$	$16\frac{3}{4}$	$3\frac{1}{5}$
× 531	$42\frac{3}{8}$	$1\frac{5}{8}$	$16\frac{1}{2}$	$2\frac{15}{16}$
× 480	$41\frac{3}{4}$	$1\frac{7}{16}$	$16\frac{3}{8}$	$2\frac{5}{8}$
× 436	$41\frac{3}{8}$	$1\frac{5}{16}$	$16\frac{1}{4}$	$2\frac{3}{8}$
× 397	41	$1\frac{1}{4}$	$16\frac{1}{8}$	$2\frac{3}{16}$
× 362	$40\frac{1}{2}$	$1\frac{1}{8}$	16	2
× 324	$40\frac{1}{8}$	1	$15\frac{7}{8}$	$\frac{13}{16}$
× 297	$39\frac{7}{8}$	$\frac{15}{16}$	$15\frac{7}{8}$	$1\frac{5}{8}$
× 277	$39\frac{3}{4}$	$\frac{13}{16}$	$15\frac{7}{8}$	$1\frac{9}{16}$
× 249	$39\frac{3}{8}$	$\frac{3}{4}$	$15\frac{3}{4}$	$1\frac{7}{16}$
× 215	39	$\frac{5}{8}$	$15\frac{3}{4}$	$1\frac{1}{4}$
× 199	$38\frac{5}{8}$	$\frac{5}{8}$	$15\frac{3}{4}$	$1\frac{1}{16}$
W 40 × 183	39	$\frac{5}{8}$	$11\frac{3}{4}$	$1\frac{1}{4}$
× 167	$38\frac{5}{8}$	$\frac{5}{8}$	$11\frac{3}{4}$	1
× 149	$38\frac{1}{4}$	$\frac{5}{8}$	$11\frac{3}{4}$	$\frac{13}{16}$
W 36 × 848	$42\frac{1}{2}$	$2\frac{1}{2}$	$18\frac{1}{8}$	$4\frac{1}{2}$
× 798	42	$2\frac{3}{8}$	18	$4\frac{5}{16}$
× 720	$41\frac{1}{4}$	$2\frac{3}{16}$	$17\frac{3}{5}$	$3\frac{7}{8}$
× 650	$40\frac{1}{2}$	2	$17\frac{5}{8}$	$3\frac{9}{16}$
× 588	$39\frac{7}{8}$	$1\frac{13}{16}$	$17\frac{3}{8}$	$3\frac{1}{4}$
× 527	$39\frac{1}{4}$	$1\frac{5}{8}$	$17\frac{1}{4}$	$2\frac{15}{16}$
× 485	$38\frac{3}{4}$	$1\frac{1}{2}$	$17\frac{1}{8}$	$2\frac{11}{16}$
× 439	$38\frac{1}{4}$	$1\frac{3}{8}$	17	$2\frac{7}{16}$
× 393	$37\frac{3}{4}$	$1\frac{1}{4}$	$16\frac{7}{8}$	$2\frac{3}{16}$
× 359	$37\frac{3}{8}$	$1\frac{1}{8}$	$16\frac{3}{4}$	2
× 328	$37\frac{1}{8}$	1	$16\frac{5}{8}$	$1\frac{7}{8}$
× 300	$36\frac{3}{4}$	$\frac{15}{16}$	$16\frac{5}{8}$	$1\frac{11}{16}$
× 280	$36\frac{1}{2}$	$\frac{7}{8}$	$16\frac{5}{8}$	$1\frac{9}{16}$
× 260	$36\frac{1}{4}$	$\frac{13}{16}$	$16\frac{1}{2}$	$1\frac{7}{16}$
× 245	$36\frac{1}{8}$	$\frac{13}{16}$	$16\frac{1}{2}$	$1\frac{3}{8}$
× 230	$35\frac{7}{8}$	$\frac{3}{4}$	$16\frac{1}{2}$	$1\frac{1}{4}$
W 36 × 256	$37\frac{3}{8}$	1	$12\frac{3}{16}$	$1\frac{3}{4}$
× 232	$37\frac{1}{8}$	$\frac{7}{8}$	$12\frac{1}{8}$	$1\frac{9}{16}$
× 210	$36\frac{3}{4}$	$\frac{13}{16}$	$12\frac{3}{16}$	$1\frac{3}{8}$
× 194	$36\frac{1}{2}$	$\frac{3}{4}$	$12\frac{1}{8}$	$1\frac{1}{4}$
× 182	$36\frac{3}{8}$	$\frac{3}{4}$	$12\frac{1}{16}$	$1\frac{3}{16}$
× 170	$36\frac{1}{8}$	$\frac{11}{16}$	12	$1\frac{1}{8}$
× 160	36	$\frac{5}{8}$	12	1
× 150	$35\frac{7}{8}$	$\frac{5}{8}$	12	$\frac{15}{16}$
× 135	$35\frac{1}{2}$	$\frac{5}{8}$	$11\frac{15}{16}$	$\frac{13}{16}$

Wide-Flange Dimensions, in (Continued)

Designation	Depth	Web thickness	Flange Width	Flange Thickness
		W-shape dimensions		
W 33 × 619	$38\frac{1}{2}$	2	$16\frac{7}{8}$	$3\frac{9}{16}$
× 567	$37\frac{7}{8}$	$1\frac{13}{16}$	$16\frac{3}{4}$	$3\frac{1}{4}$
× 515	$37\frac{3}{8}$	$1\frac{5}{8}$	$16\frac{5}{8}$	3
× 468	$36\frac{3}{4}$	$1\frac{1}{2}$	$16\frac{1}{2}$	$2\frac{3}{4}$
× 424	$36\frac{3}{8}$	$1\frac{3}{8}$	$16\frac{3}{8}$	$2\frac{1}{2}$
× 387	36	$1\frac{1}{4}$	$16\frac{1}{4}$	$2\frac{1}{4}$
× 354	$35\frac{1}{2}$	$1\frac{13}{16}$	$16\frac{1}{8}$	$2\frac{1}{16}$
× 318	$35\frac{1}{8}$	$1\frac{1}{16}$	16	$1\frac{7}{8}$
× 291	$34\frac{7}{8}$	1	$15\frac{7}{8}$	$1\frac{3}{4}$
× 263	$34\frac{1}{2}$	$\frac{7}{8}$	$15\frac{3}{4}$	$1\frac{9}{16}$
× 241	$34\frac{1}{8}$	$\frac{13}{16}$	$15\frac{7}{8}$	$1\frac{3}{8}$
× 221	$33\frac{7}{8}$	$\frac{3}{4}$	$15\frac{3}{4}$	$1\frac{1}{4}$
× 201	$33\frac{3}{8}$	$\frac{11}{16}$	$15\frac{3}{4}$	$1\frac{1}{8}$
W 33 × 169	$33\frac{3}{8}$	$\frac{11}{16}$	$11\frac{1}{2}$	$1\frac{1}{4}$
× 152	$33\frac{1}{2}$	$\frac{5}{8}$	$11\frac{5}{8}$	$1\frac{1}{16}$
× 141	$33\frac{1}{4}$	$\frac{5}{8}$	$11\frac{1}{2}$	$\frac{15}{16}$
× 130	$33\frac{1}{8}$	$\frac{9}{16}$	$11\frac{1}{2}$	$\frac{7}{8}$
× 118	$32\frac{7}{8}$	$\frac{9}{16}$	$11\frac{1}{2}$	$\frac{3}{4}$
W 30 × 581	$35\frac{3}{8}$	2	$16\frac{1}{4}$	$3\frac{9}{16}$
× 526	$34\frac{3}{4}$	$1\frac{13}{16}$	16	$3\frac{1}{4}$
× 477	$34\frac{1}{4}$	$1\frac{5}{8}$	$15\frac{7}{8}$	3
× 433	$33\frac{5}{8}$	$1\frac{1}{2}$	$15\frac{3}{4}$	$2\frac{11}{16}$
× 391	$33\frac{1}{4}$	$1\frac{3}{8}$	$5\frac{5}{8}$	$2\frac{7}{8}$
× 357	$32\frac{3}{4}$	$1\frac{1}{4}$	$15\frac{1}{2}$	$2\frac{1}{4}$
× 326	$32\frac{3}{8}$	$1\frac{1}{8}$	$15\frac{3}{8}$	$2\frac{1}{16}$
× 292	32	1	$15\frac{1}{4}$	$1\frac{7}{8}$
× 261	$31\frac{5}{8}$	$\frac{15}{16}$	$15\frac{1}{8}$	$1\frac{5}{8}$
× 235	$31\frac{1}{4}$	$\frac{13}{16}$	15	$1\frac{1}{2}$
× 211	31	$\frac{3}{4}$	$15\frac{1}{8}$	$1\frac{5}{16}$
× 191	$30\frac{5}{8}$	$\frac{11}{16}$	15	$1\frac{3}{16}$
× 173	$30\frac{1}{2}$	$\frac{5}{8}$	15	$1\frac{1}{16}$
W 30 × 148	$30\frac{5}{8}$	$\frac{5}{8}$	$10\frac{1}{2}$	$1\frac{3}{16}$
× 132	$30\frac{1}{4}$	$\frac{5}{8}$	$10\frac{1}{2}$	1
× 124	$30\frac{1}{8}$	$\frac{9}{16}$	$10\frac{1}{2}$	$\frac{15}{16}$
× 116	30	$\frac{9}{16}$	$10\frac{1}{2}$	$\frac{7}{8}$
× 108	$29\frac{7}{8}$	$\frac{9}{16}$	$10\frac{1}{2}$	$\frac{3}{4}$
× 99	$29\frac{5}{8}$	$\frac{1}{2}$	$10\frac{1}{2}$	$\frac{11}{16}$
× 90	$29\frac{1}{2}$	$\frac{1}{2}$	$10\frac{3}{8}$	$\frac{9}{16}$

Designation	Depth	Web thickness	Flange Width	Flange Thickness
		W-shape dimensions		
W 27 × 539	$32\frac{1}{2}$	2	$15\frac{1}{4}$	$3\frac{9}{16}$
× 494	32	$1\frac{13}{16}$	$15\frac{1}{8}$	$3\frac{1}{4}$
× 448	$31\frac{3}{8}$	$1\frac{5}{8}$	15	3
× 407	$30\frac{7}{8}$	$1\frac{1}{2}$	$14\frac{3}{4}$	$2\frac{3}{4}$
× 368	$30\frac{3}{8}$	$1\frac{3}{8}$	$14\frac{5}{8}$	$2\frac{1}{2}$
× 336	30	$1\frac{1}{4}$	$14\frac{1}{2}$	$2\frac{1}{4}$
× 307	$29\frac{5}{8}$	$1\frac{3}{16}$	$14\frac{1}{2}$	$2\frac{1}{16}$
× 281	$29\frac{1}{4}$	$1\frac{1}{16}$	$14\frac{3}{8}$	$1\frac{15}{16}$
× 258	29	1	$14\frac{1}{4}$	$1\frac{3}{4}$
× 235	$28\frac{5}{8}$	$\frac{15}{16}$	$14\frac{1}{4}$	$1\frac{5}{8}$
× 217	$28\frac{3}{8}$	$\frac{13}{16}$	$14\frac{1}{8}$	$1\frac{1}{2}$
× 194	$28\frac{1}{8}$	$\frac{3}{4}$	14	$1\frac{5}{16}$
× 178	$27\frac{3}{4}$	$\frac{3}{4}$	$14\frac{1}{8}$	$1\frac{3}{16}$
× 161	$27\frac{5}{8}$	$\frac{11}{16}$	14	$1\frac{1}{16}$
× 146	$27\frac{3}{8}$	$\frac{5}{8}$	14	1
W 27 × 129	$27\frac{5}{8}$	$\frac{5}{8}$	10	$1\frac{1}{8}$
× 114	$27\frac{1}{4}$	$\frac{9}{16}$	$10\frac{1}{8}$	$\frac{15}{16}$
× 102	$27\frac{1}{8}$	$\frac{1}{2}$	10	$\frac{13}{16}$
× 94	$26\frac{7}{8}$	$\frac{1}{2}$	10	$\frac{3}{4}$
× 84	$26\frac{3}{4}$	$\frac{7}{16}$	10	$\frac{5}{8}$
W 24 × 492	$29\frac{5}{8}$	2	$14\frac{1}{8}$	$3\frac{9}{16}$
× 450	$29\frac{1}{8}$	$1\frac{13}{16}$	14	$3\frac{1}{4}$
× 408	$28\frac{1}{2}$	$1\frac{5}{8}$	$13\frac{3}{4}$	3
× 370	28	$1\frac{1}{2}$	$13\frac{5}{8}$	$2\frac{3}{4}$
× 335	$27\frac{1}{2}$	$1\frac{3}{8}$	$13\frac{1}{2}$	$2\frac{1}{2}$
× 306	$27\frac{1}{8}$	$1\frac{1}{4}$	$13\frac{3}{8}$	$2\frac{1}{4}$
× 279	$26\frac{3}{4}$	$1\frac{3}{16}$	$13\frac{1}{4}$	$2\frac{1}{16}$
× 250	$26\frac{3}{8}$	$1\frac{1}{16}$	$13\frac{1}{8}$	$1\frac{7}{8}$
× 229	26	1	$13\frac{1}{8}$	$1\frac{3}{4}$
× 207	$25\frac{3}{4}$	$\frac{7}{8}$	13	$1\frac{9}{16}$
× 192	$25\frac{1}{2}$	$\frac{13}{16}$	13	$1\frac{7}{16}$
× 176	$25\frac{1}{4}$	$\frac{3}{4}$	$12\frac{7}{8}$	$1\frac{5}{16}$
× 162	25	$\frac{11}{16}$	13	$1\frac{1}{4}$
× 146	$24\frac{3}{4}$	$\frac{5}{8}$	$12\frac{7}{8}$	$1\frac{1}{16}$
× 131	$24\frac{1}{2}$	$\frac{5}{8}$	$12\frac{7}{8}$	$\frac{15}{16}$
× 117	$24\frac{1}{4}$	$\frac{9}{16}$	$12\frac{3}{4}$	$\frac{7}{8}$
× 104	24	$\frac{1}{2}$	$12\frac{3}{4}$	$\frac{3}{4}$

Wide-Flange Dimensions, in (*Continued*)

Designation	Depth	Web thickness	Width	Thickness
			Flange	
		W-shape dimensions		
W 24 × 103	$24\frac{1}{2}$	$\frac{9}{16}$	9	1
× 94	$24\frac{1}{4}$	$\frac{1}{2}$	$9\frac{1}{8}$	$\frac{7}{8}$
× 84	$24\frac{1}{8}$	$\frac{1}{2}$	9	$\frac{3}{4}$
× 76	$23\frac{7}{8}$	$\frac{7}{16}$	9	$\frac{11}{16}$
× 68	$23\frac{3}{4}$	$\frac{7}{16}$	9	$\frac{9}{16}$
W 24 × 62	$23\frac{3}{4}$	$\frac{7}{16}$	7	$\frac{9}{16}$
× 55	$23\frac{5}{8}$	$\frac{3}{8}$	7	$\frac{1}{2}$
W 21 × 402	26	$1\frac{3}{4}$	$13\frac{3}{8}$	$3\frac{1}{8}$
× 364	$25\frac{1}{2}$	$1\frac{9}{16}$	$13\frac{3}{4}$	$2\frac{7}{8}$
× 333	25	$1\frac{7}{16}$	$13\frac{1}{8}$	$2\frac{5}{8}$
× 300	$24\frac{1}{2}$	$1\frac{5}{16}$	13	$2\frac{3}{8}$
× 275	$24\frac{1}{8}$	$1\frac{1}{4}$	$12\frac{7}{8}$	$2\frac{3}{16}$
× 248	$23\frac{3}{4}$	$1\frac{1}{8}$	$12\frac{3}{4}$	2
× 223	$23\frac{3}{8}$	1	$12\frac{5}{8}$	$1\frac{13}{16}$
× 201	23	$\frac{15}{16}$	$12\frac{5}{8}$	$1\frac{5}{8}$
× 182	$22\frac{3}{4}$	$\frac{13}{16}$	$12\frac{1}{2}$	$1\frac{1}{2}$
× 166	$22\frac{1}{2}$	$\frac{3}{4}$	$12\frac{3}{8}$	$1\frac{3}{8}$
× 147	22	$\frac{3}{4}$	$12\frac{1}{2}$	$1\frac{1}{8}$
× 132	$21\frac{7}{8}$	$\frac{5}{8}$	$12\frac{1}{2}$	$1\frac{1}{16}$
× 122	$21\frac{5}{8}$	$\frac{5}{8}$	$12\frac{3}{8}$	$\frac{15}{16}$
× 111	$21\frac{1}{2}$	$\frac{9}{16}$	$12\frac{3}{8}$	$\frac{7}{8}$
× 101	$21\frac{3}{8}$	$\frac{1}{2}$	$12\frac{1}{4}$	$\frac{13}{16}$
W 21 × 93	$21\frac{5}{8}$	$\frac{9}{16}$	$8\frac{3}{8}$	$\frac{15}{16}$
× 83	$21\frac{3}{8}$	$\frac{1}{2}$	$8\frac{3}{8}$	$\frac{13}{16}$
× 73	$21\frac{1}{4}$	$\frac{7}{16}$	$8\frac{1}{4}$	$\frac{3}{4}$
× 68	$21\frac{1}{8}$	$\frac{7}{16}$	$8\frac{1}{4}$	$\frac{11}{16}$
× 62	21	$\frac{3}{8}$	$8\frac{1}{4}$	$\frac{5}{8}$
W 21 × 57	21	$\frac{3}{8}$	$6\frac{1}{2}$	$\frac{5}{8}$
× 50	$20\frac{7}{8}$	$\frac{3}{8}$	$6\frac{1}{2}$	$\frac{9}{16}$
× 44	$20\frac{5}{8}$	$\frac{3}{8}$	$6\frac{1}{2}$	$\frac{7}{16}$
W 18 × 311	$22\frac{3}{8}$	$1\frac{1}{2}$	12	$2\frac{3}{4}$
× 283	$21\frac{7}{8}$	$1\frac{3}{8}$	$11\frac{7}{8}$	$2\frac{1}{2}$
× 258	$21\frac{1}{2}$	$1\frac{1}{4}$	$11\frac{3}{4}$	$2\frac{5}{16}$
× 234	21	$1\frac{3}{16}$	$11\frac{5}{8}$	$2\frac{1}{8}$
× 211	$20\frac{5}{8}$	$1\frac{1}{16}$	$11\frac{1}{2}$	$1\frac{15}{16}$
× 192	$20\frac{3}{8}$	1	$11\frac{1}{2}$	$1\frac{3}{4}$
× 175	20	$\frac{7}{8}$	$11\frac{3}{8}$	$1\frac{9}{16}$
× 158	$10\frac{3}{4}$	$\frac{13}{16}$	$11\frac{1}{4}$	$1\frac{7}{16}$
× 143	$19\frac{1}{2}$	$\frac{3}{4}$	$11\frac{1}{4}$	$1\frac{5}{16}$
× 130	$19\frac{1}{4}$	$\frac{11}{16}$	$11\frac{1}{8}$	$1\frac{3}{16}$
W 18 × 119	19	$\frac{5}{8}$	$11\frac{1}{4}$	$1\frac{1}{16}$
× 106	$18\frac{3}{4}$	$\frac{9}{16}$	$11\frac{1}{4}$	$\frac{15}{16}$
× 97	$18\frac{5}{8}$	$\frac{9}{16}$	$11\frac{1}{8}$	$\frac{7}{8}$
× 86	$18\frac{3}{8}$	$\frac{1}{2}$	$11\frac{1}{8}$	$\frac{3}{4}$
× 76	$18\frac{1}{4}$	$\frac{7}{16}$	11	$\frac{11}{16}$

Designation	Depth	Web thickness	Flange	
			Width	Thickness
W-shape dimensions				
W 18 × 71	$18\frac{1}{2}$	$\frac{1}{2}$	$7\frac{5}{8}$	$\frac{13}{16}$
× 65	$18\frac{3}{8}$	$\frac{7}{16}$	$7\frac{5}{8}$	$\frac{3}{4}$
× 60	$18\frac{1}{4}$	$\frac{7}{16}$	$7\frac{1}{2}$	$\frac{11}{16}$
× 55	$18\frac{1}{8}$	$\frac{3}{8}$	$7\frac{1}{2}$	$\frac{5}{8}$
× 50	18	$\frac{3}{8}$	$7\frac{1}{2}$	$\frac{9}{16}$
W 18 × 46	18	$\frac{3}{8}$	6	$\frac{5}{8}$
× 40	$17\frac{7}{8}$	$\frac{5}{16}$	6	$\frac{1}{2}$
× 35	$17\frac{3}{4}$	$\frac{5}{16}$	6	$\frac{7}{16}$
W 16 × 100	17	$\frac{9}{16}$	$10\frac{3}{8}$	1
× 89	$16\frac{3}{4}$	$\frac{1}{2}$	$10\frac{3}{8}$	$\frac{7}{8}$
× 77	$16\frac{1}{2}$	$\frac{7}{16}$	$10\frac{1}{4}$	$\frac{3}{4}$
× 67	$16\frac{3}{8}$	$\frac{3}{8}$	$10\frac{1}{4}$	$\frac{11}{16}$
W 16 × 57	$16\frac{3}{8}$	$\frac{7}{16}$	$7\frac{1}{8}$	$\frac{11}{16}$
× 50	$16\frac{1}{4}$	$\frac{3}{8}$	$7\frac{1}{8}$	$\frac{5}{8}$
× 45	$16\frac{1}{8}$	$\frac{3}{8}$	7	$\frac{9}{16}$
× 40	16	$\frac{5}{16}$	7	$\frac{1}{2}$
× 36	$15\frac{7}{8}$	$\frac{5}{16}$	7	$\frac{7}{16}$
W 16 × 31	$15\frac{7}{8}$	$\frac{1}{4}$	$5\frac{1}{2}$	$\frac{7}{16}$
× 26	$15\frac{3}{4}$	$\frac{1}{4}$	$5\frac{1}{2}$	$\frac{3}{8}$
W 14 × 730	$22\frac{3}{8}$	$3\frac{1}{16}$	$17\frac{7}{8}$	$4\frac{15}{16}$
× 665	$21\frac{5}{8}$	$2\frac{13}{16}$	$17\frac{5}{8}$	$4\frac{1}{2}$
× 605	$20\frac{7}{8}$	$2\frac{5}{8}$	$17\frac{3}{8}$	$4\frac{3}{16}$
× 550	$20\frac{1}{4}$	$2\frac{3}{8}$	$17\frac{1}{4}$	$3\frac{13}{16}$
× 500	$19\frac{6}{8}$	$2\frac{3}{16}$	17	$3\frac{1}{2}$
× 455	19	2	$16\frac{7}{8}$	$3\frac{3}{16}$
W 14 × 426	$18\frac{5}{8}$	$1\frac{7}{8}$	$16\frac{3}{4}$	$3\frac{3}{16}$
× 398	$18\frac{1}{4}$	$1\frac{3}{4}$	$16\frac{5}{8}$	$2\frac{7}{8}$
× 370	$17\frac{7}{8}$	$1\frac{5}{8}$	$16\frac{1}{2}$	$2\frac{11}{16}$
× 342	$17\frac{1}{2}$	$1\frac{9}{16}$	$16\frac{3}{8}$	$2\frac{1}{2}$
× 311	$17\frac{1}{8}$	$1\frac{7}{16}$	$16\frac{1}{4}$	$1\frac{1}{4}$
× 283	$16\frac{3}{4}$	$1\frac{5}{16}$	$16\frac{1}{8}$	$2\frac{1}{16}$
× 257	$16\frac{3}{8}$	$1\frac{3}{16}$	16	$1\frac{7}{8}$
× 233	16	$1\frac{1}{16}$	$15\frac{7}{8}$	$1\frac{3}{4}$
× 211	$15\frac{3}{4}$	1	$15\frac{3}{4}$	$1\frac{9}{16}$
× 193	$15\frac{1}{2}$	$\frac{7}{8}$	$15\frac{3}{4}$	$1\frac{7}{16}$
× 176	$15\frac{1}{4}$	$\frac{13}{16}$	$15\frac{5}{8}$	$1\frac{5}{16}$
× 159	15	$\frac{3}{4}$	$15\frac{5}{8}$	$1\frac{3}{16}$
× 145	$14\frac{3}{4}$	$\frac{11}{16}$	$15\frac{1}{2}$	$1\frac{1}{16}$
W 14 × 132	$14\frac{5}{8}$	$\frac{5}{8}$	$14\frac{3}{4}$	1
× 120	$14\frac{1}{2}$	$\frac{9}{16}$	$14\frac{5}{8}$	$\frac{15}{16}$
× 109	$14\frac{3}{8}$	$\frac{1}{2}$	$14\frac{5}{8}$	$\frac{7}{8}$
× 99	$14\frac{1}{8}$	$\frac{1}{2}$	$14\frac{5}{8}$	$\frac{3}{4}$
× 90	14	$\frac{7}{16}$	$14\frac{1}{2}$	$\frac{11}{16}$
W 14 × 82	$14\frac{1}{4}$	$\frac{1}{2}$	$10\frac{1}{8}$	$\frac{7}{8}$
× 74	$14\frac{1}{8}$	$\frac{7}{16}$	$10\frac{1}{8}$	$\frac{13}{16}$
× 68	14	7	10	$\frac{3}{4}$
× 61	$13\frac{7}{8}$	16	10	$\frac{5}{8}$

Wide-Flange Dimensions, in (*Continued*)

Designation	Depth	Web thickness	Flange Width	Flange Thickness
W-shape dimensions				
W 14 × 53	$13\frac{7}{8}$	$\frac{3}{8}$	8	$\frac{11}{16}$
× 48	$13\frac{3}{4}$	$\frac{3}{8}$	8	$\frac{5}{8}$
× 43	$13\frac{5}{8}$	$\frac{5}{16}$	8	$\frac{1}{2}$
W 14 × 38	$14\frac{1}{8}$	$\frac{5}{16}$	$6\frac{3}{4}$	$\frac{1}{2}$
× 34	14	$\frac{5}{16}$	$6\frac{3}{4}$	$\frac{7}{16}$
× 30	$13\frac{7}{8}$	$\frac{1}{4}$	$6\frac{3}{4}$	$\frac{3}{8}$
W 14 × 26	$13\frac{7}{8}$	$\frac{1}{4}$	5	$\frac{7}{16}$
× 22	$13\frac{3}{4}$	$\frac{1}{4}$	5	$\frac{5}{16}$
W 12 × 336	$16\frac{7}{8}$	$1\frac{3}{4}$	$13\frac{3}{8}$	$2\frac{15}{16}$
× 305	$16\frac{3}{8}$	$1\frac{5}{8}$	$13\frac{1}{4}$	$2\frac{11}{16}$
× 279	$15\frac{7}{8}$	$1\frac{1}{2}$	$13\frac{1}{8}$	$2\frac{1}{2}$
× 252	$15\frac{3}{8}$	$1\frac{3}{8}$	13	$2\frac{1}{4}$
× 230	15	$1\frac{5}{16}$	$12\frac{7}{8}$	$2\frac{1}{16}$
× 210	$14\frac{3}{4}$	$1\frac{3}{16}$	$12\frac{3}{4}$	$1\frac{7}{8}$
× 190	$14\frac{3}{8}$	$1\frac{1}{16}$	$12\frac{5}{8}$	$1\frac{3}{4}$
× 170	14	$\frac{15}{16}$	$12\frac{5}{8}$	$1\frac{9}{16}$
× 152	$13\frac{3}{4}$	$\frac{7}{8}$	$12\frac{1}{2}$	$1\frac{3}{8}$
× 136	$13\frac{3}{8}$	$\frac{13}{16}$	$12\frac{3}{8}$	$1\frac{1}{4}$
× 120	$13\frac{1}{8}$	$\frac{11}{16}$	$12\frac{3}{8}$	$1\frac{1}{8}$
× 106	$12\frac{7}{8}$	$\frac{5}{8}$	$12\frac{1}{4}$	1
× 96	$12\frac{3}{4}$	$\frac{9}{16}$	$12\frac{1}{8}$	$\frac{7}{8}$
× 87	$12\frac{1}{2}$	$\frac{1}{2}$	$12\frac{1}{8}$	$\frac{13}{16}$
× 79	$12\frac{3}{8}$	$\frac{1}{2}$	$12\frac{1}{8}$	$\frac{3}{4}$
× 72	$12\frac{1}{4}$	$\frac{7}{16}$	12	$\frac{11}{16}$
× 65	$12\frac{1}{8}$	$\frac{3}{8}$	12	$\frac{5}{8}$
W 12 × 58	$12\frac{1}{4}$	$\frac{3}{8}$	10	$\frac{5}{8}$
× 53	12	$\frac{3}{8}$	10	$\frac{9}{16}$
W 12 × 50	$12\frac{1}{4}$	$\frac{3}{8}$	$8\frac{1}{8}$	$\frac{5}{8}$
× 45	12	$\frac{5}{16}$	8	$\frac{9}{16}$
× 40	12	$\frac{5}{16}$	8	$\frac{1}{2}$
W 12 × 35	$12\frac{1}{2}$	$\frac{5}{16}$	$6\frac{1}{2}$	$\frac{1}{2}$
× 30	$12\frac{3}{8}$	$\frac{1}{4}$	$6\frac{1}{2}$	$\frac{7}{16}$
× 26	$12\frac{1}{4}$	$\frac{1}{4}$	$6\frac{1}{2}$	$\frac{3}{8}$
W 12 × 22	$12\frac{1}{4}$	$\frac{1}{4}$	4	$\frac{7}{16}$
× 19	$12\frac{1}{8}$	$\frac{1}{4}$	4	$\frac{3}{8}$
× 16	12	$\frac{1}{4}$	4	$\frac{1}{4}$
× 14	$11\frac{7}{8}$	$\frac{3}{16}$	4	$\frac{1}{4}$
W 10 × 112	$11\frac{3}{8}$	$\frac{3}{4}$	$10\frac{3}{8}$	$1\frac{1}{4}$
× 100	$11\frac{1}{8}$	$\frac{11}{16}$	$10\frac{3}{8}$	$1\frac{1}{8}$
× 88	$10\frac{7}{8}$	$\frac{5}{8}$	$10\frac{1}{4}$	1
× 77	$10\frac{5}{8}$	$\frac{1}{2}$	$10\frac{1}{4}$	$\frac{7}{8}$
× 68	$10\frac{3}{8}$	$\frac{1}{2}$	$10\frac{1}{8}$	$\frac{3}{4}$
× 60	$10\frac{1}{4}$	$\frac{7}{16}$	$10\frac{1}{8}$	$\frac{11}{16}$
× 54	$10\frac{1}{8}$	$\frac{3}{8}$	10	$\frac{5}{8}$
× 49	10	$\frac{5}{16}$	10	$\frac{9}{16}$

Designation	Depth	Web thickness	Flange	
			Width	Thickness
W-shape dimensions				
W 10 × 45	$10\frac{1}{8}$	$\frac{3}{8}$	8	$\frac{5}{8}$
× 39	$9\frac{7}{8}$	$\frac{5}{16}$	8	$\frac{1}{2}$
× 33	$9\frac{3}{4}$	$\frac{5}{16}$	8	$\frac{7}{16}$
W 10 × 30	$10\frac{1}{2}$	$\frac{5}{16}$	$5\frac{3}{4}$	$\frac{1}{2}$
× 26	$10\frac{3}{8}$	$\frac{1}{4}$	$5\frac{3}{4}$	$\frac{7}{16}$
× 22	$10\frac{1}{8}$	$\frac{1}{4}$	$5\frac{3}{4}$	$\frac{3}{8}$
W 10 × 19	$10\frac{1}{4}$	$\frac{1}{4}$	4	$\frac{3}{8}$
× 17	$10\frac{1}{8}$	$\frac{1}{4}$	4	$\frac{5}{16}$
× 15	10	$\frac{1}{4}$	4	$\frac{1}{4}$
× 12	$9\frac{7}{8}$	$\frac{3}{16}$	4	$\frac{3}{16}$
W 8 × 67	9	$\frac{9}{16}$	$8\frac{1}{4}$	$\frac{15}{16}$
× 58	$8\frac{3}{4}$	$\frac{1}{2}$	$8\frac{1}{4}$	$\frac{13}{16}$
× 48	$8\frac{1}{2}$	$\frac{3}{8}$	$8\frac{1}{8}$	$\frac{11}{16}$
× 40	$8\frac{1}{4}$	$\frac{3}{8}$	$8\frac{1}{8}$	$\frac{9}{16}$
× 35	$8\frac{1}{8}$	$\frac{5}{16}$	8	$\frac{1}{2}$
× 31	8	$\frac{5}{16}$	8	$\frac{7}{16}$
× 28	8	$\frac{5}{16}$	$6\frac{1}{2}$	$\frac{7}{16}$
× 24	$7\frac{7}{8}$	$\frac{1}{4}$	$6\frac{1}{2}$	$\frac{3}{8}$
× 21	$8\frac{1}{4}$	$\frac{1}{4}$	$5\frac{1}{4}$	$\frac{3}{8}$
× 18	$8\frac{1}{8}$	$\frac{1}{4}$	$5\frac{1}{4}$	$\frac{5}{16}$
× 15	$8\frac{1}{8}$	$\frac{1}{4}$	4	$\frac{5}{16}$
× 13	8	$\frac{1}{4}$	4	$\frac{1}{4}$
× 10	$7\frac{7}{8}$	$\frac{3}{16}$	4	$\frac{3}{16}$
W 6 × 25	$6\frac{3}{8}$	$\frac{5}{16}$	$6\frac{1}{8}$	$\frac{7}{16}$
× 20	$6\frac{1}{4}$	$\frac{1}{4}$	6	$\frac{3}{8}$
× 15	6	$\frac{1}{4}$	6	$\frac{1}{4}$
× 16	$6\frac{1}{4}$	$\frac{1}{4}$	4	$\frac{3}{8}$
× 12	6	$\frac{1}{4}$	4	$\frac{1}{4}$
× 9	$5\frac{7}{8}$	$\frac{3}{16}$	4	$\frac{3}{16}$
W 5 × 19	$5\frac{1}{8}$	$\frac{1}{4}$	5	$\frac{7}{16}$
× 16	5	$\frac{1}{4}$	5	$\frac{3}{8}$
W 4 × 13	$4\frac{1}{8}$	$\frac{1}{4}$	4	$\frac{3}{8}$
M-shape dimensions				
M 14 × 18	14	$\frac{3}{16}$	4	$\frac{1}{4}$
M 12 × 11.8	12	$\frac{3}{16}$	$3\frac{1}{8}$	$\frac{1}{4}$
M 12 × 10.8	12	$\frac{3}{16}$	$3\frac{1}{8}$	$\frac{1}{4}$
M 12 × 10	12	$\frac{3}{16}$	$3\frac{1}{4}$	$\frac{3}{16}$
M 10 × 9	10	$\frac{3}{16}$	$2\frac{3}{4}$	$\frac{3}{16}$
M 10 × 8	10	$\frac{3}{16}$	$2\frac{3}{4}$	$\frac{3}{16}$
M 10 × 7.5	10	$\frac{1}{8}$	$2\frac{3}{4}$	$\frac{3}{16}$
M 8 × 6.5	8	$\frac{1}{8}$	$2\frac{1}{4}$	$\frac{3}{16}$
M 6 × 4.4	6	$\frac{1}{8}$	$1\frac{7}{8}$	$\frac{3}{16}$
M 5 × 18.9	5	$\frac{5}{16}$	5	$\frac{7}{16}$

Wide-Flange Dimensions, in (*Continued*)

Designation	Depth	Web thickness	Flange Width	Flange Thickness
		S-shape dimensions		
S 24 × 121	$24\frac{1}{2}$	$\frac{13}{16}$	8	$1\frac{1}{16}$
× 106	$24\frac{1}{2}$	$\frac{5}{8}$	$7\frac{7}{8}$	$1\frac{1}{16}$
S 24 × 100	24	$\frac{3}{4}$	$7\frac{1}{4}$	$\frac{7}{8}$
× 90	24	$\frac{5}{8}$	$7\frac{1}{8}$	$\frac{7}{8}$
× 80	24	$\frac{1}{2}$	7	$\frac{7}{8}$
S 20 × 96	$20\frac{1}{4}$	$\frac{13}{16}$	$7\frac{1}{4}$	$\frac{15}{16}$
× 86	$20\frac{1}{4}$	$\frac{11}{16}$	7	$\frac{15}{16}$
S 20 × 75	20	$\frac{5}{8}$	$6\frac{3}{8}$	$\frac{13}{16}$
× 66	20	$\frac{1}{2}$	$6\frac{1}{4}$	$\frac{13}{16}$
S 18 × 70	18	$\frac{11}{16}$	$6\frac{1}{4}$	$\frac{11}{16}$
× 54.7	18	$\frac{7}{16}$	6	$\frac{11}{16}$
S 15 × 50	15	$\frac{9}{16}$	$5\frac{5}{8}$	$\frac{5}{8}$
× 42.9	15	$\frac{7}{16}$	$5\frac{1}{2}$	$\frac{5}{8}$
S 12 × 50	12	$\frac{11}{16}$	$5\frac{1}{2}$	$\frac{11}{16}$
× 40.8	12	$\frac{7}{16}$	$5\frac{1}{4}$	$\frac{11}{16}$
S 12 × 35	12	$\frac{7}{16}$	$5\frac{1}{8}$	$\frac{9}{16}$
× 25.4	12	$\frac{3}{8}$	5	$\frac{9}{16}$
S 10 × 35	10	$\frac{5}{8}$	5	$\frac{1}{2}$
× 25.4	10	$\frac{5}{16}$	$4\frac{5}{8}$	$\frac{1}{2}$
S 8 × 23	8	$\frac{7}{16}$	$4\frac{1}{8}$	$\frac{7}{16}$
× 18.4	8	$\frac{1}{4}$	4	$\frac{7}{16}$
S 7 × 20	7	$\frac{7}{16}$	$3\frac{7}{8}$	$\frac{3}{8}$
× 15.3	7	$\frac{1}{4}$	$3\frac{5}{8}$	$\frac{3}{8}$
S 6 × 17.25	6	$\frac{7}{16}$	$3\frac{5}{8}$	$\frac{3}{8}$
× 12.5	6	$\frac{1}{4}$	$3\frac{3}{8}$	$\frac{3}{8}$
S 5 × 14.75	5	$\frac{1}{2}$	$3\frac{1}{4}$	$\frac{5}{16}$
× 10	5	$\frac{3}{16}$	3	$\frac{5}{16}$
S 4 × 9.5	4	$\frac{5}{16}$	$2\frac{3}{4}$	$\frac{5}{16}$
× 7.7	4	$\frac{3}{16}$	$2\frac{5}{8}$	$\frac{5}{16}$
S 3 × 7.5	3	$\frac{3}{8}$	$2\frac{1}{2}$	$\frac{1}{4}$
× 5.7	3	$\frac{3}{16}$	$2\frac{3}{8}$	$\frac{1}{4}$
		HP-shape dimensions		
HP 14 × 117	$14\frac{1}{2}$	$\frac{13}{16}$	$14\frac{7}{8}$	$\frac{13}{16}$
× 102	14	$\frac{11}{16}$	$14\frac{3}{4}$	$\frac{11}{16}$
× 89	$13\frac{7}{8}$	$\frac{5}{8}$	$14\frac{3}{4}$	$\frac{5}{8}$
× 73	$13\frac{5}{8}$	$\frac{1}{2}$	$14\frac{5}{8}$	$\frac{1}{2}$
HP 13 × 100	$13\frac{1}{8}$	$\frac{3}{4}$	$13\frac{1}{4}$	$\frac{3}{4}$
× 87	13	$\frac{11}{16}$	$13\frac{1}{8}$	$\frac{11}{16}$
× 73	$12\frac{3}{4}$	$\frac{9}{16}$	13	$\frac{9}{16}$
× 60	$12\frac{1}{2}$	$\frac{7}{16}$	$12\frac{7}{8}$	$\frac{7}{16}$

Designation	Depth	Web thickness	Flange Width	Flange Thickness
		HP-shape dimensions		
HP 12 × 84	$12\frac{1}{4}$	$\frac{11}{16}$	$12\frac{1}{4}$	$\frac{11}{16}$
× 74	$12\frac{1}{8}$	$\frac{5}{8}$	$12\frac{1}{4}$	$\frac{1}{2}$
× 63	12	$\frac{1}{2}$	$12\frac{1}{8}$	$\frac{1}{2}$
× 53	$11\frac{3}{4}$	$\frac{7}{16}$	12	$\frac{7}{16}$
HP 10 × 57	10	$\frac{9}{16}$	$10\frac{1}{4}$	$\frac{9}{16}$
× 42	$9\frac{3}{4}$	$\frac{7}{16}$	$10\frac{1}{8}$	$\frac{7}{16}$
HP 8 × 36	8	$\frac{7}{16}$	$8\frac{1}{8}$	$\frac{7}{16}$
		Channel iron—American Standard dimensions		
C 15 × 50	15	$\frac{11}{16}$	$3\frac{3}{4}$	$\frac{5}{8}$
× 40	15	$\frac{1}{2}$	$3\frac{1}{2}$	$\frac{5}{8}$
× 33.9	15	$\frac{3}{8}$	$3\frac{3}{8}$	$\frac{5}{8}$
C 12 × 30	12	$\frac{1}{2}$	$3\frac{1}{8}$	$\frac{1}{2}$
× 25	12	$\frac{3}{8}$	3	$\frac{1}{2}$
× 20.7	12	$\frac{5}{16}$	3	$\frac{1}{2}$
C 10 × 30	10	$\frac{11}{16}$	3	$\frac{7}{16}$
× 25	10	$\frac{1}{2}$	$2\frac{7}{8}$	$\frac{7}{16}$
× 20	10	$\frac{3}{8}$	$2\frac{3}{4}$	$\frac{7}{16}$
× 15.3	10	$\frac{1}{4}$	$2\frac{5}{8}$	$\frac{7}{16}$
C 9 × 20	9	$\frac{7}{16}$	$2\frac{5}{8}$	$\frac{7}{16}$
× 15	9	$\frac{5}{16}$	$2\frac{1}{2}$	$\frac{7}{16}$
× 13.4	9	$\frac{1}{4}$	$2\frac{3}{8}$	$\frac{7}{16}$
C 8 × 18.75	8	$\frac{1}{2}$	$2\frac{1}{2}$	$\frac{3}{8}$
× 13.75	8	$\frac{5}{16}$	$2\frac{3}{8}$	$\frac{3}{8}$
× 11.5	8	$\frac{1}{4}$	$2\frac{1}{4}$	$\frac{3}{8}$
C 7 × 14.75	7	$\frac{7}{16}$	$2\frac{1}{4}$	$\frac{3}{8}$
× 12.25	7	$\frac{5}{16}$	$1\frac{1}{4}$	$\frac{3}{8}$
× 9.8	7	$\frac{3}{16}$	$1\frac{1}{8}$	$\frac{3}{8}$
C 6 × 13	6	$\frac{7}{16}$	$2\frac{1}{8}$	$\frac{5}{16}$
× 10.5	6	$\frac{5}{16}$	2	$\frac{5}{16}$
× 8.2	6	$\frac{3}{16}$	$1\frac{7}{8}$	$\frac{5}{16}$
C 5 × 9	5	$\frac{5}{16}$	$1\frac{7}{8}$	$\frac{5}{16}$
× 6.7	5	$\frac{3}{16}$	$1\frac{3}{4}$	$\frac{5}{16}$
C 4 × 7.25	4	$\frac{5}{16}$	$1\frac{3}{4}$	$\frac{5}{16}$
× 5.4	4	$\frac{3}{16}$	$1\frac{5}{8}$	$\frac{5}{16}$
C 3 × 6	3	$\frac{3}{8}$	$1\frac{5}{8}$	$\frac{1}{4}$
× 5	3	$\frac{1}{4}$	$1\frac{1}{2}$	$\frac{1}{4}$
× 4.1	3	$\frac{3}{16}$	$1\frac{3}{8}$	$\frac{1}{4}$
		Channel iron—miscellaneous dimensions		
MC 18 × 58	18	$\frac{11}{16}$	$4\frac{1}{4}$	$\frac{5}{8}$
× 51.9	18	$\frac{5}{8}$	$4\frac{1}{8}$	$\frac{5}{8}$
× 45.8	18	$\frac{1}{2}$	4	$\frac{5}{8}$
× 42.7	18	$\frac{7}{16}$	4	$\frac{5}{8}$
MC 13 × 50	13	$\frac{13}{16}$	$4\frac{3}{8}$	$\frac{5}{8}$
× 40	13	$\frac{9}{16}$	$4\frac{1}{8}$	$\frac{5}{8}$
× 35	13	$\frac{7}{16}$	$4\frac{1}{8}$	$\frac{5}{8}$
× 31.8	13	$\frac{3}{8}$	4	$\frac{5}{8}$

Wide-Flange Dimensions, in (Continued)

Designation	Depth	Web thickness	Flange Width	Flange Thickness
Channel-iron—miscellaneous dimensions				
MC 12 × 50	12	$^{13}/_{16}$	$4^1/_8$	$^{11}/_{16}$
× 45	12	$^{11}/_{16}$	4	$^{11}/_{16}$
× 40	12	$^9/_{16}$	$3^7/_8$	$^{11}/_{16}$
× 35	12	$^7/_{16}$	$3^3/_4$	$^{11}/_{16}$
× 31	12	$^3/_8$	$3^5/_8$	$^{11}/_{16}$
MC 12 × 10.6	12	$^3/_{16}$	$1^1/_2$	$^5/_{16}$
MC 10 × 41.1	10	$^{13}/_{16}$	$4^3/_8$	$^9/_{16}$
× 33.6	10	$^9/_{16}$	$4^1/_8$	$^9/_{16}$
× 28.5	10	$^7/_{16}$	4	$^9/_{16}$
MC 10 × 25	10	$^3/_8$	$3^3/_8$	$^9/_{16}$
× 22	10	$^5/_{16}$	$3^3/_8$	$^9/_{16}$
MC 10 × 8.4	10	$^3/_{16}$	$1^1/_2$	$^1/_4$
MC 10 × 6.5	10	$^1/_8$	$1^1/_8$	$^3/_{16}$
MC 9 × 25.4	9	$^7/_{16}$	$3^1/_2$	$^9/_{16}$
× 23.9	9	$^3/_8$	$3^1/_2$	$^9/_{16}$
MC 8 × 22.8	8	$^7/_{16}$	$3^1/_2$	$^1/_2$
× 21.4	8	$^3/_8$	$3^1/_2$	$^1/_2$
MC 8 × 20	8	$^3/_8$	3	$^1/_2$
× 18.7	8	$^3/_8$	3	$^1/_2$
MC 8 × 8.5	8	$^3/_{16}$	$1^7/_8$	$^5/_{16}$
MC 7 × 22.7	7	$^1/_2$	$3^5/_8$	$^1/_2$
× 19.1	7	$^3/_8$	$3^1/_2$	$^1/_2$
MC 6 × 18	6	$^3/_8$	$3^1/_2$	$^1/_2$
× 15.3	6	$^5/_{16}$	$3^1/_2$	$^3/_8$
MC 6 × 16.3	6	$^3/_8$	3	$^1/_2$
× 15.1	6	$^5/_{16}$	3	$^1/_2$
MC 6 × 12	6	$^5/_{16}$	$2^1/_2$	$^3/_8$

Pipe Dimensions—Standard Weight, in

Nominal diameter	Outside diameter	Inside diameter	Wall thickness	Weight per foot
½	0.840	0.622	0.109	0.85
¾	1.050	0.824	0.113	1.13
1	1.315	1.049	0.133	1.68
1¼	1.660	1.380	0.140	2.27
1½	1.900	1.610	0.145	2.72
2	2.375	2.067	0.154	3.65
2½	2.875	2.469	0.203	5.79
3	3.500	3.068	0.216	7.58
3½	4.000	3.548	0.226	9.11
4	4.500	4.026	0.237	10.79
5	5.563	5.047	0.258	14.62
6	6.625	6.065	0.280	18.97
8	8.625	7.981	0.322	28.55
10	10.750	10.020	0.365	40.48
12	12.750	12.000	0.375	49.56

Pipe Dimensions—Extra Strong, in

Nominal diameter	Outside diameter	Inside diameter	Wall thickness	Weight per foot
½	0.840	0.546	0.147	1.09
¾	1.050	0.742	0.154	1.47
1	1.315	0.957	0.179	2.17
1¼	1.660	1.278	0.191	3.00
1½	1.900	1.500	0.200	3.63
2	2.375	1.939	0.218	5.02
2½	2.875	2.323	0.276	7.66
3	3.500	2.900	0.300	10.25
3½	4.000	3.364	0.318	12.50
4	4.500	3.826	0.337	14.98
5	5.563	4.813	0.375	20.78
6	6.625	5.761	0.432	28.57
8	8.625	7.625	0.500	43.39
10	10.750	9.750	0.500	54.74
12	12.750	11.750	0.500	65.42

Pipe Dimensions—Double Extra Strong, in

Nominal diameter	Outside diameter	Inside diameter	Wall thickness	Weight per foot
2	2.375	1.503	0.436	9.03
2½	2.875	1.771	0.552	13.69
3	3.500	2.300	0.600	18.58
4	4.500	3.152	0.674	27.54
5	5.563	4.063	0.750	38.55
6	6.625	4.897	0.864	53.16
8	8.625	6.875	0.875	72.42

Structural Tubing Dimensions (Square and Rectangular)

Nominal size, in	Wall thickness, in	Weight per foot
	Square tubing dimensions	
16 × 16	⅝	127.37
	½	103.30
	⅜	78.52
	⁵⁄₁₆	65.87
14 × 14	⅝	110.36
	½	89.68
	⅜	68.31
	⁵⁄₁₆	57.36
12 × 12	⅝	93.34
	½	76.07
	⅜	58.10
	⁵⁄₁₆	48.86
	¼	39.43
	³⁄₁₆	29.84
10 × 10	⅝	76.33
	⁹⁄₁₆	69.48
	½	62.46
	⅜	47.90
	⁵⁄₁₆	40.35
	¼	32.63
	³⁄₁₆	24.73
9 × 9	⅝	67.82
	⁹⁄₁₆	61.83
	½	55.66
	⅜	42.79
	⁵⁄₁₆	36.10
	¼	29.23
	³⁄₁₆	22.18
8 × 8	⅝	59.32
	⁹⁄₁₆	54.17
	½	48.85
	⅜	37.69
	⁵⁄₁₆	31.84
	¼	25.82
	³⁄₁₆	19.63
7 × 7	⁹⁄₁₆	46.51
	½	42.05
	⅜	32.58
	⁵⁄₁₆	27.59
	¼	22.42
	³⁄₁₆	17.08
6 × 6	⁹⁄₁₆	38.86
	½	35.24
	⅜	27.48
	⁵⁄₁₆	23.34
	¼	19.02
	³⁄₁₆	14.53

Nominal size, in	Wall thickness, in	Weight per foot
Square tubing dimensions		
5 × 5	1/2	28.43
	3/8	22.37
	5/16	19.08
	1/4	15.62
	3/16	11.97
4.5 × 4.5	1/4	13.91
	3/16	10.70
4 × 4	1/2	21.63
	3/8	17.27
	5/16	14.83
	1/4	12.21
	3/16	9.42
3.5 × 3.5	5/16	12.70
	1/4	10.51
	3/16	8.15
3 × 3	5/16	10.58
	1/4	8.81
	3/16	6.87
2.5 × 2.5	5/16	8.45
	1/4	7.11
	3/16	5.59
2 × 2	5/16	6.32
	1/4	5.41
	3/16	4.32
Rectangular tubing dimensions		
20 × 12	1/2	103.30
	3/8	78.52
	5/16	65.87
20 × 8	1/2	89.68
	3/8	68.31
	5/16	57.36
20 × 4	1/2	76.07
	3/8	58.10
	5/16	48.86
18 × 6	1/2	76.07
	3/8	58.10
	5/16	48.86
16 × 12	5/8	110.36
	1/2	89.68
	3/8	68.31
	5/16	57.36
16 × 8	1/2	76.07
	3/8	58.10
	5/16	48.86

Structural Tubing Dimensions (Square and Rectangular) (*Continued*)

Nominal size, in	Wall thickness, in	Weight per foot
	Rectangular tubing dimensions	
16 × 4	$\frac{1}{2}$	62.46
	$\frac{3}{8}$	47.90
	$\frac{5}{16}$	40.35
14 × 10	$\frac{5}{8}$	93.34
	$\frac{1}{2}$	76.07
	$\frac{3}{8}$	58.10
	$\frac{5}{16}$	48.86
14 × 16	$\frac{1}{2}$	62.46
	$\frac{3}{8}$	47.90
	$\frac{5}{16}$	40.35
	$\frac{1}{4}$	32.63
14 × 4	$\frac{1}{2}$	55.66
	$\frac{3}{8}$	42.79
	$\frac{5}{16}$	36.10
	$\frac{1}{4}$	29.23
12 × 8	$\frac{5}{8}$	76.33
	$\frac{9}{16}$	69.48
	$\frac{1}{2}$	62.46
	$\frac{3}{8}$	47.90
	$\frac{5}{16}$	40.35
	$\frac{1}{4}$	32.63
	$\frac{3}{16}$	24.73
12 × 6	$\frac{5}{8}$	67.82
	$\frac{9}{16}$	61.83
	$\frac{1}{2}$	55.66
	$\frac{3}{8}$	42.79
	$\frac{5}{16}$	36.10
	$\frac{1}{4}$	29.23
	$\frac{3}{16}$	22.18
12 × 4	$\frac{5}{8}$	59.32
	$\frac{9}{16}$	54.17
	$\frac{1}{2}$	48.85
	$\frac{3}{8}$	37.69
	$\frac{5}{16}$	31.84
	$\frac{1}{4}$	25.82
	$\frac{3}{16}$	19.63
12 × 2	$\frac{1}{4}$	22.42
	$\frac{3}{16}$	17.08
10 × 8	$\frac{5}{8}$	67.82
	$\frac{9}{16}$	61.83
	$\frac{1}{2}$	55.66
	$\frac{3}{8}$	42.79
	$\frac{5}{16}$	36.10
	$\frac{1}{4}$	29.23
	$\frac{3}{16}$	22.18

Nominal size, in	Wall thickness, in	Weight per foot
	Square tubing dimensions	
10 × 6	$\frac{5}{8}$	59.32
	$\frac{9}{16}$	54.17
	$\frac{1}{2}$	48.85
	$\frac{3}{8}$	37.69
	$\frac{5}{16}$	31.84
	$\frac{1}{4}$	25.82
	$\frac{3}{16}$	19.63
10 × 5	$\frac{5}{8}$	55.06
	$\frac{9}{16}$	50.34
	$\frac{1}{2}$	45.45
	$\frac{3}{8}$	35.13
	$\frac{5}{16}$	29.72
	$\frac{1}{4}$	24.12
	$\frac{3}{16}$	18.35
10 × 4	$\frac{9}{16}$	46.51
	$\frac{1}{2}$	42.05
	$\frac{3}{8}$	32.58
	$\frac{5}{16}$	27.59
	$\frac{1}{4}$	22.42
	$\frac{3}{16}$	17.08
10 × 2	$\frac{3}{8}$	27.48
	$\frac{5}{16}$	23.34
	$\frac{1}{4}$	19.02
	$\frac{3}{16}$	14.53
9 × 7	$\frac{5}{8}$	59.32
	$\frac{9}{16}$	54.17
	$\frac{1}{2}$	48.85
	$\frac{3}{8}$	37.69
	$\frac{5}{16}$	31.84
	$\frac{1}{4}$	25.82
	$\frac{3}{16}$	19.63
9 × 6	$\frac{5}{8}$	55.06
	$\frac{9}{16}$	50.34
	$\frac{1}{2}$	45.45
	$\frac{3}{8}$	35.13
	$\frac{5}{16}$	29.72
	$\frac{1}{4}$	24.12
	$\frac{3}{16}$	18.35
9 × 5	$\frac{9}{16}$	46.51
	$\frac{1}{2}$	42.05
	$\frac{3}{8}$	32.58
	$\frac{5}{16}$	27.59
	$\frac{1}{4}$	22.42
	$\frac{3}{16}$	17.08

Structural Tubing Dimensions (Square and Rectangular) (*Continued*)

Nominal size, in	Wall thickness, in	Weight per foot
	Rectangular tubing dimensions	
9 × 3	½	35.24
	⅜	27.48
	⁵⁄₁₆	23.34
	¼	19.02
	³⁄₁₆	14.53
8 × 6	⁹⁄₁₆	46.51
	½	42.05
	⅜	32.58
	⁵⁄₁₆	27.59
	¼	22.42
	³⁄₁₆	17.08
8 × 4	⁹⁄₁₆	38.86
	½	35.24
	⅜	27.48
	⁵⁄₁₆	23.34
	¼	19.02
	³⁄₁₆	14.53
8 × 3	½	31.84
	⅜	24.93
	⁵⁄₁₆	21.21
	¼	17.32
	³⁄₁₆	13.25
8 × 2	⅜	22.37
	⁵⁄₁₆	19.08
	¼	15.62
	³⁄₁₆	11.97
7 × 5	½	35.24
	⅜	27.48
	⁵⁄₁₆	23.34
	¼	19.02
	³⁄₁₆	14.53
7 × 4	½	31.84
	⅜	24.93
	⁵⁄₁₆	21.21
	¼	17.32
	³⁄₁₆	13.25

Nominal size, in	Wall thickness, in	Weight per foot
	Rectangular tubing dimensions	
7 × 3	1/2	28.43
	3/8	22.37
	5/16	19.08
	1/4	15.62
	3/16	11.97
7 × 2	1/4	13.91
	3/16	10.70
6 × 5	1/2	31.84
	3/8	24.93
	5/16	21.21
	1/4	17.32
	3/16	13.25
6 × 4	1/2	28.43
	3/8	22.37
	5/16	19.08
	1/4	15.62
	3/16	11.97
6 × 3	3/8	19.82
	5/16	16.96
	1/4	13.91
	3/16	10.70
6 × 2	3/8	17.27
	5/16	14.83
	1/4	12.21
	3/16	9.42
5 × 4	3/8	19.82
	5/16	16.96
	1/4	13.91
	3/16	10.70
5 × 3	1/2	21.63
	3/8	17.27
	5/16	14.83
	1/4	12.21
	3/16	9.42
5 × 2	5/16	12.70
	1/4	10.51
	3/16	8.15
4 × 3	5/16	12.70
	1/4	10.51
	3/16	8.15
4 × 2	5/16	10.58
	1/4	8.81
	3/16	6.87
3.5 × 3.5	1/4	8.81
	3/16	6.87
3 × 2	1/4	7.11
	3/16	5.59

Sheet Steel Thickness

Gage numbers and equivalent thicknesses, hot-rolled and cold-rolled sheet			Gage numbers and equivalent thicknesses, galvanized sheet		
Manufacturers' standard gage number	Thickness equivalent		Galvanized sheet gage number	Thickness equivalent	
	in	mm		in	mm
3	0.2391	6.073			
4	0.2242	5.695	8	0.1681	4.270
5	0.2092	5.314	9	0.1532	3.891
6	0.1943	4.935	10	0.1382	3.510
7	0.1793	4.554	11	0.1233	3.132
8	0.1644	4.176	12	0.1084	2.753
9	0.1495	3.800	13	0.0934	2.372
10	0.1345	3.416	14	0.0785	1.993
11	0.1196	3.038	15	0.0710	1.803
12	0.1046	2.657	16	0.0635	1.613
13	0.0897	2.278	17	0.0575	1.460
14	0.0747	1.900	18	0.0516	1.311
15	0.0673	1.709	19	0.0456	1.158
16	0.0598	1.519	20	0.0396	1.006
17	0.0538	1.366	21	0.0366	0.930
18	0.0478	1.214	22	0.0336	0.853
19	0.0418	1.062	23	0.0306	0.777
20	0.0359	0.912	24	0.0276	0.701
21	0.0329	0.836	25	0.0247	0.627
22	0.0299	0.759	26	0.0217	0.551
23	0.0269	0.660	27	0.0202	0.513
24	0.0239	0.607	28	0.0187	0.475
25	0.0209	0.531	29	0.0172	0.437
26	0.0179	0.455	30	0.0157	0.399
27	0.0164	0.417	31	0.0142	0.361
28	0.0149	0.378	32	0.0134	0.340

This table is for information only. This product is commonly specified to decimal thickness, not to gage number.

B

Welding Symbols

Welding Symbols
Standard Location of Elements of a Welding Symbol

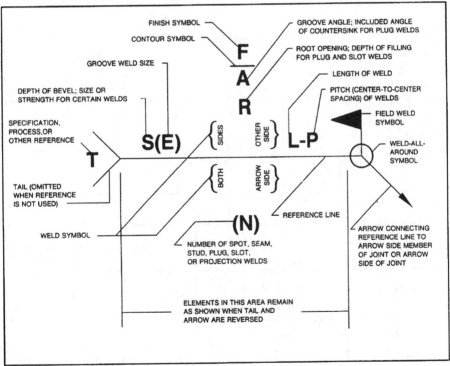

Supplementary Symbols

WELD ALL AROUND	FIELD WELD	MELT THROUGH	CONSUMABLE INSERT (SQUARE)	BACKING OR SPACER (RECTANGLE)	CONTOUR		
					FLUSH OR FLAT	CONVEX	CONCAVE

Welding Symbols

GROOVE							
SQUARE	SCARF	V	BEVEL	U	J	FLARE-V	FLARE-BEVEL

FILLET	PLUG OR SLOT	STUD	SPOT OR PROJECTION	SEAM	BACK OR BACKING	SURFACING	FLANGE	
							EDGE	CORNER

NOTE: THE REFERENCE LINE IS SHOWN DASHED FOR ILLUSTRATIVE PURPOSES.

American Welding Society

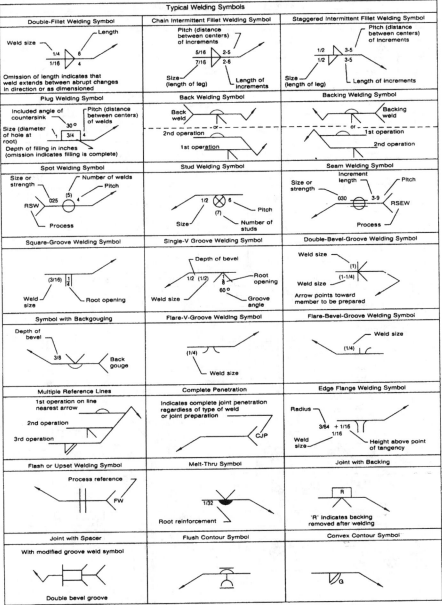

Typical Welding Symbols

* It should be understood that these charts are intended only as shop aids. The only complete and official presentation of the standard welding symbols is in A2.4.

American Welding Society

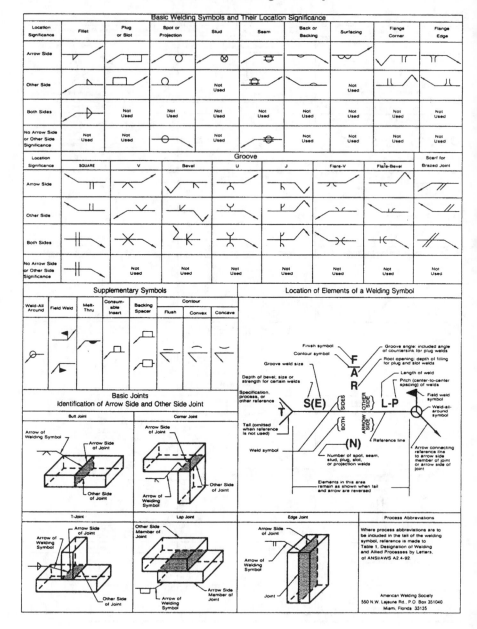

C

SI Metric Conversion Table*

Some Conversion Factors, between U.S. Customary and SI Metric Units, Useful in Structural-Steel Design

	To convert	To	Multiply by
Forces	kip force	kN	4.448
	lb	N	4.448
	kN	kip	0.2248
Stresses	ksi	MPa (that is, N/mm^2)	6.895
	psi	MPa	0.006895
	MPa	ksi	0.1450
	MPa	psi	145.0
Moments	ft-kip	$kN \cdot m$	1.356
	$kN \cdot m$	ft-kip	0.7376
Uniform loading	kip/ft	kN/m	14.59
	kN/m	kip/ft	0.06852
	kip/ft^2	kN/m^2	47.88
	psf	N/m^2	47.88
	kN/m^2	kip/ft^2	0.02089

Note: For proper use of SI, see "Standard for Metric Practice (ASTM E380)," American Society for Testing and Materials, Philadelphia. Also see "Standard Practice for the Use of Metric (SI) Units in Building Design and Construction" (Committee E-6 Supplement to E380) (ANSI/ASTM E621), American Society for Testing and Materials, Philadelphia.

Basis of conversions (ASTM E380): 1 in = 25.4 mm; 1 lb force = 4.448 221 615 260 5 N.

*From *Steel Design Handbook: LRFD Method.*

Basic SI Units Relating to Structural Steel Design

Quantity	Unit	Symbol
Length	Meter	m
Mass	Kilogram	kg
Time	Second	s

Derived SI Units Relating to Structural Steel Design

Quantity	Unit	Symbol	Formula
Force	Newton	N	$kg \cdot m/s^2$
Pressure, stress	Pascal	Pa	N/m^2
Energy, or work	Joule	J	$N \cdot m$

Nomenclature*

A	cross-sectional area, in^2
A_B	loaded area of concrete, in^2
A_b	nominal body area of a fastener, in^2
A_c	area of concrete, in^2
A_c	area of concrete slab within effective width, in^2
A_D	area of an upset rod based on the major diameter of its threads, in^2
A_e	effective net area, in^2
A_f	area of flange, in^2
A_{fe}	effective tension flange area, in^2
A_{fg}	gross area of flange, in^2
A_{fn}	net area of flange, in^2
A_g	gross area, in^2
A_{gt}	gross area subject to tension, in^2
A_{gv}	gross area subject to shear, in^2
A_n	net area, in^2
A_{nt}	net area subject to tension, in^2
A_{nv}	net area subject to shear, in^2
A_{pb}	projected bearing area, in^2
A_r	area of reinforcing bars, in^2
A_s	area of steel cross section, in^2
A_{sc}	cross-sectional area of stud shear connector, in^2
A_{sf}	shear area on the failure path, in^2

*From *Steel Design Handbook: LRFD Method.*

A_w	web area, in^2
A_1	area of steel bearing concentrically on a concrete support, in^2
A_2	total cross-sectional area of a concrete support, in^2
B	factor for bending stress in tees and double angles
B	factor for bending stress in web-tapered members
B_1, B_2	factors used in determining M_u for combined bending and axial forces when first-order analysis is employed
C_{PG}	plate-girder coefficient
C_b	bending coefficient dependent on moment gradient
C_m	coefficient applied to bending term in interaction formula for prismatic members and dependent on column curvature caused by applied moments
C_m'	coefficient applied to bending term in interaction formula for tapered members and dependent on axial stress at the small end of the member
C_p	ponding flexibility coefficient for primary member in a flat roof
C_s	ponding flexibility coefficient for secondary member in a flat roof
C_v	ratio of "critical" web stress, according to linear buckling theory, to the shear yield stress of web material
C_w	warping constant, in^2
D	outside diameter of circular hollow section, in
D	dead load due to the weight of the structural elements and permanent features on the structure
D	factor dependent on the type of transverse stiffeners used in a plate girder
E	modulus of elasticity of steel ($E = 29,000$ ksi)
E	earthquake load
E_c	modulus of elasticity of concrete, ksi
E_m	modified modulus of elasticity, ksi
F_{BM}	nominal strength of the base material to be welded, ksi
F_{EXX}	classification number of weld metal (minimum specified strength), ksi
F_L	smaller of $(F_{yf} F_r)$ or F_{yw}, ksi
F_{by}	flexural stress for tapered members
F_{cr}	critical stress, ksi
$F_{crft}, F_{cry}, F_{crz}$	flexural-torsional buckling stresses for double-angle and tee-shaped compression members, ksi

F_e	elastic buckling stress, ksi
F_{ex}	elastic flexural buckling stress about the major axis, ksi
F_{ey}	elastic flexural buckling stress about the minor axis, ksi
F_{ez}	elastic torsional buckling stress, ksi
F_{my}	modified yield stress for composite columns, ksi
F_n	nominal shear rupture strength, ksi
F_r	compressive residual stress in flange (10 ksi for rolled, 16.5 ksi for welded), ksi
F_{sy}	stress for tapered members, ksi
F_u	specified minimum tensile strength of the type of steel being used, ksi
F_w	nominal strength of the weld electrode material, ksi
F_{wy}	stress for tapered members, ksi
F_y	specified minimum yield stress of the type of steel being used, ksi; *yield stress* denotes either the specified minimum yield point (for those steels that have a yield point) or specified yield strength (for those steels that do not have a yield point)
F_{yf}	specified minimum yield stress of the flange, ksi
F_{yr}	specified minimum yield stress of reinforcing bars, ksi
F_{yst}	specified minimum yield stress of the stiffener materials, ksi
F_{yw}	specified minimum yield stress of the web, ksi
G	shear modulus of elasticity of steel, ksi ($G = 11,200$)
H	horizontal force, kips
H	flexural constant
H_s	length of stud connector after welding, in
I	moment of inertia, in^4
I_d	moment of inertia of the steel deck supported on secondary members, in^4
I_p	moment of inertia of primary members, in^4
I_s	moment of inertia of secondary members, in^4
I_{st}	moment of inertia of a transverse stiffener, in^4
I_{yc}	moment of inertia about y axis referred to compression flange, or if reverse curvature bending referred to smaller flange, in^4
J	torsional constant for a section, in^4
K	effective length factor for prismatic member
K_z	effective length factor for torsional buckling

K_γ	effective length factor for a tapered member
L	story height, in
L	length of connection in the direction of loading, in
L	live load due to occupancy and movable equipment
L_b	laterally unbraced length; length between points which are either braced against lateral displacement of compression flange or braced against twist of the cross section, in
L_c	length of channel shear connector, in
L_e	edge distance, in
L_p	limiting laterally unbraced length for full plastic bending capacity, uniform moment case ($C_b = 1.0$), in
L_p	column spacing in direction of girder, ft
L_{pd}	limiting laterally unbraced length for plastic analysis, in
L_r	limiting laterally unbraced length for inelastic lateral-torsional buckling, in
L_r	roof live load
L_s	column spacing perpendicular to direction of girder, ft
M_A	absolute value of moment at quarter point of the unbraced beam segment, kip-in
M_B	absolute value of moment at centerline of the unbraced beam segment, kip-in
M_C	absolute value of moment at three-quarter point of the unbraced beam segment, kip-in
M_{cr}	elastic buckling moment, kip-in
M_{lt}	required flexural strength in member due to lateral frame translation only, kip-in
M_{max}	absolute value of maximum moment in the unbraced beam segment, kip-in
M_n	nominal flexural strength, kip-in
M'_{nx}, M'_{ny}	flexural strength for use in alternate interaction equations for combined bending and axial force, kip-in or kip-ft as indicated
M_{nt}	required flexural strength in member assuming there is no lateral translation of the frame, kip-in
M_p	plastic bending moment, kip-in
M'_p	moment for use in alternate interaction equations for combined bending and axial force, kip-in
M_r	limiting buckling moment, M_{cr}, when $\lambda = \lambda_r$ and $C_b = 1.0$, kip-in
M_u	required flexural strength, kip-in

M_y	moment corresponding to onset of yielding at the extreme fiber from an elastic stress distribution ($= F_y S$ for homogeneous sections), kip-in
M_1	smaller moment at end of unbraced length of beam or beam-column, kip-in
M_2	larger moment at end of unbraced length of beam or beam-column, kip-in
N	length of bearing, in
N_r	number of stud connectors in one rib at a beam intersection
P_{e1}, P_{e2}	elastic Euler buckling load for braced and unbraced frame, respectively, kips
P_n	nominal axial strength (tension or compression), kips
P_p	bearing load on concrete, kips
P_u	required axial strength (tension or compression), kips
P_y	yield strength, kips
Q	full reduction factor for slender compression elements
Q_a	reduction factor for slender stiffened compression elements
Q_n	nominal strength of one stud shear connector, kips
Q_s	reduction factor for slender unstiffened compression elements
R	load due to initial rainwater or ice exclusive of the ponding contribution
R_{PG}	plate girder bending strength reduction factor
R_e	hybrid girder factor
R_n	nominal strength
R_v	web shear strength, kips
S	elastic section modulus, in^3
S	spacing of secondary members, ft
S	snow load
S_x'	elastic section modulus of larger end of tapered member about its major axis, in^3
S_{eff}	effective section modulus about major axis, in^3
S_{xt}, S_{xc}	elastic section modulus referred to tension and compression flanges, respectively, in^3
T	tension force due to service loads, kips
T_b	specified pretension load in high-strength bolt, kips
T_u	required tensile strength due to factored loads, kips

U	reduction coefficient, used in calculating effective net area
V_n	nominal shear strength, kips
V_u	required shear strength, kips
W	wind load
X_1	beam buckling factor
X_2	beam buckling factor
Z	plastic section modulus, in^3
a	clear distance between transverse stiffeners, in
a	distance between connectors in a built-up member, in
a	shortest distance from edge of pin hole to edge of member measured parallel to direction of force, in
a_r	ratio of web area to compression flange area
a'	weld length, in
b	compression element width, in
b_e	reduced effective width for slender compression elements, in
b_{eff}	effective edge distance, in
b_f	flange width, in
c_1, c_2, c_3	numerical coefficients
d	nominal fastener diameter, in
d	overall depth of member, in
d	pin diameter, in
d	roller diameter, in
d_L	depth at larger end of unbraced tapered segment, in
d_b	beam depth, in
d_c	column depth, in
d_o	depth at smaller end of unbraced tapered segment, in
e	base of natural logarithm = 2.71828…
f	computed compressive stress in the stiffened element, ksi
f_{b1}	smallest computed bending stress at one end of a tapered segment, ksi
f_{b2}	largest computed bending stress at one end of a tapered segment, ksi
f'_c	specified compressive strength of concrete, ksi
f_o	stress due to $1.2D+1.2R$, ksi
f_{un}	required normal stress, ksi
f_{uv}	required shear stress, ksi
f_v	required shear stress due to factored loads in bolts or rivets, ksi

g	transverse center-to-center spacing (gage) between fastener gage line, in
h	clear distance between flanges less the fillet or corner radius for rolled shapes; and for built-up sections, the distance between adjacent lines of fasteners or the clear distance between flanges when welds are used, in
h	distance between centroids of individual components perpendicular to the member axis of buckling, in
h_c	twice the distance from the centroid to the following: the inside face of the compression flange less the fillet or corner radius, for rolled shapes; the nearest line of fasteners at the compression flange or the inside faces of the compression flange when welds are used, for built-up sections, in
h_r	nominal rib height, in
h_s	factor used for web-tapered members
h_w	factor used for web-tapered members
j	factor for minimum moment of inertia for a transverse stiffener
k	distance from outer face of flange to web toe of fillet, in
k_v	web plate buckling coefficient
l	laterally unbraced length of member at the point of load, in
l	length of bearing, in
l	length of connection in the direction of loading, in
l	length of weld, in
m	ratio of web to flange yield stress or critical stress in hybrid beams
r	governing radius of gyration, in
r_{To}	for the smaller end of a tapered member, the radius of gyration, considering only the compression flange plus one-third of the compression web area, taken about an axis in the plane of the web, in
r_i	minimum radius of gyration of individual component in a built-up member, in
r_{ib}	radius of gyration of individual component relative to centroidal axis parallel to member axis of buckling, in
r_m	radius of gyration of the steel shape, pipe, or tubing in composite columns; for steel shapes it may not be less than 0.3 times the overall thickness of the composite section, in
$\bar{r}_o$	polar radius of gyration about the shear center, in
r_{ox}, r_{oy}	radius of gyration about x and y axes at the smaller end of a tapered member, respectively, in

r_x, r_y	radius of gyration about x and y axes, respectively, in
r_{yc}	radius of gyration about y axis referred to compression flange, or if reverse curvature bending, referred to smaller flange, in
s	longitudinal center-to-center spacing (pitch) of any two consecutive holes
t	thickness of connected part, in
t_f	flange thickness, in
t_f	flange thickness of channel shear connector, in
t_w	web thickness of channel shear connector, in
t_w	web thickness, in
w	plate width; distance between welds, in
w	unit weight of concrete, lb/ft^3
w_r	average width of concrete rib or haunch, in
x	subscript relating symbol to strong-axis bending
x_o, y_o	coordinates of the shear center with respect to the centroid, inh
$\bar{x}$	connection eccentricity, in
y	subscript relating symbol to weak-axis bending
z	distance from the smaller end of tapered member used, for the variation in depth, in
α	separation ratio for built-up compression members = $h/2r_{ib}$
Δ_{oh}	translation deflection of the story under consideration, in
γ	depth tapering ratio; subscript for tapered members
ζ	exponent for alternate beam-column interaction equation
η	exponent for alternate beam-column interaction equation
λ_c	column slenderness parameters
λ_e	equivalent slenderness parameter
λ_{eff}	effective slenderness ratio
λ_p	limiting slenderness parameter for compact element
λ_r	limiting slenderness parameter for noncompact element
ϕ	resistance factor
ϕ_b	resistance factor for flexure
ϕ_c	resistance factor for compression
ϕ_c	resistance factor for axially loaded composite columns
ϕ_{sf}	resistance factor for shear on the failure path
ϕ_t	resistance factor for tension
ϕ_v	resistance factor for shear

Index

Quality control (*see* Inspection and quality control)

Radiographic testing (RT), 427–428
Reduced beam section (RBS) for special moment frames:
  shape of, 308–310
  size of, 311–313
Reinforced concrete columns with steel beams, joints for, 473–485
Reinforcement of axial force connections, 167–170
Reinforcing ribs in RBS connection design, 311–312
Required strengths in connection design, 37
Resistance welding, 201, 211
Reuse, bolt, 413–414
Rib-plate connections, 304–305
Rigid composite member connections, 449
Rigid frame and connections for trusses, 99–106
Rivets, 1–2
Roof beam details, 332
Roof deck attachment details, 345
Roof deck attachment pattern details, 346
Roof joist bearing details, 362
Roof opening details, 341
Roof-truss splices, fully tensioned bolts for, 3
Root openings in welding symbols, 23–24
Rotation:
  in composite member connections, 449
  in CW-BB PR connections, 255
  in PR connections classification, 243
  in steel/steel-concrete coupling beam-wall connections, 460
Rotation tolerance in bolt installation, 403
Rotational-capacity tests, 399–400
Run way girder details, 356
Rust on bolts, 410

S-shape dimensions, 531
Sampling techniques for fastener testing, 400
SAW (submerged arc welding) process, 195–197
  electrodes for, 420
  filler metals for, 27
Screws in steel deck connections, 433–436

SCWB requirements for cover-plated connection column face, 307–308
Segregation-induced centerline cracking, 186–187
Seismic design of connections:
  eccentrically braced frames, 325
  Northridge earthquake damage, 325–326
  requirements for, 292–294
  special concentrically braced frames, 313–315
    beam-to-column connections, 324
    brace-to-gusset connection, 317, 323
    example, 315–323
    force distribution, 316–317
    force level, 316
    gusset-plate design, 323
    gusset-plate-to-beam-and-column connection, 323–324
  special issues in, 287–292
  special moment frames, 294–295
    cover-plated connections, 301–303, 305–308
    deformation capacities, 298
    geometry determination, 311–313
    haunched connections, 303–304
    load capacities, 295–297
    RBS shape for, 308–310
    recent developments, 298–299
    strengthened connections, 300–301
    toughened connections, 299–300
    vertical rib-plate connections, 304–305
    weakened connections, 308
Self-drilling screws, 433–434
Self-shielded flux core, 193–195, 207
Semiautomatic welding, 196
Semirigid composite member connections, 449
Service level forces in CB-BB PR connections, 270
Service load slip capacity in CW-BB PR connections, 255
Serviceability of composite member connections, 450
Shear-bearing joints, washers for, 412
Shear connections, 140
  beam, 330
  beam shear splices, 156–159
  in beam-to-wall connections, 456
  economic considerations in, 38
  framed connections, 140–143
  outrigger beam-to-wall, 466–473

ABOUT THE EDITOR

Akbar R. Tamboli is a senior project engineer with CUH2A in Princeton, New Jersey, a well-recognized company specializing in architecture, engineering, and planning. He is a former vice president and project manager for Cantor-Seinuk Group PC, Consulting Engineers, in New York City, and was the principal consulting engineer on a number of noteworthy projects, including Morgan Guaranty Bank Headquarters at 60 Wall Street and Salomon Brothers World Headquarters, Seven World Trade Center.

How to Use the Accompanying FabriCAD CD-ROM Disk

System Requirements

Pentium

Graphics card of 800 × 600 or higher resolution

100 MB free on hard disk

32 MB of RAM

Windows 95/Windows 98/Windows NT 4.x

Installation

1. Place the FabriCAD CD-ROM in your CD-ROM drive.

2. Windows 95/98/NT: In Start menu, choose Run.

3. Type d:\Fabricd (if your CD-ROM drive is not drive D, type the appropriate letter instead)

4. Choose OK.

5. Follow the instructions on the screen.

Viewing the Online Manuals

The following manuals are included on the FabriCAD CD-ROM:

FabriCAD Reference Manual

Visual Draw Reference Manual

To view the online manuals:

In the Start menu, choose Program.

Select FabriCAD from the list of applications.

Select the desired manual.

Technical Support

For questions about FabriCAD, contract Research Engineers, Inc. at:

Phone: (714) 974-2500

Fax: (714) 974-4771

E-mail: info@reiusa.com

Interactive LRFD examples can be downloaded from Research Engineers website: www.reiusa.com

SOFTWARE AND INFORMATION LICENSE

The software and information on this diskette (collectively referred to as the "Product") are the property of The McGraw-Hill Companies, Inc. ("McGraw-Hill") and are protected by both United States copyright law and international copyright treaty provision. You must treat this Product just like a book, except that you may copy it into a computer to be used and you may make archival copies of the Products for the sole purpose of backing up our software and protecting your investment from loss.

By saying "just like a book," McGraw-Hill means, for example, that the Product may be used by any number of people and may be freely moved from one computer location to another, so long as there is no possibility of the Product (or any part of the Product) being used at one location or on one computer while it is being used at another. Just as a book cannot be read by two different people in two different places at the same time, neither can the Product be used by two different people in two different places at the same time (unless, of course, McGraw-Hill's rights are being violated).

McGraw-Hill reserves the right to alter or modify the contents of the Product at any time.

This agreement is effective until terminated. The Agreement will terminate automatically without notice if you fail to comply with any provisions of this Agreement. In the event of termination by reason of your breach, you will destroy or erase all copies of the Product installed on any computer system or made for backup purposes and shall expunge the Product from your data storage facilities.

LIMITED WARRANTY

McGraw-Hill warrants the physical diskette(s) enclosed herein to be free of defects in materials and workmanship for a period of sixty days from the purchase date. If McGraw-Hill receives written notification within the warranty period of defects in materials or workmanship, and such notification is determined by McGraw-Hill to be correct, McGraw-Hill will replace the defective diskette(s). Send request to:

Customer Service
McGraw-Hill
Gahanna Industrial Park
860 Taylor Station Road
Blacklick, OH 43004-9615

The entire and exclusive liability and remedy for breach of this Limited Warranty shall be limited to replacement of defective diskette(s) and shall not include or extend to any claim for or right to cover any other damages, including but not limited to, loss of profit, data, or use of the software, or special, incidental, or consequential damages or other similar claims, even if McGraw-Hill has been specifically advised as to the possibility of such damages. In no event will McGraw-Hill's liability for any damages to you or any other person ever exceed the lower of suggested list price or actual price paid for the license to use the Product, regardless of any form of the claim.

THE McGRAW-HILL COMPANIES, INC. SPECIFICALLY DISCLAIMS ALL OTHER WARRANTIES, EXPRESS OR IMPLIED, INCLUDING BUT NOT LIMITED TO, ANY IMPLIED WARRANTY OF MERCHANTABILITY OR FITNESS FOR A PARTICULAR PURPOSE. Specifically, McGraw-Hill makes no representation or warranty that the Product is fit for any particular purpose and any implied warranty of merchantability is limited to the sixty day duration of the Limited Warranty covering the physical diskette(s) only (and not the software or information) and is otherwise expressly and specifically disclaimed.

This Limited Warranty gives you specific legal rights; you may have others which may vary from state to state. Some states do not allow the exclusion of incidental or consequential damages, or the limitation on how long an implied warranty lasts, so some of the above may not apply to you.

This Agreement constitutes the entire agreement between the parties relating to use of the Product. The terms of any purchase order shall have no effect on the terms of this Agreement. Failure of McGraw-Hill to insist at any time on strict compliance with this Agreement shall not constitute a waiver of any rights under this Agreement. This Agreement shall be construed and governed in accordance with the laws of New York. If any provision of this Agreement is held to be contrary to law, that provision will be enforced to the maximum extent permissible and the remaining provisions will remain in force and effect.